Animal Geographies

Animal Geographies

Place, Politics, and Identity in the Nature-Culture Borderlands

Edited by
JENNIFER WOLCH and JODY EMEL

V

VERSO

London • New York

First published by Verso 1998
© in the collection Jennifer Wolch and Jody Emel 1998
© in individual contributions the contributors 1998

The rights of Jennifer Wolch and Jody Emel to be identified
as the editors of this work have been asserted by them in
accordance with the Copyright, Designs and Patents Act 1988

Verso
UK: 6 Meard Street, London W1V 3HR
USA: 180 Varick Street, New York NY 10014–4606

Verso is the imprint of New Left Books

ISBN 1–85984–831–1
ISBN 1–85984–137–6 (pbk)

British Library Cataloguing in Publication Data
A catalogue record for this book is available from the British Library

Library of Congress Cataloging-in-Publication Data
A catalog record for this book is available from the Library of Congress

Typeset by SetSystems Ltd, Saffron Waldon
Printed by Biddles Ltd, Guildford and King's Lynn

Contents

List of Figures

List of Tables

Preface

> We need another and a wiser and perhaps more mystical
> concept of animals ... We patronize them for their incom-
> pleteness, for their tragic fate of having taken form so far
> below ourselves. And therein we err and err greatly. For the
> animal shall not be measured by man. In a world older and
> more complete than ours they move finished and complete,
> gifted with extensions of the senses we have lost or never
> attained, living by voices we shall never hear. They are not
> brethren, they are not underlings; they are other nations,
> caught with ourselves in the net of life and time, fellow
> prisoners of the splendor and travail of the earth.[1]

The plight of animals worldwide has never been more serious than it is
today. Each year, by the billions, animals are killed in factory farms;
poisoned by toxic pollutants and waste; driven from their homes by logging,
mining, agriculture, and urbanization; dissected, re-engineered, and used
as spare body-parts; and kept in captivity and servitude to be discarded as
soon as their utility to people has waned. This reality is mostly obscured by
the progressive elimination of animals from everyday human experience,
and by the creation of a thin veneer of civility surrounding human-animal
relations, embodied largely by language tricks, isolation of death camps,
and food preparation routines that artfully disguise the true origins of flesh-
food. Despite the efforts made to minimize human awareness of animal lives
and fates, however, the brutality of human domination over the animal
world and the catastrophic consequences of such dominionism are everywhere evident.

The premise of *Animal Geographies* is that animals have been so indispensable to the structure of human affairs and so tied up with our visions of
progress and the good life that we have been unable to (even try to) fully
see them. Their very centrality prompted us to simply look away and to
ignore their fates. But human practices now threaten the animal world and
the entire global environment as never before. Our own futures are on the
line too. Hence we have an intellectual responsibility as well as an ethical
duty to consider the lives of animals closely. Over the past two decades, a
willingness to take animals seriously appeared both within academe and

beyond. This concern with human-animal relations—or what we term "the animal question"—has expanded dramatically in the last few years, leading to radical reconceptualizations of the nature of human-animal relations. *Animal Geographies* seeks to contribute to this fundamental rethinking of animals and to suggest how, by looking through geographical lenses, we may be able to bring animals into clearer focus and back into our understanding of social life.

This goal reflects not only intellectual but also political commitments that we want to acknowledge. The conservation of natural species and traditional cultures for their intrinsic value has often been associated with the extreme right. Accelerating environmental degradation, along with the rise of environmental and animal rights movements, however, has made concern for nature a legitimate focus of the political left as well. But animals were never part of the modernist ideology of progress for either the right or the left, except as commodities. And despite the fundamental questioning of the ideology of progress that is now underway, which seeks to revalue those who were marginalized by it, the issue of human-animal relations has still not been embraced by the left. Some strands of feminism and postmodern environmentalism have been sympathetic to animal rights and welfare, but most of those on the left interpret a focus on animals as an attempt to regain some pre-modern, spiritual or animistic stance. Animal activism also has been dismissed as just one more form of commodification: "saving lobsters, spraying furs—is a symptom of the pervasive commodification of nature as much as a visceral response to it. It is as much about making its advocates feel good about themselves and their world as it is about lobsters."[2] Such worries are not entirely unjustified,[3] but the result of such misgivings is that animals are nowhere in most leftist texts; in fact, "nature" is rarely let out of its two-dimensional black box.

Is it possible to construct a democratic, non-racist, non-sexist politics that embraces an animated and embodied nature? Surely there must be spaces in which both animals and humans have authority, but they are neither in the dictatorial fascism of some forms of deep ecology or sociobiology, nor in the smug authoritarianism of anthropocentric humanism. *Our political project is the creation of many forms of shared space.* It will not be an easy task. Animals may be the last group to be brought into the circle of morality and subjectivity; no other group has been admitted without bloodshed and strife. As the circle enlarges to include all sentient beings, issues of costs and benefits become much more difficult to decide. If suspicions of misanthropy and a hatred of modernity on the part of some animal rights activists and other animal protectors are not altogether unfounded, huge doses of reflexivity are called for on all sides. With barriers to emotional engagement eroding, battles over material and cultural goods increasing, and global capitalism expanding, democratic communication and decision-making will be absolutely necessary to maintain any sort of social, economic, and ecological balance. For us, building a progressive politics for the twenty-first century means combining critical analysis with a commitment to inclusive, caring, and democratic campaigns for a justice capable of embracing both people and animals.

Geography and the Animal Question

This book demonstrates that taking geographical approaches to the animal question, or the issue of human-animal relations, will generate rich and provocative ideas. Geography boasts a long tradition of inquiry into relationships between nature and society, and the ways in which natural resources and conditions (such as climate, soils, and so on) and human cultural practices shape one another. This is one rationale for considering the animal question through geographical lenses. But geography has also emerged as crucial to critiques of modernity and contemporary social theory. In contrast to modernist social theory and its obsession with time, feminists and postmodern social theorists have turned to geographical theory to guide their understanding. This book takes advantage of both nature/society traditions in geography and geographical social theory to shed light on animals as central agents in the constitution of space and place, and all that that entails.

Nature-society traditions in geography

In the twentieth century, the nature-culture relations tradition in geography was perhaps best epitomized by Carl Sauer and the Berkeley school of cultural geography. The Berkeley school focused on human impacts on the landscape to reveal the co-evolution of environments and cultures in places and regions, and thus to understand the morphology of the cultural landscape. While animals did not figure as subjects in Sauerian geographies, studies of domestication and diffusion of animal husbandry, the cultural and economic role of animals in agrarian societies, and the environmental changes attendant upon agriculture- and livestock-based lifeways revealed the importance of certain animals to cultural practices and environmental conditions.

By the 1960s, Sauerian cultural geography had receded as a dominant approach in the field, criticized for its reliance on a super-organic theory of culture, naive empiricism, and obsessive orientation to the dry details of material artifacts and place facts.[4] It satisfied neither those searching for explanations of landscapes based on an understanding of social processes, nor those seeking greater scientific legitimacy for the discipline.[5] Cultural geography had become a backwater, and its central questions about human/ environment relations receded from view as well.

The reassertion of space in social theory

Within the larger academy, cultural geography, along with other branches of the field, had long been marginalized as "merely" descriptive and ideographic.[6] But during the 1970s and 1980s, as modernist beliefs in historicism and foundationalism gave way to the recognition of context and difference, time made way for space. Geography seemingly re-enchanted the social sciences and humanities, and treatises on space sprouted up in an astonishing variety of fields concerned with theorizing society.

Why space? At first glance, space appears decidedly unproblematic: space is simply where things happen. But Henri Lefebvre explicated an enormous range of spaces, the social dynamics that produce them, and their implications for the constitution of social life, concluding that space is never simply a stage for human action, and never "innocent" in terms of its role in shaping human affairs.[7] Similarly, Michel Foucault denied the innocence of space, pointing to the profound ways in which knowledge and power were linked to the uses of space and place—to isolate and exclude, to segregate and thus manage social difference.[9] And the arrival of what Fredric Jameson termed "postmodern hyperspace," created by an "alarming disjunction point between the body and its built environment" and "the incapacity of our minds ... to map the great global multinational and decentered communications network in which we find ourselves caught as individual subjects,"[9] brought questions of space to the forefront of postmodern theories of social change.

The reassertion of space in social theory occurred in a number of social science fields. As John Berger presciently claimed, "(p)rophesy now involves geographical rather than historical projection; it is space, not time, that hides consequences from us."[10] The most prominent example, perhaps, was the sociology of Anthony Giddens, whose theory of structuration was grounded in geographic notions of region, locale, and time-space routine (drawing on Swedish geographer Torsten Haägerstrand's time-geography).[11] Understandings of both the object and tasks of social theory itself were revolutionized; now the challenge was to understand social practices and evolution, not only through time but over space and within the context of particular places.

As social theory tackled the problematics of space, geography itself was transformed in the late 1980s and early 1990s. Influential books on geography and social theory (such as Ed Soja's *Postmodern Geographies*,[12] and Susan Hanson and Geraldine Pratt's *Gender, Work, and Space*),[13] led the way to a renaissance of interest in the role of space in shaping social processes.[14] There was also an increasing focus on place—the unique, empirical manifestations of socio-spatial processes as they are played out in particular locales and regions. Any particular place or locale could be viewed as a distillation of past, present, and emergent processes that constitute, constrain, and mediate social organization. By the mid 1990s, the task of what Derek Gregory termed a "geographical imagination" was to unravel those time-space horizons and interpret landscape to understand the creation of place.[15]

Rediscovering nature
Despite the efflorescence of geographical work on space and place, geographers working in the social theory tradition were slow to attend to questions

of how nature and environment were implicated in spatial processes or the constitution of place. But because of the emergence of feminism, postmodernism, and the "interpretive turn" throughout the social sciences, and the flowering of the social theory tradition within geography itself, cultural geography was reinvigorated. Employing the metaphor of "landscape as text," a "new" cultural geography drew upon poststructural theory (especially Foucault and Derrida) to tease out the social forces that created landscapes, lent them power, and engendered social conflict. Seeing landscapes as culturally produced texts or icons revealed "the inherent instability of meaning, fragmentation or absence of integrity, lack of authorial control, polyvocality and unresolvable social contradictions" inherent in places.[16] Cultural geographers were thus drawn to issues of how landscapes related to personal subjectivity, cultural identity and conflict, and systems of symbolic meaning.

How was the environment theorized within this new cultural geographic discourse? Paralleling the wholesale rejection of Sauerian cultural geography, the entire idea of a nature separate from human culture was essentially abandoned as hopelessly naive and outdated. But in so completely denaturalizing nature and treating geographic places as cultural productions, the agency of nature and especially animals was denied.[17] This "writing out" of nature by cultural geographers catalyzed a lively debate with scholars engaged in the "new" environmental history (influenced, ironically, by Sauer and the Berkeley school), who were at pains to demonstrate nature's agency. Their project was to understand how culture-nature boundaries were drawn and redrawn over time as a result of both cultural processes, ecological features and agency, and their interactions.

As this debate unfolded, some extreme positions were established at both the relativist and realist ends of the spectrum. But along with ideas from social theory, especially powerful feminist arguments about the socially constructed and situated character of everyday life and knowledge,[18] the landscape debate stimulated a resurgence of interest in nature-society relations and a wide range of studies capitalizing on the intellectual ferment left in its wake. Contemporary geographical work on nature-society relations, fully informed by social theory, philosophy, and cultural studies, thus provides solid ground on which to base new thinking about animal-human relations.

Inside *Animal Geographies*

Geographers—like most other intellectuals in the academy—have tended to deal with nature in a black-box manner. Aside from arguments about ecological realism versus cultural constructionism, nature has remained a largely undifferentiated concept, its constituent parts rarely theorized separately. To read most geographical texts, one might never know that nature was populated by sentient creatures; animals have simply been confined within the black box, or conflated with ecological and production systems.

We think the time has come to let animals out of the box, to add body and other forms of presence. *Animal Geographies* seeks to harness the multiple perspectives of social theory and the insights of geographical work on nature and culture to initiate a lively and satisfying discourse on human-animal relations. The chapters in this book will demonstrate this potential and enrich dialogues about people and animals that are continuing across the social sciences and humanities. They have been organized around four nested themes central to contemporary thinking about people and animals and about how a project of coexistence can be pursued (see below).

The first part of the book explores human-animal relations at the level of the individual, considering in particular the links between animal subjectivity and the formation of human identities. Human-animal relations at this level form the basis for the evolution of "borderland" communities, in which humans and animals share space, however uneasily. The possibilities of such communities are treated in the second part of the book. Borderland communities, and more generally human and animal populations in any place-time, exist within the context of an encompassing political economy in which animal bodies play an enormous role. Some of the ways animals relate to political-economic structures and processes at both local and global scales are taken up in the book's third section. All aspects of human-animal relations—individual and structural, local and global—are subject to examination from an ethical perspective, and hence the final part of the book addresses the question of animals and the moral landscape.

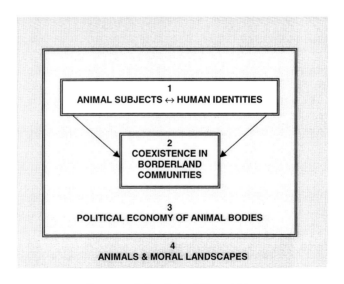

Framework for *Animal Geographies*

Animal subjects/human identities

Largely unremarked in discourse about individual identities is the role that animals play in identity construction. Part I of this book seeks to redress this gap, by exploring the diverse ways in which animals shape the formation of human personal identity. Loren Eisley, Paul Shepard, and many others have argued that humans as a species define themselves in relation to animals.[19] Here, however, we focus on the formation of the more subtle identities that people adopt or have ascribed to them by others. These range from identities linked to particular places to "race"/ethnic and gender identities.

Using the case of the Adelaide Zoo, for example, Anderson (chapter 2) shows how zoo practices served to naturalize the imposition of colonial rule and the oppression of indigenous peoples, to reinforce gendered and racialized underpinnings of human-nonhuman boundaries, and to consolidate and legitimate Australian colonial identities. Philo (chapter 3) explores the emergence of a distinctive urban identity in Victorian Britain, associated with standards of civility, public decency, and sexual license, and norms of compassion that stood in contrast to rural stereotypes. Live meat markets and urban slaughterhouses violated these emergent standards and norms, and thus were ultimately excised from the city, a move reinforcing urban identities defined in opposition to a countryside populated by beastly people and animals.

Turning toward the formation of racialized identities, Elder, Wolch, and Emel (chapter 4) show that culturally specific practices involving harm to animals help construct the human-animal divide and allocate particular racial/cultural groups to either civilization or savagery. In the US context, animal practices are being employed to racialize and devalue immigrants by dominant groups seeking to protect national identity in the face of growing population diversity linked to globalization. Emel (chapter 5) then turns to ecofeminism to understand how the construction of masculinity in the American West during the late nineteenth and early twentieth centuries was associated with wolf eradication efforts. Dominant representations emphasized the wolf's so-called savagery, lack of mercy, unfair habit of pack hunting, and cowardice—all of which contravened norms of masculinity in the American frontier. These sorts of images were not only devastating to the wolf and other animals, but were analogous to practices that perpetuated racism and sadism in the treatment of humans.

Negotiating the human/animal borderlands

In part II we move from questions of subjectivities and identities to the places in which people and animals confront the realities of coexistence on an everyday basis. Traditional nature/culture dualisms have led to the creation of mutually exclusive spaces and places for wild animals (pristine wildernesses) and humans (cities and towns). But there remain extensive, permeable border zones of metropolitan regions inhabited by both people and animals. This inquiry takes up the possibilities of such zones of potential coexistence and examines cases of negotiation/struggle over sharing space which reveal how representations of both animals and people reflect the interspecific balance of borderlands power.

In chapter 6 Wolch suggests that the creation of a "zoöpolis" in which people and animals coexist might serve as a method for re-establishing networks of care, friendship, and solidarity between people and animals. She outlines a framework to describe the development of a trans-species urban theory and practice that could extend our understandings of human relations with animals in the city, and holds out hope that the ideal of zoöpolis can create political space for movements intent on peace between the species.

Gullo, Lassiter, and Wolch (chapter 7) then consider the specific case of relations between people and mountain lions during the 1980s and 1990s in California, where urbanization-driven increases in human-cougar inter-actions along with scientific discord over cougar ecology stimulated a renegotiation of state-level cougar management. Gullo et al. portray the political and technical difficulties of large predator management in human/animal borderlands but also suggest the potential for mutual learning and coexistence achieved through education and behavioral modification of both people and cougars. Similarly, Michel (chapter 8) considers the urban-wildlands boundary. She represents the politics of eagle conservation in Southern California as one entrenched not only in urban growth rhetoric, real estate interests, and scientific environmentalism, but also shaped by community activists such as eagle rehabilitators and wildlife educators who contest scientific and utilitarian visions of nature. She shows how their work, imbued with gendered notions of motherhood and family, and an ethic of care, constitutes a personal politics of both animal and human reproduction that attempts to assert the agency of wildlife in defining pathways to human-animal coexistence.

The political economy of animal bodies

In part III we turn from our consideration of human-animal borderlands to how animals figure in the ongoing globalization of production. In certain contexts, animals have become central symbols in battles over the disappear-ance of local production regimes and associated ways of life threatened by globalization. For example, Proctor (chapter 9) considers how one animal—the northern spotted owl—simultaneously became a symbol of the nation's biodiversity and wilderness patrimony, and a vehicle for internationalization of the forest products industry. He argues that the political force of the owl emanated from its powerful symbolism of a particular moral landscape, but under pressures of timber-industry restructuring and flight, residents of the Northwest linked owl protection to the loss of timber jobs, the death of timber-dependent communities, and the elimination of a traditional rural way of life.

More directly, animal bodies themselves are key to an emerging world food order based on flesh-food, with enormous political-economic, social, and environmental consequences. Robbins (chapter 10) argues that the emergence of India as an actor in the internationalizing livestock sector intensified long-standing political struggles between ethno-religious groups. He shows how transformations in international meat demand in the desert state of Rajasthan have catalyzed shifts in agricultural production along with changes in caste identity and notions of property, fundamentally altering the bargain between people and domesticated animal species. Conservative Hindu politicians have also used the increasing presence of meat in the economy and diet as evidence of a decline in morality, to generate ideological support for economic isolation and anti-Muslim sentiment. Internationalization is also influencing livestock practices in the West, as Ufkes (chapter 11) shows for the US Midwest. Here, a new wave of agro-industrialization driven by changes in consumer demand and meatpacking practices now centers around lean meat production and involves alterations in genetics, feeding regimes, facilities construction, and management practices "down on the farm." Ufkes argues that demand for value-added meats for worldwide consumption has resulted in increased scale/standardization of production, while demand for leaner hogs on the part of affluent consumers has spurred development of an array of new commercial inputs—lean genetics, partitioning agents, medications—designed to refashion the interior geography of the pig for profit.

Animals and the moral landscape
In the final portion of the book, we explore legal and ethical approaches to human-animal relations in which justice for both people and animals is paramount. Wescoat (chapter 12), for instance, maintains that little attention has been given to animals' access to water in different cultural and legal contexts, with the "right of thirst" in Islamic law constituting an important exception. In a comparative analysis of Pakistan and the American West, he shows that while in Pakistan the right of thirst is not legally required, this religious imperative has stimulated the emergence of animal rights and welfare activities. In the American West, the right of thirst asserts a moral imperative which reveals western water law as morally inadequate, and thus provides an example of how duties to provide water for animals might be expanded.

Finally, Lynn (chapter 13) compares geographical ethics—or geoethics—to other ethical traditions, and argues that the concept of "geographical community" is valuable as a means to encompass ethical questions involving people, animals, and the rest of nature. Lynn makes a case for including animals in the moral community, and proposes four normative principles to guide the conduct of human-animal relations and resolve the moral dilemmas inherent in sharing space with animals.

Taken together, the essays of *Animal Geographies* serve to frame debates about animals and their centrality to individual identity, social and political life, economic organization and dynamics, and the moral choices we face about how to relate to animals and nature. We hope that the book will

motivate an ongoing discourse on these questions; the fate of both humans
and animals "caught in the splendor and travail of the earth" surely lie in
the balance.

Notes

1. H. Beston, *The Outermost House*, Boston: Houghton Mifflin, 1928, pp. 19–20.
2. Neil Smith, "The Production of Nature," in George Robertson, Melinda Mash, Lisa
Tickner, Jon Bird, Barry Curtis, and Tim Putnam, eds, *Future Natural: Nature/Science/
Culture*, London and New York: Routledge, 1996, pp. 35–54.
3. Deep ecology, for example, has serious problems with modern civilization. Luc Ferry,
in *The New Ecological Order* (trs. Carol Volk, Chicago, Ill. and London: University of
Chicago Press, 1995, p. 90), argues that deep ecology "continually hesitates between
conservative romantic themes and progressive ... anticapitalist ones," so much so that
"one can say that some of deep ecology's roots lie in Nazism, while its branches extend far
into the distant reaches of the cultural left."
4. James Duncan, *The City as Text: The Politics of Landscape Interpretation in the Kandyan
Kingdom*, New York: Cambridge University Press, 1990; Peter Jackson, *Maps of Meaning: An
Introduction to Cultural Geography*, London: Unwin Hyman, 1989.
5. David Demeritt, "The Nature of Metaphors in Cultural Geography and Environmental History," *Progress in Human Geography* 12, 1994, pp. 163–85.
6. Edward Soja, *Postmodern Geographies: The Reassertion of Space in Critical Social Theory*,
London: Verso Press, 1989.
7. Henri Lefebvre, *The Production of Space*, Oxford and Cambridge, Mass.: Blackwell,
1991.
8. Michel Foucault, *Discipline and Punish: The Birth of the Prison*, New York: Pantheon,
1977; and Foucault, *Madness and Civilization: A History of Insanity in the Age of Reason*, New
York: Pantheon, 1965.
9. Fredric Jameson, *Postmodernism, or The Cultural Logic of Late Capitalism*, Durham, N.C.:
Duke University Press, 1991, p. 44.
10. John Berger, *The Look of Things*, New York: Viking 1974, p. 40.
11. Anthony Giddens, *The Constitution of Society: Outline of the Theory of Structuration*,
Cambridge: Polity Press, 1984.
12. Soja, *Postmodern Geographies.*
13. Susan Hanson and Geraldine Pratt, *Gender, Work, and Space*, London and New York:
Routledge, 1995.
14. Michael Dear, "The Postmodern Challenge: Reconstructing Human Geography,"
Transactions, Institute of British Geographers, new series 13, 1988, pp. 262–74.
15. Derek Gregory, *Geographical Imaginations*, Oxford and New York: Blackwell, 1994.
16. Trevor J. Barnes and James S. Duncan, "Introduction," in T. J. Barnes and J. S.
Duncan, eds, *Writing Worlds: Discourse, Text and Metaphor in the Representation of Landscape*,
New York: Routledge, 1992, p. 7.
17. Demeritt, "The Nature of Metaphors."
18. See Donna Haraway, *Simians, Cyborgs, and Women*, New York: Routledge, 1991.
19. Loren Eisley, *The Unexpected Universe*, New York: Harcourt, Brace and World, 1964;
and Paul Shepard, *The Others: How Animals Made Us Human*, Washington, D.C.: Island
Press, 1995.

Acknowledgements

We are grateful to both the people and institutions who gave us various forms of support in the production of this book. First and foremost, we would like to thank Gerry Pratt, Michael Dear, the *Society & Space* editorial team, and Pion Ltd, publishers of the journal, for the encouragement and opportunity to launch our work on animals, and for permitting us to reprint, in modified form, articles first published in *Society & Space*. Kay Anderson was one of the first to voice her interest and convince us that we should embark on the project, which we greatly appreciate; we are also grateful to the Royal Geographical Society for allowing Kay's article on the Adelaide Zoo, first published in *Transactions of the Institute of British Geographers*, to be reprinted in modified form here. Roger Keil was an early and keen promoter of the *Animal Geographies* project, and his intellectual support was vital. We want to thank Roger for his contributions to Jennifer's chapter on the idea of zoöpolis, which first appeared in an issue of *Capitalism, Nature, Socialism* that he co-edited, and which Guilford Press has generously allowed us to reprint here.

Mike Davis, who himself has long explored the relationships between nature and culture, enthusiastically embraced our endeavor and brought us to Verso, where Mike Sprinker deftly maneuvered the book through to publication and provided a myriad of suggestions along the way that enhanced the final product. Verso, most notably Jane Hindle, was invaluable in smoothing out the inevitable wrinkles in the production process.

Jennifer's University of Southern California colleagues (especially Laura Pulido, Doug Sherman, and David Sloane) and students (Tom Gaines, Andrea Gullo, Unna Lassiter, Lynn Whitley, and Andrew Straw); Southern California Edison colleague Kathleen West; UCLA colleagues Dana Cuff, Susanna Hecht, and Stephanie Pincetl; and Center for Advanced Study in the Behavioral Sciences colleagues (especially Lynn Gale, Christine Williams, and Alison Wylie) all generously embraced this effort and offered insightful and much appreciated advice. Both Bill Shaw of the University of Arizona and Ray Sauvajot of the National Park Service were extremely supportive and helpful at earlier stages. At Clark University, Jody's colleagues Sally Deutsch and Cynthia Enloe, along with students Rob Krueger, Bruce Bratley, Suzi Moser, and Pat Benjamin provided enthusiastic support at critical moments. Lydia Savage of the University of Southern Maine and

Joni Seager and Glen Elder at the University of Vermont lent their encouragement from the beginning and thoughtful critique at the end.

Finally, we would also like to acknowledge the Center for Advanced Study in the Behavioral Sciences and the National Science Foundation, who provided Jennifer with a Center Fellowship and superior administrative support, enabling her to complete work on the book, and the University of Southern California for granting her the leave necessary to take up this fellowship.

Jennifer Wolch, La Honda, California
Jody Emel, Worcester, Massachusetts

Witnessing the Animal Moment

Jody Emel and Jennifer Wolch

Why Animals in Social Theory? And Why Now?

In a 1928 article entitled "The Culture of Canines," sociologist Read Bain made the case for a serious "animal sociology."[1] He suggested that, along with other nonhumans, dogs possessed a distinct culture that was a result of dog-dog as well as dog-human interaction and socialization processes. Canine behaviors such as responding to a whistle or giving a paw when greeting a human were "not unlearned instinctive responses, nor are they individual habits, but they are common to practically all civilized dogs in America . . . resulting from the acquisition of culture traits."[2] He described visiting white friends in Texas whose terrier, although friendly to white children, barked and snarled at African American children passing their home. Observing the approbation (stroking, patting) with which the dog's white mistress responded to this aggressive behavior, he raised the possibility of "sectional" canine culture, and noted:

> I suppose the dog was no more oblivious to the import of these tonal and motor gestures than I was . . . I wondered if this might be a case of canine "race prejudice." Upon inquiry, I discovered several people who had observed similar white [sic] canine responses to Negroes. If this is true, it would seem to be a clear case of canine race prejudice, a culture trait acquired by all dogs socially responsive to that particular culture trait of their white masters.[3]

Chiding his fellow social theorists for failing to consider animals, Bain suggested that "the persistent attempt to set human phenomena distinctly and widely apart from all other natural phenomena is a hang-over of theological teleology, an instance of organic ego-centrism, a type of wishful aggrandizement and self-glorification" that belonged "in the realm of valuation, not in the realm of science."[4] He predicted that "the denial of culture of subhuman [sic] animals is probably a phase of anthropocentrism."[5]

At the end of the twentieth century, Bain's prediction has come to pass.

The multiple and nefarious linkages between human and nonhuman animals have become provocative and of growing, serious concern to American and European social theorists. Why are animal-human relationships suddenly so topical and central to social theory? What political and intellectual purposes are served by studies of the "animal question"?

In this introductory chapter, we provide some initial answers to these fundamental questions. Theoretical debates and positions are deeply engrained in the environmental, material, and political circumstances of time-place. We therefore begin, in the chapter's first part, by considering economic contexts surrounding the rise of the animal question in social theory. Economic globalization, industrialization, and environmental destruction on a world scale have stimulated a politics of resistance based on vast animal suffering, loss of wilderness, and fears about the "end of nature" which we detail in part two. This politics of resistance has multiple sites of contestation, and the movement's branches address problems of animals both as species and as individual beings. It has both inspired and been influenced by specific theoretical challenges to modernist epistemologies. In the final section, we argue that the resulting shifts in social thought have had the effect of suddenly bringing animals into sharper theoretical focus. Feminism and postmodernism, in particular, have undermined the beliefs that defined modernism, rendering the boundaries between humans and animals erected by intellectuals much more fluid and contestable. Freed from these theoretical tethers, we are now led, inevitably, to the animal question.

1. The Animal Economy: Environmental and Ethical Dilemmas

Over the past two decades, the animal economy has become simultaneously both more intensive and more extensive. More profits are squeezed out of each animal life, more quickly, while the reach of animal-based industries has grown to include most of the developing world. In the sections that follow, we consider the animal economy, its environmental impacts, and the ethical dilemmas it raises.

A major part of this economy involves traditional uses of animals, as clothing and especially food. Due to the globalization of the animal food economy, animal-linked food production is growing rapidly as western meat consumption norms spread worldwide. The rise of factory farming, which creates pervasive environmental problems and poses profound moral choices, is one result. Meat-driven agricultural practices with devastating implications for environmental quality and habitat loss are another. In addition, however, more general models for economic development have put enormous pressure on old and new lands, eliminating or degrading spaces critical to wildlife populations and species and calling the models' logic into question. And the wild animal trade, some of it involving smuggling and poaching, is big business that kills hundreds of thousands of animals each year, bringing some species to the brink of extinction.

Animals are also central to biomedical research, and its spin-off, biotechnology. Through biotechnology, animals in some transgenically altered form become living commodities, the new products of biomedical research and bioengineering enterprises stimulated by heavy investments from private and public entities convinced of their profitable future. More traditional biomedical and product research, as well as firms producing animal-based "biologicals," continue to use huge numbers of animals for the development and production of medicines, vaccines, and consumer products. The ethical implications of animal-based biomedical labs, product research, and biotechnology have loomed increasingly large, especially as the potential consequences of reconfiguring animal bodies shaped over millions of millennia become apparent.

1.1. Globalization and the world diet
Globalization has augmented dramatically the circulation of animal bodies (whole and in parts) as western food norms and development strategies together create a "world diet" predicated on grain-fed animal proteins for the rich and starvation and food insecurity for the poor. The average meat-eating person in the US consumes 112 kilograms/year in beef, pork, mutton, lamb, and poultry; this consumer also directly or indirectly consumes fish as well as crustaceans and mollusks, silkworms, horses, goats, turkey, pigs, geese and/or ducks, mice, rabbits, and rats. The world average meat consumption is 32 kilograms/year, with consumption steadily rising. Global meat production has quadrupled since 1950 (population has doubled), and worldwide cheese consumption has doubled since 1970.[6]

Demand for meat has stimulated a profound shift in grain production, as animal feed grains now account for almost 40 percent of all grain production, and many poor nations which used to be grain exporters are now net importers. Demand for meat among urban affluent consumers has skyrocketed; by 1981 the Food and Agriculture Organization estimated that 75 percent of Third World grain imports went to feed animals.[7] This process, which Mexican agronomist David Barkin terms *ganaderización* or livestockization, has threatened food security throughout the Third World.[8]

Some of these animal products are produced and consumed locally, but many others arrive from far-flung spots spanning the globe, for example, the "world steer" or "global steer."[9] Worldwide in 1990, 64 million tons of pork, 48 million tons of beef, and 34 million tons of poultry were "produced"—the top producers being the US, China, the European Community, and the countries of the former USSR. The aggregate size of the globalized animal economy is enormous: the world trade in cows alone (their flesh, skin, organs, and hooves) employs 200 million people and involves approximately 1.3 billion cattle, who take up almost a quarter of the earth's landmass. In the US alone, the cow trade is a $36 billion industry.[10]

Factory farms have multiplied to meet the expanding demand for meat in an increasingly competitive internationalizing livestock market.

Agroindustrial production of animals is energy intensive: one kilogram of US pork not only requires 6.9 kilograms of grain and several thousand liters of water, but 30,000 kilocalories—the equivalent of about 4 liters of gasoline.[11] Factory farming also has created its own set of environmental damages, such as heightened greenhouse gas levels (with farm animals accounting for 15–20 percent of all global methane emissions) and ground-water pollution (especially nitrates). Some European countries are so bogged down with manure that they have been deemed "manure surplus" nations by the European Community.

The environmental and potential public health disasters associated with factory farming have become big news. During the 1996 US presidential campaign, for example, contenders for the Republican presidential nomination faced a barrage of angry constituents in pork-belt states where huge hog factories are decimating traditional smaller scale operations. Furious neighbors have been left behind to face the air and water pollution from the vast open cesspools of animal waste mixed with drugs and food additives flowing from factory farm operations. Factory farms have also undermined the popular image of the farm homestead; as news of factory farming techniques has spread, such farms have been transformed in the popular consciousness into animal concentration camps. In this regard, the 1995 British mad cow crisis served not only to awaken the public about the health threats associated with factory farms, but to project images of bioengineered cannibalism—feeding herbivorous animals their rendered, sanitized, and granulated brethren down on the factory farm—into living rooms around the world.

1.2. Economic development and habitat loss

Wildlife loss and extinction resulting from industrialization and models of "development" predicated on massive exploitation of land and natural resources have been another context for public alarm, debate, and action over animals. Biodiversity loss is expected to escalate as countries join a "race to the bottom" in environmental protection in an effort to secure the economic gains of trade. Land-use intensification and frontier expansion, development strategies employed throughout the world, have generally entailed the subdual or removal of existing peoples and the elimination or control of animal populations. In European Russia and Siberia, the US, Australia, South Africa, India, Brazil, China, and other parts of the world, the outcome has been the same. "Explorers" and prospectors led the way for commodity extractors; settlers plowed grasslands or semi-arid lands, drained wetlands, built dams, and cleared forests. The forests of Brazil and Indonesia, as well as the northern lands of the Arctic Circle, are perhaps the last land frontiers. The major ecosystems in the Indo-Malayan realm are estimated to have lost 70 percent of their original vegetation and 30 percent of the region's coral reefs are considered degraded.[12] The US has lost over 50 percent of its coastal and freshwater wetlands and many parts of Europe have lost nearly all wetlands. Chad, Cameroon, Niger, Bangladesh, India,

Thailand, and Vietnam have lost more than 80 percent of their freshwater wetlands.[13] Overall, nearly as much land was converted from natural habitat to agriculture from 1960 to 1980 as had been converted prior to 1960 in sub-Saharan Africa, Latin America, South Asia, and South-East Asia.[14]

Historically, in some areas like the southern High Plains of the United States or most of the British Isles and the Mediterranean, nearly all of the landscape was altered following human colonization. Accompanying such development was an ecological cleansing in which most large mammals (except for deer, elk, and other select ungulates) were eliminated by hunting, poisoning, trapping, and other means of making the land secure for farming, ranching, and urban settlement. The litany of "last killed" animals in these European-settled lands is extensive and provocative (for example, the last Arizona grizzly was killed in 1939; the last wolf was killed in Great Britain by 1509; the last quagga (lesser-striped zebra) died in the Amsterdam zoo in 1883; the last Carolina parakeet died in the Cincinnati zoo in 1914). Rates of animal loss have accelerated rapidly. Amazon deforestation and the coincident projections of species elimination, for example, soared in the 1980s; in 1986, Simberloff estimated that 66 percent of plant species and 69 percent of Amazon birds would be lost by the year 2000, given current estimates of deforestation.[15]

Hunting for ivory, fur, hides, and other animal commodities also puts tremendous pressure on animal populations, and at various times during the nineteenth and twentieth centuries has caused the near extinction of numerous species from North American beaver to South African elephants. Fishing and whaling, in conjunction with pollution, have destroyed several fisheries and caused a reduction in the numbers of many other species due to current net technology. Prior to the passage of the US Marine Mammal Protection Act of 1972, the US tuna fishing industry was killing over 360,000 dolphins annually (85 percent of the total) in the eastern Pacific.[16] Whale populations were decimated by commercial whaling. Only the minke whale, the smallest species, survives in commercial numbers today. Blue whale populations have never recovered, even after twenty-five years of protection. At the height of the "taking" of blue whales, 30,000 animals were taken in one season alone.[17]

Introduction of industrial chemicals has augmented the potential and actual spatial scales of destruction. Acid rain, which poisons water and kills forests on a regional scale, alteration of atmospheric chemistry, and ozone depletion increase the threat to wildlife. The very bases of the world's food chains—for example, phytoplankton and zooplankton—may be affected. Just which biome boundaries will remain where they are currently is highly uncertain, but the outlook for wildlife in the face of renewed and continued human pressure upon the land from such shifts is onerous. Some estimates place the number of extinctions at 20 percent of all species by the year 2000.[18]

1.3. The wild animal trade

Hunting, trading, and raising wild animals for circuses, laboratories, pets, trophies, sport, and other uses is big business. The international trade in live wild animals and body parts alone is estimated at from $7–8 billion a year. Most of the business is legal, but about one quarter to one third depends upon poaching and smuggling, usually across borders.[19]

The number of animals (including fish) involved in the trade is estimated at nearly half a billion annually. Some 50,000 primates a year are on the market, as well as 6–7 million live birds and about 350 million tropical fish. Furs, leather, and ivory are also traded in huge quantities. Tusk ivory from an estimated 70,000 African elephants is on the market annually, plus some 10 million reptile skins and 15 million pelts.[20]

The effects of the trade include placing some animal species in danger of extinction and causing hardship to those that are used as pets, circus animals, and laboratory experiments. African elephants, horned oryxs, Kemp's Ridley sea turtles, and northern bald ibis are some of the most threatened and endangered species. Perhaps the most endangered species are the black rhino, which has been reduced by 95 percent since 1970, and the northern white rhino, of which there are only twenty left. Asian and African rhino horn sells at $1,000 an ounce or more for medicines and aphrodisiacs. Traffic in the pelts of endangered feline species such as jaguars, ocelots, and pumas is contributing to the gradual disappearance of these animals from many countries, notably Mexico.

Animals do not fare well in the trade circuits. An estimated 50 to 80 percent of the live animals die en route to their destinations. Animals maintained in captivity may thrive in some cases, but in others do not. Elephants in the wild may live more than seventy years; in circuses, their average life is reduced to fourteen years because of stress, traveling in circus boxcars, and being stabled in unsuitable quarters.[21] Between 4,000 and 5,000 chimpanzees are incarcerated around the world. For every infant that survives a year at the final destination, ten die in transit or on arrival, or are killed in the wild by poachers.[22] Many animals do not reproduce in captivity, a problem besieging many captive breeding programs.

The major buyers of live animals are in the US, Japan, and Europe. The US is the largest importer of live primates; other big primate importers include Canada, Japan, France, and the Netherlands.[23] The Netherlands is the largest importer of live birds, and Japanese dealers purchase the biggest supplies of reptile skins. Main suppliers are located in Latin America, Africa, Asia, and the Middle East. Indonesia is the largest exporter of live primates; Senegal is the largest exporter of live birds; and China is the largest exporter of reptile skins. The UK and the US are also primary exporters of reptile skins. The United Arab Emirates, Taiwan, Paraguay, Bolivia, Yemen, Laos, Myanmar, Vietnam, Hong Kong, Taiwan, and China are important middlemen for illegal wildlife trade, as was Amsterdam until the Netherlands joined the Convention on the International Trade in Endangered Species of Flora and Fauna (CITES) in 1984. Hunters and cartels in French Guiana, Bolivia, and Paraguay are also movers of illegal birds and other animals.[24]

CITES has, since 1973, done a great deal to mitigate the threats to animals in trade. It has been signed by over ninety countries. Nevertheless, many animals are not covered by the convention, and even countries that have signed are unable to effectively police the illegal trade. Poaching is a problem throughout the world. In US parks, poaching is at an all-time high. An estimated three thousand American black bears are taken illegally every year, primarily for bear paws and bear gall bladders (one gall bladder may fetch as much as $64,000 for Asian medicinal purposes). Current prices for ready-to-mount bighorn sheep can go as high as $10,000; grizzlies can go for $25,000. Snake poaching is a multimillion-dollar industry in the US alone, and, as we write, reptile-skin shoes, belts, and purses are high-fashion again in the spring of 1996.[25]

Recreational hunting also contributes to the animal economy. Hunting-based tourism continues to earn profits, and not just in "traditional" countries in Africa and Asia where colonial sport hunting gained its highest acclaim. "Canned hunts" as they are called, can now be found in many states of the US. In Texas alone there are some five hundred ranches where exotic animals are bred for killing. With either bows or high-powered rifles, "hunters" from America and elsewhere can obtain African game hunt trophies without having to endure the hardships of going to a Third World country. A perhaps less exotic example of this type of hunting, again in the US, is the boom in prairie dog safaris. The skill it takes to down them is the attraction. These "boutique hunts" are not inexpensive, however. Just like the exotic animal hunting, which can cost up to $35,000 or more per animal, prairie dog hunting requires rifles that can easily cost $5,000. And one need not go to the prairies to be outfitted: Dog-town Varmint Supplies of Newport Beach, California, offers everything from hunting clothes and high-powered rifle scopes to the services of gun technicians who will answer questions by telephone.

1.4. Biotechnology and the transgenic animal kingdom
Biomedical research laboratories, product testing companies, and drug manufacturers have long used animals for experimentation, testing, and production purposes. Exact numbers of animals who die as a result of these uses are difficult to determine. Estimates of world use of laboratory animals in the 1970s ranged between 100 and 225 million; in the US, the figure during the mid 1980s is estimated by Rowan to have been about 71 million.[26]

Despite an increasing emphasis on replacement, refinement, and reduction of animal use in traditional biomedical, product testing, and pharmaceutical production labs—through in vitro methods, for example—biotechnology is a fresh arena for animal experimentation and production. Indeed, biotechnology has become a major growth area in advanced industrial economies and the hot new promise of international development planners. The rise of biotechnology has upped the ante on animals to support healthy living and beautiful human bodies, and, generally, to produce more for less. Given its potential to become a bigger earner than traditional sectors such as the chemical industry, virtually every developed country and many developing countries have identified leadership in biotechnology as a national goal.

Animal biotechnology companies were expected to reach annual revenues of $150 million from US sales and close to $500 million worldwide in 1996.[27] Current developments, such as transgenic animals, including the patented mice bred for specific predispositions to cancer, add new twists to the old debates about animal welfare versus human health. In agriculture, genetically altered rhizobia (designed to enhance nitrogen fixation) have already been added to millions of acres of farm land. Farm animals are targets for bovine growth hormones and human gene transplants that promote faster maturation and more body weight with less feed. Transgenic chickens and pigs are expected to inundate the market by 2015, earning billions of dollars in the US market alone.[28]

Arguing that bioengineering improves on rather than mimicks nature, its proponents claim it will remedy the failures and inefficiencies of industrial agriculture.[29] New plants and animals will be created that are more resistant to the old pests, diseases, and stresses. These new forms of life, created by transplanting genetic material, are and will be "owned" by their engineers and corporate funders—a development that has already caused a furor among people in a number of countries. In addition, the ethics of "creating" new animals is under serious scrutiny: should humans be so quick to tinker with the results of the evolutionary process, and to what end? And, looming in the background, how long until the androids arrive?

2. Animal Politics: Sites and Social Movements

The threats of massive environmental degradation and species extinction, and the commodification of billions of animals as the economy goes global, have led to a turbulent politics surrounding animals. Animal-related issues have increasingly found their way onto the public agenda, and as a result the state now plays a major role in protecting animals from suffering, minimizing species loss, and balancing economic and environmental objectives. Yet the globalization of the economy and changes in the international division of labor since the 1970s have produced or coincided with a substantial reduction in the state's control of national affairs in both economic and noneconomic realms.[30]

Free trade agreements like the General Agreement on Trade and Tariffs (GATT) may be used to further weaken the ability of activists to promote animal protection through state governance. The dismantlement of dolphin protection by way of the US Marine Mammal Protection Act of 1972 is a case in point. The act required tuna fleets to adopt dolphin protection programs; embargoes on imports of tuna caught by foreign fleets (from

Mexico, Venezuela, and Vanuatu) were established until those fleets came into compliance. Mexico charged that the US was protecting US fishers rather than dolphins, and a GATT panel found in favor of Mexico. The Clinton administration, with the support of Greenpeace and other activist groups, quietly retreated from the dolphin protection issue. Friends of Animals, the Sea Shepherds, and other groups are opposed to the resultant changes in the act because of the impact this will have on dolphins and the precedent set for erosion of animal protection in a globalized economy.

In general, revelations about the scale of habitat loss and endangerment, and animal death and suffering, along with the unwillingness or inability of the state to stop or effectively regulate the slaughter, have catalyzed a wave of social movements. Although some have long histories (especially wilderness conservation groups), such movements can be characterized overall as "new" social movements, which address broad quality of life issues rather than purely economic concerns (such as wages, worker protection, workplace discrimination, and so on). Like other social movements, the animal movement comprises a broad spectrum of organizations that range from large bureaucratic institutions to small-scale informal collectives, and have varied political orientations and causes: wildlife and wildlife conservation organizations, animal protection societies, animal rights groups, wildlife habitat restoration projects, farm animal protective leagues, wildlands and forest protection campaigns, and animal rescue operations. Despite this enormous diversity, to some extent all animal social-movement organizations contest the harmful human treatment of animals, and the destruction and degradation of their habitats.

In the sections that follow, we consider the contested sites of animal politics. Then, we focus on two domains into which much of the organized activism has tended to fall: wildlife and wilderness protection groups within the larger environmental movement, whose primary aims are habitat and species conservation; and animal protection groups concerned with the protection of individual animals or classes of animals.

2.1. The contested sites of animal activism
The sites of animal politics range from Western to non-Western, local to global. In non-Western localities, debate sometimes revolves around domestic animals, such as cattle in India, whose religious status simultaneously protects them from slaughter and exposes them to the depredations of hunger and disease. More often, political conflict centers on how local subsistence economies can coexist with wild animals protected for their ability to attract rich western ecotourists. Conflicts also arise between resource managers and subsistence dwellers living on or near newly created bioreserves, whose traditional practices are suddenly redefined as illegal poaching, or who are removed from their homelands to make way for bioreserve establishment. And battles erupt between Western conservation organizations and national governments intent upon cashing in on their unfungible wildlife resources, including whales and elephants.[31] In northern

Sulawesi in Indonesia, for example, the World Bank and the World Wildlife Fund (now called Worldwide Fund for Nature) sponsored the establishment of a reserve (destined to become a national park) which entailed the eviction of some seven hundred families from the area, many of whom were indigenous Mongoneow who had been forced into the highlands because of pressure from the resettlement and migration of other Indonesians.

Local conflicts in the US are often over how and where to manage local wildlife and how to prevent species endangerment in the context of urbanization-driven habitat loss and fragmentation. Urban wildlife, especially large predators in a suburban setting, presents a particularly troublesome and delicate problem. Bears, coyotes, deer, moose, and alligators—to name just a few—are expanding their ranges and experiencing human encroachment upon existing ranges. Eliminating or shooting such animals is not the foregone conclusion it once was, given the zoophile spirit that infuses our contemporary culture. The result is often a pitched battle between pro-hunting forces, wildlife management agencies, ordinary residents, and animal rights activists to determine how "problem" wildlife and developments will be handled.

At the regional and national levels too, the US norm is chronic political conflict between environmentalists and animal advocates, land management agencies, pro-growth factions, or industry interests. The battles are over wildlife conservation, land management practices, wildlife management techniques, and the impacts of resource extraction and land development on animal habitat. Some high-profile examples include the sharpshooting of mustangs in the American West and the spotted-owl controversy in the Pacific Northwest. Continual battles occur between animal welfare and rights lobbies on the one hand, and livestock interests, rodeo and circus groups, bioengineering companies, and pharmaceutical interests on the other; these wars tend to be waged in Statehouses and Congress, since legislation governing animal welfare, livestock transport, product and drug testing, and patents is written there.

Animal politics also rage in the international arena, where struggles revolve around efforts to protect endangered species and eliminate smuggling and poaching. For example, the ivory-trade ban, initiated by placing the African elephant on the Appendix I listing at the 1989 CITES meeting, is another extremely controversial issue. Undertaken to protect elephant populations, which were declining during the 1980s, the ban has received considerable criticism (and support) from Westerners and non-Westerners. One of the criticisms is that Western countries, through the World Bank and nongovernmental organizations, were directing land use and wildlife policy against the interests of the people living alongside the elephants. Consequently, more efforts are being made to share tourist dollars with people living around parks, to ensure that people are not moved off their lands and to generally try to find people-friendly and socially equitable ways to protect biodiversity.

2.2. *Protecting wilderness and wildlife species*

One of the most sustained political efforts around animals has been the battle to conserve wildlife and prevent species endangerment. The attempt to conserve and protect wildlife species has a long and complicated history, beginning well before the twentieth century. Bird protection leagues and hunting societies in Europe and the US were among the first conservation groups to concern themselves with the protection of wild animals. The International Union for the Conservation of Nature and the World Wildlife Fund, for example, grew out of British and French interest in maintaining species within their empires. Such groups were prominent in politics during the late nineteenth and early twentieth centuries, but after achieving some successes in establishing reservations and protective legislation declined in membership and public visibility.

In the 1960s and 1970s a second wave of the environmental movement was catalyzed by the chemical insecticide killing of birds, mammals, fishes, and other forms of wildlife. Existing groups—such as the US National Wildlife Federation and the Audubon Society, the Italian League for the Protection of Birds, the French Society for the Protection of Nature, and the Royal Society for the Protection of Birds—were reinvigorated and many new wildlife-oriented organizations—Greenpeace, Friends of the Earth, Mouvement Ecologique, and Robinwood—also sprang up to protect animals and wilderness areas.

These new groups, along with the older conservation societies, are characterized by white, middle- and upper-class memberships and white male leaderships. They tend to run the political gamut from left to right. In general, the organizations are interested in "conservation," "ecology," or a mixture of the two. The ecology groups have been most sympathetic to the ideas of the new left, while the conservation groups tend to split between green, new left, and several other ideological orientations.[32]

These groups differ in their political strategies as well. Practices of the wilderness and wildlife preservation groups range from the more staid letter-writing campaigns, to educational programs and lobbying, to illegal direct actions. One large grassroots organization, Earth First!, includes "ecotage" among its major tactics. Ecotage involves blockades, taking over equipment, sitting in trees scheduled to be cut down, pulling up survey spikes, cutting down billboards, doing damage to logging equipment and bulldozers, and destroying traps. Australian Earth First! members have buried themselves up to their necks and chained themselves to logs in front of bulldozers. Both Greenpeace and the Canadian Sea Shepherd Conservation Society, funded in part by the Fund for Animals and the Royal Society for the Prevention of Cruelty to Animals, have also taken direct action and rammed whaling and factory-fishing boats. These dramatic direct actions have succeeded in bringing international media and government attention to a number of issues, including sealing, illegal whaling, scientific abuse of whales, and dolphin slaughter. They have also resulted in these more militant groups (including Germany's Robinwood) being targeted by governments as dangerous and subjected to a variety of covert and official investigations.

More traditional groups—Friends of the Earth (with offices throughout the world), the US Defenders of Wildlife, and the British-founded World-wide Fund for Nature, for example—have opted for political lobbying and public relations in their campaigns to protect wild animals. The nonmilitant groups have pursued a more litigious route toward ensuring some measure of protection for endangered species and habitats.

Pressures from these environmental organizations and the public led many countries to pass wildlife protection legislation in the 1970s. In the US, for example, Congress passed the Endangered Species Act in 1966, amending and strengthening it in 1973. The act is based on the assumptions that each life-form may prove valuable in ways not yet measurable and that each one is entitled to exist for its own sake. In 1973 the list of threatened and endangered species numbered 109; now the total is over 900 (more than 1400 counting foreign species) and some 3700 officially recognized candidates await review. While this legislation has been a boon to animal protection, critics claim that due to lack of funding the act has never been implemented as it was intended. During the first eighteen years of the act, annual funding for the program averaged $39 million—"about enough to build a mile of urban interstate highway, or about 16 cents per year from every taxpayer."[33]

With the slowdown in growth experienced by many of the industrialized nations in the late 1980s and early 1990s, the environmental impulse was stifled by governments that veered off to the far right. In Britain, Germany, and the US, in particular, wildlife protection legislation and other environmental regulations were scrutinized for the brakes they put on private development. These sentiments echoed those of farmers, developers, and timber industry groups that had for years tried to reverse regulation. Oppositional groups, like the Wise Use Movement in the US with its anti-government and anti-elite philosophy, gained some prominence through their own high-profile campaigns. Anti-environment backlash came from the populist left as well. In response, both local and global conservation groups, such as the Worldwide Fund for Nature, the International Union for the Conservation of Nature, and especially Greenpeace, have tried to become more "grassroots" in their ideologies and programs. Community-based conservation is now the word on the street; however, neo-liberal approaches that rely on the market (which means only animals that can pay their own way can stay) are also in the ascendant.

2.3. In defense of individual animals and their rights
The other major arm of the animal social movement is dedicated to animal protection, specifically the protection of individual animal lives, the reduc-tion of animal suffering, and, in some cases, the "liberation" of captive animals. Unlike wildlife conservation and habitat protection efforts, this branch of animal-oriented activism did not originate as part of the broader

environmental movement (although firm linkages now exist). Rather, animal welfare groups as a part of the animal protection movement emerged along with early abolition, suffrage, and especially social welfare and child protective societies (in fact, the Society for the Prevention of Cruelty to Animals predated, and was used as a model for, child protective associations of the late nineteenth century). In the 1820s, for example, British women led campaigns opposed to vivisection. By the 1870s opposition to vivisection and other forms of animal suffering stimulated the growth of organizations including the British Union for the Abolition of Vivisection, the Victoria Street Society, and the Royal Society for the Protection of Children and Animals. Such organizations expanded rapidly throughout the US, Britain, and other countries of the industrialized West, becoming powerful national bodies with political influence and venerable local institutions attracting charitable donations and volunteers. Later, during the second half of the twentieth century, the protection movement expanded to include animal rights organizations. The efforts of these groups, who sought the liberation of captive animals and extended rights as well as protection for animals, were often modeled on the civil rights campaigns, women's liberation movements, and other progressive political struggles of the 1960s and 1970s.

Animal protection organizations range widely in terms of philosophy. Traditional animal welfare-oriented societies tend to predicate their activism on beliefs in the virtue of human kindness and the evil of suffering (human or animal), while animal rights groups are steeped in ethics supporting the intrinsic value of animal subjects and their rights to equal consideration. Some of these latter groups seek legal standing for animals and oppose many conventional animal practices that involve captivity (including pet keeping and animal-based entertainments). Traditionally, most animal protection groups were oriented toward domestic animals and animals hunted for sport or trapped for fur. Over time, however, animal protection organizations have also become deeply involved in the protection of wild animal lives.

The animal welfare movement is one of the biggest coalitions of activists in the West and has spread to many other areas of the world. In Britain, there are hundreds of pro-animal groups, and the Royal Society for the Prevention of Cruelty to Animals alone has five hundred thousand supporters. In the US, while traditional groups such as the Humane Society, the Anti-Vivisection League, and the Society for the Prevention of Cruelty to Animals (SPCA) are still active, among the new groups that have sprung up are People for the Ethical Treatment of Animals (PETA, with a membership of around three hundred thousand), the Gorilla Foundation, the Humane Farming Association, Farm Animal Reform, Alliance for Animals, Citizens to End Animal Suffering and Exploitation (CEASE), Trans-Species Unlimited, the Digit Fund, and many more. In other places around the world, animal welfare/rights organizations are active, for example, the Philippines' National Society for the Protection of Animals; South Korea's Animal

Figure 1.1 Rescue workers from a farm animal advocacy group (Farm
Sanctuary) try to help a downed cow dying in the slaughterhouse yard

Photograph by Anacleto Rapping, © 1995, *Los Angeles Times*, and reprinted by permission

Protection Society; and Japan's Animal Welfare Society. Similar organizations exist in the Ukraine, the Czech Republic, Mexico, India, Russia, and
many, many more locales.

Today's animal protection activism has expanded its focus from the initial
targets of anti-cruelty campaigns (pets, working animals, and animals killed
in blood sports) mounted by the humane societies of the nineteenth century
and still waged by many traditional groups today. In the US, animal
experimentation and fur-wearing have become important targets of cam-
paigning, while for British groups, factory farming and the trans-shipment
of animals have loomed large. Groups in other countries illustrate a range
of concerns, from trying to solve the problem of whales caught in fishers'
nets (Republic of Korea), to anti-vivisection (Japan and the Czech Republic), to ending dog poisoning in urban areas (Peru and Portugal).

The emergence of concern and activism around so-called food animals
reveals the erosion of lines that historically divided the animal world into
those worth protecting because they were seen as either part of nature
(wildlife) or the human community (pets), and those not worth protecting
because they were neither (farm animals) and constituted sources of profit
and value. The status of commodified domestic animals such as cattle,
sheep, pigs, and chickens, once excluded from spheres of moral concern
and legal protections, is being re-evaluated (see Figure 1.1). And because of
the environmental damage inherent in large-scale factory farming, campaigns around farm animal welfare and farming practices are increasingly

waged by coalitions of greens and wildlife conservation groups on the one hand, and animal welfare/rights organizations on the other.

Like some wildlife conservation organizations, many of the animal protection groups engage in civil disobedience as well as educational campaigns and animal rescue/sanctuary work; some even engage in violence. The Animal Liberation Front, Last Chance for Animals, People for the Ethical Treatment of Animals, Band of Mercy, and True Friends are some of the groups considered "terrorist" by the US Federal Bureau of Investigation and New Scotland Yard.[34] Most pro-animal activists, however, disclaim violence although the scientific community and the media have tended to lump all activists together as extremists.

In the US and Britain, the vast majority of animal protection activists are middle-class women. Feminists have a long history of association with animal welfare and anti-vivisection societies.[35] Recognizing similar sources of oppression for both women and animals, ecofeminists have attempted to walk a thin line between not romanticizing nature or animals and yet refusing a reductionist reason in such considerations. Though generally considered a middle-class affair, animal protection activism has had its share of working-class collaborators. The Brown Dog riot of the nineteenth century in Britain involved trade unionists and others fighting medical students in the streets of Battersea to preserve a statue erected in memory of a terrier that was vivisected. According to Carol Lansbury and Matt Cartmill, the Brown Dog riot symbolized "the power that doctors and other men of wealth and influence exerted over the poor."[36] Recently, Ted Benton and Simon Redfearn argued that the Brightlingsea protests of 1995 in resistance to live exports of animals could be seen as an opportunity for left politics. Attempting to do something, even something so apparently moderate as these campaigns have tried to do, brings protesters up against the power of capital and the full force of the law.[37] With few exceptions, however, the left in the US and in Britain has remained wary and dismissive of animal protection activism.

Animal protection activists have achieved some measure of success. Several countries, including the US and Great Britain, passed stronger regulations for animal care in research and on the farm during the 1980s. The cosmetics companies Revlon and Avon agreed to stop the Draize test in 1989. After PETA's Heartbreak of America campaign in 1991, General Motors agreed to stop the world's last animal test-crash experiments. By some reports, fur sales decreased during the 1980s but have risen again in the 1990s. Scandinavian Airlines Systems and British Airways have refused to ship animals destined for laboratory experiments. Nevertheless, the battles still rage in the face of continued vast animal suffering and death.

3. Critiques of Modernism: Opening Theoretical Space for Animals

Modernity and modern social theory reached their zeniths in the West during the twentieth century. Western modernity as a historical epoch was characterized by rapid developments in science, machine technology, and modes of industrial production that together led to unprecedented living standards, dependence upon inanimate energy sources, the political system of the nation-states, and the rise of massive bureaucracies. Modernism as a set of values for human behavior emphasized rationalism, individualism, and humanism, while modernist norms for social order were predicated upon the possibility of liberal democracy and secular culture. Modernization (or "development") thus entailed following a path of social and technological change that maximized these values and norms. By the 1970s, however, the legacies of modernity and modernist ways of knowing were under severe attack. Critics argued that the achievements of modernity rested on race, class, and gender domination, colonialism and imperialism, anthropocentrism and the destruction of nature. The legitimacy of modernism as a system of thought was simultaneously undermined. Given some of the more horrific of modernism's "achievements"—namely, the twentieth century's "age of camps," to use Zygmunt Bauman's term in reference to Auschwitz and the Gulag—criticism of the modernist project unfolded with exceptional force and passion, if not always with pristine clarity.

Many strands of this critique—feminist, multicultural, and postmodern—have created spaces for reconsidering animals. It is not so much any one particular theorization that has set out to produce these opportunities, but rather discourses about several general themes have converged to make a consideration of animals appropriate if not inevitable. In the sections that follow, we consider some of the most important critiques of modernism with crucial implications for thinking about animals. These critiques have aimed to challenge the hegemony of the material as foundational to social life and, by extension, the assumed irrelevance of everyday cultural forms such as play, advertisements, and fictionalization as carriers of profound social meanings; question the notion of the unitary subject and expose intertwined dualisms, which together have served to silence a multiplicity of different voices, foster human conflict, and engender environmental degradation; and unveil aspects of the modernist vision of social progress as inimical to global safety and security. We treat each of these elements of the critique in turn below.

3.1. Reasserting the relevance of cultural forms
Traditionally, animals have been dismissed as too down home, too trivial, too close to nature for most serious intellectual authorities to consider, even the most avant-garde. In a reflection on animals and children's literature, for example, Ursula LeGuin claimed that

> if you want to clear a room of derrideans, mention Beatrix Potter without sneering. . . . In literature as in real life, women, children, and animals are the obscure matter upon which Civilization erects itself, phallologically. . . . If Man vs. Nature is the name of the game, no wonder the team players kick

out all these non-men who won't learn the rules and run around the cricket pitch squeaking and barking and chirping![38]

Feminism and postmodernism, however, have managed to defy the old lines of reasoning regarding the privileged position of the material in explanations of social life and organization. What can and cannot be discussed seriously is now an open question. Postmodernism especially, with its emphasis on culture, has challenged the somberness of modernism and flung open the doors to deliberation on subjects ranging from gangsta rap and high fashion to world creation and the meaning of life. The feminist emphasis on the significance of the everyday, and the rents in modernity's materialist fabric produced by the "cultural turn," have allowed us to see and examine the rules imposed upon everyone and every life occasion. A denial of universalism lets us hear a choir of voices that sing no single melody, not even a harmonic chord; all sorts of folks now have their say, including those whose "pet" peeves and extended families include animals of myriad kinds.

Once one starts looking around the cultural landscape, the animal is everywhere. Comic strips like the Far Side and Calvin and Hobbes use animals to show us in good-humored fashion how silly, naive, and contradictory we are. Cat mysteries are hot sellers. Even a superficial review of folk traditions illustrates the prominence of animal teachers (and bad guys) in every culture. The Berenstain Bears, Winnie the Pooh, Barney, Paddington Bear, Brer Fox and Brer Rabbit teach children morality, kindness, good manners, and self-respect. That animals are deeply ensconced in children's folk culture is a reality not lost upon toymakers, particularly those who manufacture soft, cuddly animal companions and children's books. Products for adults are equally susceptible to animal-based marketing campaigns, however. Black rhinos and vervet monkeys sell Nissan Pathfinders. Wolves and black panthers sell Jeeps, and an alligator with frogs on its tail walks off a pier with a case of Budweiser. A quick look at the movie industry's production of animal movies for family consumption illustrates a growing trend. Disney's *101 Dalmatians* is the seventh-highest-grossing film of all time. Animals, it would seem, *are* serious business.

The legitimation of animals and human-animal interactions as appropriate subjects for scholarly investigation has led to a variety of feminist and postmodern critiques of the use of animals as cultural forms and the problems they pose for social relations; even the cultural politics of toys, children's culture, and TV marketing for children are now matters for serious study.[39] Critiques of pet keeping and the anthropomorphizing of animals in cartoons, movies, books, and other media are also serious scholarship.[40] Matt Cartmill's analysis of Bambi and the men involved in writing the book (Siegmund Salzmann, a Jewish intellectual born in Budapest) and translating it (Whittaker Chambers, a Communist Party member who later became Nixon's star witness in the Alger Hiss case) is a brilliant example of an "animal story" in which humanized animals reveal the "truth" about human nature.[41] The point is that "we know by now, or ought to know, that what gets us off as entertainment is rarely simple and never innocent."[42] As works like those by Cartmill and Kline reveal, cultural forms including literature, film, and other types of entertainment exert a potent impact on the formation of ideologies of human-animal relations among both children and adults alike.

3.2. Decentering the subject and debunking dualisms

Writers such as Carol Adams and Steven Baker suggest that the decentering of the human subject is a major opportunity to see animals and humans differently. Adams, for instance, has argued that "the tumbling away of a unitary subject opens up space for discussing other-than-human subjects."[43] Baker, in his book *Picturing the Beast: Animals, Identity and Representation*, writes that "the decentering of the human subject opens up a valuable conceptual space for shifting the animal out of the cultural margins. It does so precisely by destabilizing that familiar clutch of entrenched stereotypes which works to maintain the illusion of human identity, centrality and superiority."[44]

The unitary subject rests on foundational notions of humanity, the subject and the citizen. Most postmodern and feminist theorists argue against the fiction of the essential man or subject, identifying this subject fiction as grounded in Kantian anthropology. Taking what could be considered an anti-humanist position, these theorists deny the existence of any such human nature or essence and see humans as an "ever-varying matrix of biological, social, and cultural determinants."[45] Foucault, in particular, denied the transcendent or universal subject, arguing that the subject is configured and reconfigured in the conjunction of discursive and nondiscursive practices.[46] Such interrogations of "man" as "subject" have resulted in a clear picture of the exclusionary and often violent operations by which such a position has been established.[47]

The idea of the unitary subject has been in part buttressed by the dualisms so characteristic and pervasive to modernist thought. The concept of dualism refers to "the construction of a devalued and sharply demarcated sphere of otherness" that stands in opposition to an essentialized and valorized sphere of identification.[48] The operation of dualisms, analyzed most extensively by feminists and Derrideans, has historically relied on exclusions and denials of dependency, instrumentalism or objectification, and homogenization or stereotyping.

As Luc Ferry explains in his splendid exploration of radical ecology, the "animal" is the first being encountered in the process of decentering the unitary subject and thus undermining modernity.[49] Why? Because the animal has stood in opposition to "man," creating the fiction that man's humanity resides in his freedom from instinct and even from history. But through the dualist lens of "constitutive otherness" we can see animals as constitutive of humans, revealing the porosity of the distinction between human freedom and animal necessity.

As the frontier between civility and barbarity, culture and nature increasingly drifts, animals bodies flank the moving line. It is upon animal bodies that the struggles for naming what is human, what lies within the grasp of human agency, what is possible are taking place. Biological science has stood sentinel at the frontier, reading the animal and reporting its findings to psychologists and other normative gatekeepers. But, as Zygmunt Bauman points out, biology occupies a "hotly contested spot, pregnant with profound political and *weltanschauliche* controversies."[50] Donna Haraway's prodigious writings on the inextricably related findings of primate science and ongoing political and social dialogues, or the mapping of human and nonhuman primate social dynamics upon one another through research, have shown the indeterminacy that accompanies efforts to construct humanity from the animal other (and vice versa).[51]

Some question the porosity of the line between human and nonhuman animals, or otherwise argue that the boundary has always been avidly policed. Yet the history of human oppression is replete with examples in which specific categories of humans have been grouped with, and treated like, animals. Marjorie Spiegel, for example, delineates the self-conscious parallels drawn by slaveholders, who modeled the lexicon of slavery and its systems of surveillance and physical restraint on those used with animal captives.[52] The British "simianized" the Irish and the Blacks; the Croatians forbade Serbs, Jews, Gypsies, and dogs to enter restaurants, parks, and means of public transport. In a new assault on the fungible border, bushmen have become the most recent exhibits in the Kagga Kamma Game Park in South Africa. *New York Times* reporter Suzanne Daley wrote that the bushpeople were rescued from the squalor of a shantytown on the edge of the Kalahari Desert: "In a country that has treated them savagely for centuries, being in what feels very much like a zoo may seem like a step up."[53] The tourists spend $7 to view these people; the bushmen receive $1.50 for the viewing. The photograph accompanying the story shows four "bush" youngsters, mostly naked, down on their hands and knees while a white boy stands over them, staring down at them as if they were *animals in a zoo* (see Figure 1.2).

3.3. Exposing modernist myths of social progress

Grand stories of progress, humanity, wealth creation, reason, and the unity of science are, among others, critical targets of postmodern, feminist, and other theorists. Distrust of these narratives is not exclusively postmodern and can be identified throughout history from at least the sixteenth century on. Horkheimer and Adorno's critique of the domination of nature in *Dialectic of Enlightenment*, published in 1947, is one example. Another is the critique of Cartesian mechanism proffered in the seventeenth century by Margaret Cavendish and Anne Finch.[54] Nevertheless, the sort of wholesale rupturing of modernity's bulk that has taken place within the last fifteen or twenty years has enabled a re-evaluation of systems of domination and the animal's place within them.

Figure 1.2   A visitor to Kagga Kamma Game Park in South Africa observes
the newest park residents

Photograph by Greg Marinovich, courtesy of *New York Times*

Generally, animals are part of the stories of progress, rationality, economic growth, and emancipation only by their eradication, sacrifices, bred
domesticity, and genetic transfiguration. One humanitarian thread woven
through the mantle of modernism had, of course, refused the Cartesianism
that denied animals affectivity, sentience, and intelligence. A number of
activist associations sought to reduce the gap that separated humans from
animals, just as many fought scientific sexism and racism. Nevertheless, the
humanitarian orientation was eclipsed by modernity's fear of death and the
ascendancy of science. Death was the ultimate challenge to the modern
ambition to transcend all limits and its refusal justified a hyperseperation of
human and nonhuman animals. In his compelling anatomy of modernism,
Zygmunt Bauman observes that "the angst bred by the inevitability of death
was spread all over the life-process, transformed by the same token into a
sequence of death-preventing actions and lived in a state of constant
vigilance against everything smacking of abnormality."[55]

Modern society throws enormous resources into death avoidance. Animal
bodies are no exception. In fact, it is stunning to consider the extent to
which animals serve to shield humans from death. The American Medical
Association likes to point to the enormous importance of animals in
medicine by observing that fifty-four of the seventy-six Nobel Prizes awarded
in physiology or medicine between 1901 and 1989 were for discoveries and
advances made through the use of experimentation on animals.[56] Neglected

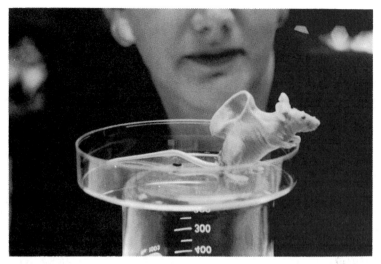

Figure 1.3   A recent experiment in "tissue engineering"
Photograph courtesy of AP/Wide World Photos

in the pro-experiment arguments are the billions of dollars earned by the drug companies (not to mention the medical industry) in selling the vaccines and other chemical substances derived from animal lives.

Biotechnology and genetic engineering promise to continue and heighten the human reliance upon animals for health and safety, with a Frankensteinian twist. In some ways this development in science and technology is a true transcendence of nature. *Time* published a story in November 1995 of a live mouse with a human ear growing out of its back (Figure 1.3), heralding the event as "the latest and most dramatic demonstration of progress in tissue engineering, a new line of research aimed at replacing body parts lost to disease, accident or, as is sometimes the case with a missing ear, a schoolyard fight."[57] Experiments creating chimeras have been done with mice, rats, monkeys, and other animals. Laws generally exist against human cloning, the creation of chimeras between human and animal embryos, and any trade or commerce involving human embryos or embryonic material. Memories of Nazi atrocities have prompted European legislators to be quite stringent on this issue, particularly in Germany.[58] But animal cloning and transgenics are not such a problem. As James O'Connor has so eloquently written, "(h)ere we enter a world in which capital docs not merely appropriate nature, then turn it into commodities that function as elements of constant and variable capital (to use Marxist categories), but rather a world in which capital remakes nature and its products biologically and physically (and politically and ideologically) in its own image."[59]

Some may argue, correctly, that humans have been always dependent upon animals. What is modern, however, is the institutionalization and bureaucratization of this dependency at a scale significantly augmented from "pre-modern" dependencies. The hallmarks of this dependency are globalized commodity chains managed by large-scale industrial institutions and government bureaucracies cutting any link between animals and meat, medicine, famine, or environmental degradation. These chains are supported by powerful narratives about the centrality of food security, health, and corporeal beauty to happiness and progress. By such spatial and functional divisions of labor, and by bureaucratization and institutionalization, the role of individual human agency in producing evil is severely fragmented, eliding individual responsibility.

The modernist insistence on cool rationality and an objectifying attitude also promotes an insensitivity or indifference to suffering that makes factory farms and animal labs possible. As several students of the Holocaust have argued, such modernist codes allowed persons like Eichmann to undertake the infliction of pain on his fellows.[60] Bauman, for instance, maintains that "emotion marks the exit from the state of indifference lived among thing-like others."[61] He suggests that the principal tool of the severance between moral guilt and the acts which entail participation in cruel deeds is adiaphorization: making certain actions, or certain objects of action, morally neutral or irrelevant—exempt from the category of phenomena suitable for moral evaluation. Artfully hidden behind factory-farm gates or research-lab doors, obscured by disembodiment and endless processing, and normalized by institutional routines and procedures, the thoroughly modern instrumental rationality that characterizes contemporary human-animal dependency has rendered animals both spatially and morally invisible.

When we do break through these surfaces, the resulting visibility is often excruciating. But actually seeing and understanding the vast extent of animal suffering and death is unavoidable if we are to transcend the invisibility of animals, engage in corrective struggles, and bear witness to the animal moment.

Acknowledgements

We are extremely grateful to Michael Dear and Joni Seager for comments on earlier versions of this chapter.

Notes

1. Read Bain, "The Culture of Canines," *Sociology and Social Research*, July–August, 1928, pp. 545–56. We thank Jessica Walsh for bringing this article to our attention.

2. Bain, "The Culture of Canines," p. 554.

3. Bain, "The Culture of Canines," p. 555.

4. Bain, "The Culture of Canines," p. 554.

5. Bain, "The Culture of Canines," p. 556.

6. Alan B. Durning and Holy B. Brough, "Taking Stock: Animal Farming and the Environment," Washington, D.C.: Worldwatch Institute Working Paper 103, 1991, p. 11.

7. Durning and Brough, "Taking Stock," pp. 14, 31.

8. David Barkin et al. *Food Crops Vs. Feed Crops: Global Substitution of Grains in Production*, Boulder, Colo.: Lynne Reinner Publishers, 1990.

9. Jeremy Rifkin, *Beyond Beef: The Rise and Fall of the Cattle Culture*, New York: Dutton, 1992.

10. Food and Agriculture Organization of the United Nations, *Production 1989 Yearbook*, vol. 43, Rome: FAOUN, 1990; Paul Ehrlich and Anne Ehrlich, *The Population Explosion*, New York: Simon and Schuster, 1990.

11. Durning and Brough, "Taking Stock," pp. 17–18.

12. Samar Singh, "The Biological Value of the Asia-Pacific Region in Biodiversity Conservation in the Asia and Pacific Region: Constraints and Opportunities," Proceedings of a regional conference 6–8 June 1994, Manila: Asian Development Bank, 1995, pp. 35–48.

13. Mostafa Tolba, *Saving Our Planet: Challenges and Hopes*, London, New York, Tokyo, Melbourne, and Madras: Chapman and Hall, 1992.

14. Robert Repetto and M. Gillis, eds, *Public Policies and the Misuse of Forest Resources*, Cambridge: Cambridge University Press, 1988.

15. Daniel Simberloff, "Are We on the Verge of a Mass Extinction in Tropical Rainforests?," in D. K. Elliott, *Dynamics of Extinction*, New York: Wiley, 1986, pp. 165–80.

16. Douglas H. Constance, Alessandro Bonanno, and William Heffernan, "Global Contested Terrain: The Case of the Tuna-Dolphin Controversy," *Agriculture and Human Values* 12, 1995, p. 23.

17. Piers Blaikie and Sally Jeanrenaud, "Biodiversity and Human Welfare," Discussion paper no 72, United Nations Research Institute for Social Development: Geneva and Norwich, 1996.

18. Robert Peters and Thomas E. Lovejoy, "Terrestrial Fauna," in B.L. Turner, William C. Clark, Robert W. Kates, John F. Richards, Jessica T. Matthews, and William B. Meyer, eds, *The Earth as Transformed by Human Action* Cambridge: Cambridge University Press, 1990, pp. 353–70.

19. Joni Seager, with Clark Reed and Peter Stott, *The New State of the Earth Atlas*, 2nd edn, New York: Simon and Schuster, 1995, p. 125.

20. Joe Kirwin, "Policing the Wildlife Trade," *Our Planet*, no. 6, vol. 4, 1994, pp. 44–5. Based on data from the 1990 Trade Records Analysis of Fauna and Flora in Commerce, developed by the Convention on International Trade in Endangered Species (CITES), Washington, D.C.

21. Ingrid E. Newkirk, "Elephants: Deprived of Dignity and Freedom, Captive Pachyderms Fight Back," *Boston Globe*, 9 December 1994.

22. Gail Vines "Planet of the Free Apes?" *New Scientist* 138 (1876), 1993, p. 39.

23. Seager, *The New State of the Earth Atlas*, p. 81.

24. Seager, *The New State of the Earth Atlas*, p. 81.

25. David Van Biema, "The Killing Fields," *Time Magazine* 1448(8): 36–37, 1994.

26. Andrew Rowan, *Animals and People Sharing the World*, Hanover, New Hampshire, and London: University Press of New England, 1984, pp. 65–71.

27. Robin Mather, *A Garden of Unearthly Delights: Bioengineering and the Future of Food*, New York, London, Melbourne, Toronto, and Auckland: Penguin Press, 1996.

28. Mather, *A Garden of Unearthly Delights*, p. 23.

29. Ann Gibbons. "Biotech's Second Generation", *Science* 256, 1992, pp. 766–8.

30. David Harvey, *The Condition of Postmodernity*, Oxford: Basil Blackwell, 1990.

31. Milton M. R. Freeman and Urs P. Krueter, eds, *Elephants and Whales: Resources for Whom?*, Basel: Gordon and Breach Science Publishers, 1994.

32. Russell Dalton, *The Green Rainbow: Environmental Groups in Western Europe*, New Haven, Conn.: Yale University Press, 1994.

33. Douglas H. Chadwick, "The Endangered Species Act," *National Geographic*, March 1995, p. 9.

34. Jerod M. Loeb, William R. Hendee, Steven J. Smith, and M. Roy Schwartz, "Human vs. Animal Rights," *The Journal of the American Medical Association* 262, 1989, pp. 2716–20.

35. Carol Lansbury, *The Old Brown Dog: Women, Workers, and Vivisection in Edwardian*

England, Madison: University of Wisconsin Press, 1985; Josephine Donovan, "Animal Rights and Feminist Theory," in Greta Gaard, ed., *Ecofeminism: Women, Animals, Nature*, Philadelphia, Pa: Temple University Press, 1993, pp. 167–94.

36. Lansbury, *The Old Brown Dog*; and Matt Cartmill, *View to a Death in the Morning: Hunting and Nature through History*, Cambridge, Mass.: Harvard University Press, 1993. Quote from Cartmill, p. 142.

37. Ted Benton and Simon Redfearn, "The Politics of Animal Rights," *New Left Review* 215, 1996, p. 57.

38. Ursula LeGuin, *Buffalo Gals and Other Animal Presences*, New York: Penguin, 1990, p. 8.

39. Stephen Kline, *Out of the Garden: Toys and Children's Culture in the Age of TV Marketing*, London and New York: Verso, 1993.

40. See, for example, Paul Shepard's *The Others: How Animals Made Us Human*, Washington, D.C.: Island Press, 1995.

41. Cartmill, *View to a Death*, pp. 172–3.

42. Fred Pfeil quoted in Cartmill, *View to a Death*, p. 161.

43. Carol J. Adams, *Neither Man Nor Beast: Feminism and the Defense of Animals*, New York: Continuum, 1995, pp. 12–13; Steve Baker, *Picturing the Beast: Animals, Identity and Representation*, Manchester and New York: Manchester University Press, 1993.

44. Baker, *Picturing the Beast*, p. 26.

45. John Johnston, "Ideology, Representation, Schizophrenia: Toward A Theory of the Postmodern Subject," in Gary Shapiro, ed., *After the Future: Postmodern Times and Places*, Albany, N.Y.: State University of New York Press, 1990, pp. 67–95.

46. Michel Foucault, *The Order of Things: An Archaeology of the Human Sciences*, New York: Vintage Books, 1973.

47. Judith Butler and Joan Scott, *Feminists Theorize the Political*, New York: Routledge, 1992.

48. Val Plumwood, *Feminism and the Mastery of Nature*, London: Routledge, 1993, p. 41.

49. Luc Ferry, *The New Ecological Order*, Chicago, Ill. and London: The University of Chicago Press, 1995, p. xxix.

50. Zygmunt Bauman, *Life in Fragments: Essays in Postmodern Morality*, Oxford and Cambridge, Mass.: Blackwell Press, 1995, p. 168.

51. Donna J. Haraway, *Simians, Cyborgs, and Women: The Reinvention of Nature*, New York: Routledge, 1991; Haraway, *Primate Visions: Gender, Race, and Nature in the World of Modern Science*, New York and London: Routledge, 1989.

52. Marjorie Spiegel, *The Dreaded Comparison: Human and Animal Slavery*, London and New York: New Society Publishers, 1988.

53. Suzanne Daley, "Endangered Bushmen Find Refuge in a Game Park," *New York Times*, International edn, 18 January 1996.

54. Donovan, "Animal Rights and Feminist Theory," p. 178.

55. Bauman, *Life in Fragments*, p. 183.

56. American Medical Association, "Animal Experimentation Benefits Human Health," in Janelle Rohr, ed., *Animal Rights: Opposing Viewpoints*, San Diego, Calif.: Greenhaven Press, 1989, pp. 59–67.

57. Anastasia Toufexis, "An Eary Tale," *Time Magazine*, 6 November 1995, p. 60.

58. As a result, BASF opened its biotechnology headquarters in Worcester, Massachusetts; Baker similarly opened a lab in Berkeley, California.

59. James O'Connor, "Is Sustainable Capitalism Possible?," in Martin O'Connor, ed., *Is Capitalism Sustainable: Political Economy and the Politics of Ecology*, New York and London: Guilford Press, 1994, p. 185.

60. Arne Johan Vetlesen, *Perception, Empathy and Judgement: An Inquiry into the Preconditions of Moral Performance*, State College, Pa.: Pennsylvania State University Press, 1984; Hannah Arendt, *The Origins of Totalitarianism*, London: André Deutsch, 1985; Bauman, *Life in Fragments*.

61. Bauman, *Life in Fragments*, p. 62.

PART I
ANIMAL SUBJECTS/HUMAN IDENTITIES

2

Animals, Science, and Spectacle in the City

Kay Anderson

Introduction

Like other spectacles of the Victorian age, the metropolitan zoo was an exhibitionary complex that gave expression to the Baconian idea that things, as well as words, could be instruments of knowledge and instruction. Through the medium of nontheatrical exhibitions—including shows, panoramas, zoos, diversions, and resorts—abstract knowledge of the printed word became concrete, the vicarious became immediate in everyday experience, the global became local. "Just by looking," stated a 1909 news report of the Adelaide Zoo in South Australia, "visitors get an object lesson in natural history, animal behaviour and world geography. An inspection of the denizens of the rows of cages and houses will give the enquiring visitor a clearer understanding of the lower orders of life than would be gained by hours of reading."[1] At the zoo, the eighteenth-century scientific ideal of observation via the artifice of distance became popularized in a particular form of visualization and sensory appeal. The (animal) objects of a more universal (zoological) knowledge were brought into proximity with, yet separateness from, human subjects. Thus the elaborate scene of the metropolitan zoo was constructed, a space in which an illusion of Nature was created and re-presented to human audiences in a cultural achievement that is the subject matter of this chapter.

Cultures of Natural History: The Case of the Metropolitan Zoo

Nature, for all its apparent remoteness and distance from humans, is in some senses at least socially constructed. In this paper I develop this argument with reference to the zoo, a space which I will examine not from a zoological or natural history vantage point, but more critically for the human perceptions and purposes that are encoded within it. The zoo is a cultural institution, I will argue, following the preliminary work of Mullan and Marvin, and Tuan,[2] one which does not reflect nature itself—as if such an unmediated thing exists—but a human adaptation of the ensemble of life forms that bears the name "nature." In terms of its changing animal composition and visual technologies, its exhibition philosophy and social function, the zoo inscribes various human strategies for domesticating, mythologizing, and aestheticizing the animal universe.

Zoos contain a highly selective array of the species of the natural world, the majority of which are never seen by people beyond the walls of zoos. Moreover, they are displayed in ways that cater to cultural demands and public expectations about animals and the world regions that exhibits are made to represent. After all, most zoos in the Western world are businesses seeking to attract fee-paying visitors—and many do. In the United States in 1993, the 154 zoos and aquariums accredited by the American Association of Zoological Parks and Aquariums were visited by a hundred million people, exceeding the combined attendance at all major-league baseball, football, and basketball games.[3]

Some zoos are more successful than others, however, at attracting visitors. And human responses to zoos are themselves wide-ranging and profoundly ambiguous. Surveys tell us that people's reactions to zoo animals typically combine excitement, fear, awe, sadness, and nostalgia, with unease about the captivity of animals.[4] These varied and contradictory responses are abundantly evident on the faces of children and adults lining enclosures at the zoo. Whatever the nature of human responses to zoos, however—and we need to learn more about them[5]—I will argue here that zoos ultimately tell us stories about boundary-making activities on the part of humans. In the most general terms, Western metropolitan zoos are spaces where humans engage in cultural self-definition against a variably constructed and opposed nature. With animals as the medium, they inscribe a cultural sense of distance from that loosely defined realm that has come to be called "nature."

The social category of nature has recently attracted critical, deconstructive study by geographers and other social scientists.[6] No longer can we speak of separate spheres of culture and nature, but rather of "hybrid spaces" of the cultural and the natural where are conjoined knowledges, products, images, and experiences of both artificial and natural derivation. Zoos—like other ensembles of natural history knowledge such as museums, gardens, parks, nurseries, and wild animal reserves[7]—sit precisely at this interface, creating, fixing, and reshaping objects of public interest and enquiry (Figure 2.1).

Cartesian Legacies: Constructing a Racialized and Gendered "Human" Subject

Historically, the nature/culture opposition has informed diverse and culturally variable practices of domination and subordination on the part of humans. The cultural sense of separation has implied no neutral relation between humans and the nonhuman world but rather entailed detailed and persistent disciplinary practices.[8] This is abundantly evident in the treatment of animals in Western cultural traditions where animals have long served human purposes.[9] Well before the erection of places called zoos, humans

Figure 2.1 The nature/
culture interface: giraffes
at Taronga Zoo in Sydney

Photograph by Peter Clayton,
courtesy of Earthfocus

kept animals in captivity for diverse purposes, chiefly as creatures of worship. By the third century BC, Romans introduced violent uses of animals in gladiatorial contests and triumphal processions. Again, during the Middle Ages, royal sports included bear-baiting and bullfighting.[10]

By the late seventeenth century, private collections of caged animals called menageries had become status symbols for their princely owners. One such collection was the Versailles menagerie, opened in 1665 when France's Louis XIV arranged a botanical garden and an enclosure for lions and elephants around his house in a pattern that is said to have inspired Bentham's Panoptican prison of the late eighteenth century.[11] In 1804, in keeping with the democratic order of post-revolutionary France, the royal animals were moved to Paris and a zoological garden at the Jardin des Plantes opened to the public.

The opening of the Paris collection coincided with the rise of scientific formulations about nonhuman nature that supported practices of animal confinement. Following the publication in 1735 of Linnaeus's influential classification of plants and animals, there developed a tradition of research and writing about animal physiology and comparative anatomy.[12] The scientific community lent legitimacy to the menagerie concepts which quickly spread throughout Europe, including England, where in 1826 the scientists Sir Joseph Banks and Sir Humphry Davy formed the research and display organization, the Zoological Society of London. By 1847 the term "zoo" had become the colloquial abbreviation for the Zoological Gardens of London, made up of animals from the royal menageries at Windsor Park and the Tower of London.[13]

The Western world's zoos evolved historically out of a much older and more general logic and desire for classification and control of the non-human world. This frame of mind, described as "rationalist," enshrined the premises and dualities of classical thought that from ancient times had progressively carved a deep philosophical rift between the spheres of the human mind and nature's apparent "mindlessness."[14] "Humans," with their capacity for thought (calculation), agency, and intentionality, had come to be rigidly set apart from nonhuman "nature," which was defined, in opposition to "human," as "irrational." The history of ideas is complex, and some classical scholars (notably Aristotle) modeled the human/nature distinction as a continuum: a hierarchy of orders of life with humans as the pinnacle and inanimate things at the base. And humans themselves were differentiated by both Plato and Aristotle into higher and lower groups, using gendered and racialized distinctions and including slaves as the most lowly.[15] But, in the course of the development of Christianity and humanism from the fifteenth century, the original rationalist dualism of human/nature began to congeal. The idea of all humanity as possessing a common nature or potential that set people apart from other orders of life was secured. Moreover, the reason/nature dualism progressively came to imply a hierarchy that pitted nature both against and *beneath* the human, who was henceforth justified in treating nature as object, as background to—and instrument of—human purposes.

What is interesting, in terms of this chapter, is the imaginative act that assimilated those thinking, sentient, intentional, and animate creatures called "animals" into the black box category of "nature." From the seventeenth century, when Descartes overlaid the ancient dualism of reason/nature with the distinction of mind and body—one which privileged the former (as the embodiment of reason and intellect) over the latter (the locus of sensation and passion)—the conceptual ground was further cleared to differentiate "human" nature from "animal" nature. While both humans and animals were believed to be capable of physical sensation, Descartes deduced that since animals lack reasoning capacity, their sensations are merely "bodily" (physical/mechanical), of which they can't be "aware" or "conscious." Henceforth, the conceptual boundaries between "animal" and "human" were increasingly chauvinistically drawn within the larger Cartesian framework of Western dualistic thought. In a boundary-making exercise of "hyperseparation,"[16] animals were not only opposed to humans, they were consigned to the already inferiorized and homogenized sphere of "dead" (unconscious) nature—that residual realm inhabited by such diverse things as plants, soils, stones, the elements, and the land.

The collapse of "animal" into "nature" repays further scrutiny in the light of feminist and post-colonial critiques of Western science and philosophy. Some feminist writers have persuasively correlated Cartesian thought with the interconnected projects of environmental, species, racial, and gender oppression, the interrogation of which clears a way to problematizing not just the category of "nature" and "animal" but, more radically, of "human." According to Plumwood,[17] these projects of oppression privileged "not a masculine identity pure and simple" but a "complex dominator identity"—a "master subject"—formed in contexts of "multiple oppressions." Plumwood argues that a master subjectivity was built into Western science and philosophy out of a web of exclusions based on dualistic thinking and a privileging of rationality that, over time, came to colonize the very conception of the human self in Western culture. The normative "human" identity became constructed out of the capacity for reason (instead, for example, of emotion, imagination, sensation, and attributes we share with other sentient creatures). In time, with the rise of weapons of mass destruction and other technology, as well as scientific enterprises, teleological conceptions of the rational human afforded it the justification to order and control other spheres of life. These included the feminine (equated with the body, the irrational, and nature), the racialized slave, the animal, and the environment in general. In contexts of power-differentiated relations, the rational (male) subject's perspective began to be set up as universal, as the generic "human" gaze around which all else turned. Indeed it set *itself* up as neutral, objective, panoramic, and all-knowing—as history's master subjectivity— when in reality it was a "partial perspective" that relied on various strategies of denial, exclusion, spatial separation, and stereotyping of women, racialized peoples, nonhuman animals, and "nature" more generally.[18]

Zoos as Mapping Projects

The arguments concerning the master subject's gaze need to be carefully specified for particular settings and for distinctive manifestations (as is attempted in the case study that follows). In the example of the zoo, it seems reasonable to argue that it is a most transparent institutional embodiment of the hierarchizing, rationalist oppositions of reason/nature, mind/body, human/animal. Colonial settings such as Australia were one of the sites through which the confidence and privilege of partial perspective was encoded and "naturalised."[19] At the zoo, the raw material of nature is crafted into an iconic representation of human capacity for order and control. Indeed, the images constructed are ones that dramatize, even glorify, this capacity for intervention in nonhuman nature (whether in the menagerie-style caging of the nineteenth century or the naturalistic spectacles of the 1990s). For zoos not only enshrine the (arbitrary) boundaries of humanity and animality, they impose their *own* boundaries *between* creatures defined as "animal": different enclosures, separate paddocks that segregate not only keeper and kept, but also (nonhuman) animal from (nonhuman) animal, birds from reptiles (Figure 2.2). Indeed, if, as Haraway (1988, 595) argues, "boundaries are drawn by mapping practices, objects

Figure 2.2 An 1898 plan of Adelaide's Zoological Gardens

don't pre-exist as such. Objects are boundary projects,"[20] then the zoo would seem to be the mapping project par excellence.

As the earlier comments about feminist deconstructions of master subjectivity suggest, the "scopic regime" that is enshrined in the nineteenth- and twentieth-century zoo is no neutral perspective.[21] In her analysis of the American Museum of Natural History, Donna Haraway argues that the experience constructed for the early-twentieth-century viewer by that museum's "visual technology" produced a particular form of "human" in relation to nature.[22] This is a historically specific type of (white) masculine that is unseen, that is *not* the spectacle but rather the privileged eye (I), the bearer of reason, the author, the knower.

At a more popular exhibitionary complex, that of the nineteenth- and twentieth-century zoo, the public was provided en masse with a forum to know rather than be known.[23] They could become subjects of knowledge (with animals as objects) and experience indirectly, off-exhibit, what it was and is to be "human," to be "self" as opposed to something "other." As this chapter seeks to demonstrate, the visual experience of the zoo has been complexly contrived. Through it, the public has been initiated into a way of seeing—that in manifestly variable ways over time—parades humans' capacity for order and command. These are the rationalist principles of a "Cartesian perspectivalism" out of which a master subjectivity was constructed at particular sites in Western societies.[24]

That this "way of seeing" enshrined distinctive Western "ways of being" in relation to nature is surely also the case. Indeed, it is interesting to reflect upon the long history of the domestication of plants and animals in European (and other) societies that had seen landscapes cleared for agricultural complexes and other human uses since Neolithic times.[25] In the context of this chapter, the theme can be usefully pressed further. For just as landscapes and animals were progressively shaped and molded with the elaboration of European colonialism and capitalism, so too were women's roles and statuses adapted ("domesticated") to the requirements of white, male-centered projects.[26] If there was a place for nature in a Western rationalist universe, so too was there a place for women, not only in the (literally) domestic sphere but more generally as instruments and servicers of (higher) male purposes. Domestication, conceived in the expanded sense, of the taming and converting of that which is "wild" for pragmatic human ends, can thus be understood as a form of conceptual and instrumental power alongside more extensively studied forms of disciplinary and coercive power. This chapter is only suggestive in opening up the vast and as yet largely unexamined topic of the historical interconnections between the domestication of animals, women, and racialized peoples.[27] But it offers glimpses—from a site where nature was converted into a supremely domesticated object—of a mode of control that surely filtered to other inferiorized categories, including those defined around sexual and racial difference.

If zoos grew out of and carried forward a rationalist perspectivalism and practice in relation to the animal universe, they also bear witness to contests and acts of resistance on the part of both women *and* men. The hints of dissent to the practice of animal captivity evident in the history of the Adelaide Zoo attest, for example, to the deeply ambiguous impulses in the human relation with animals in Western societies. Certainly the idea of Animal that has been constructed in zoos has a more complex relationship to human society than simply as "other."[28] The inconsistent variety of human responses to animals at the zoo that I noted earlier—including excitement, pleasure, wonder, distaste, guilt, nostalgia—reflect precisely the fragility of the cultural hold of reason as the master source of "human" identity in Western societies. Indeed, the tensions in part explain the adaptations in zoo philosophy and design, some early forms of which are charted in this chapter. Zoos thus provide an arena to investigate the ways "humans" have not only defined but also struggled with their own complex relationship to animals and nature.[29]

The argument concerning the mutual construction of culture and nature at the zoo has a further component. For if zoos hold up mirrors to cultural practices toward nature in general and animals in particular, it is also the case that zoos are urban institutions. Not only did the zoo bring nature into Western popular culture, it also imported it to the city. This seems particularly significant historically and conceptually. For if, from the late nineteenth century, cities came to be read as monuments to peoples' capacity for progress and order, then it follows that zoos—where nature was introduced to the metropolis and converted into domesticated spectacle— represented the ultimate triumph of modern man [sic] over nature, of city over country, of reason over nature's apparent wildness and chaos. That this process was accompanied by nostalgia for lost natures and for the animals who were progressively removed from everyday life is evident in the ambivalent human responses to nature that persist to this day.[30] Zoos tell us something, then, not only about the making of Western popular culture but also the complex construction of metropolitan cultures and identities—of what it was, and is, to be a city-dweller.

Animal Otherness: Methodological Issues

It was during a sabbatical in London—where against the backdrop of the financial district's buildings, I watched the necks of giraffes rising high above the walls of Regent's Park Zoo—that I first grew curious about zoos. En route to the tube station each day I encountered a range of groups waving placards whose slogans told of an unfolding cultural contest. First, there were the animal rights activists, intent on rendering the zoo an endangered species itself. Then there were sympathizers moved to "defend this romantic park at the heart of urbanised, domesticated life."[31] Finally, the London Zoo director himself spoke out in defense of the zoo, highlighting its role in what he called "the breeding and conserving of endangered species." The London experience prompted my first critical reflections about zoos but it was only a methodological start, as the following comments attempt to elaborate.

Returning to Australia I decided to venture to the zoo in Adelaide, the capital city of South Australia, where I read the annual reports of the Royal Zoological Society of South Australia (RZSSA) from its inception in 1878 (as the South Australian Acclimatization Society) through to the present. My interest in the Adelaide example lay less in conducting a semiotic or textual analysis that might decode—as a deconstructive exercise—the representational strategies implemented at the zoo. Nor was my primary purpose to read the symbolism or iconography of that zoo's landscape for what it might say about human attitudes to animals. My narrative device was no universalised Western mind-set and praxis, no invariant human/animal ontology, no abstracted model of "us" and "them." Rather my objective was to write a critical account of a particular institution's historical emergence, social production, and cultural transformation.

The following account of the Adelaide Zoo speaks to general themes concerning the social construction of animal difference and nature, its externalization and commodification. The analysis also connects the Adelaide Zoo to trends in zoo philosophy, exhibition, and design established elsewhere in the Western world. But as part of my methodological and writing strategy, such global processes are brought into view through an insistently local focus.[32] If the zoo is a "space," the Adelaide Zoo is a "place" which tells us about its own framing contexts of colonialism to which must be added: an Imperial network of animal trading, the confluence of nineteenth-century ideas surrounding race, gender, and empire, the growth of extractive and consumer capitalism over the twentieth century, the role of science/knowledge in everyday life and the spectacle, and changing animal import and quarantine policies. Such were some of the contexts out of which general zoo forms in British settler societies were created, but with specific manifestations in Australia and the state of South Australia. By examining a particular zoo, then, my purpose is to shift the analytical focus from the discursive construction of animal otherness to the material production (and form) of human-animal relations at a specific site.

In the story that follows, I chart the mutable technologies for fashioning a distinctive ("human") experience of animals and nature during two periods in the history of the Adelaide Zoo. The processes are traced from the period of menagerie-style caging (1878–1930) through the Great Depression and Second World War when the zoo became a fairground. During both periods, the connections of the zoo to the making of not only "human," but also colonial and metropolitan identities are described. Within the space constraints of a chapter, I also hint at the traffic of ideas between formal and informal knowledge regimes—science and spectacle— at that zoo. Omitted in the chapter is discussion of the post-Second World War period when the Adelaide Zoo became "modernized", and thenceforth its evolution to the present, when a form of ecological theater in the fanciful

World of Primates continued to craft the means for the popular experience of animals in South Australia.[33]

Adelaide's "Pleasure Garden": 1878–1930

In the mid nineteenth century, zoological garden construction rapidly expanded in many "New World" societies, including Australia. Prominent citizens in the colonies of New South Wales, Victoria, Tasmania, and South Australia felt moved by civic pride and political pressure to establish zoos in the nascent capital cities. Adding a zoo—as well as a museum, library, and art gallery—appears to have been part of the process of converting an impersonal array of buildings and houses into a "city," of affording a sense of permanence, wealth, and metropolitan identity that mapped places locally and regionally.

In South Australia, the impulse to add a zoo to the new city of Adelaide led to the formation in July 1878 of the South Australian Acclimatization Society (SAAS). Its officers included the governor of South Australia, Sir W.F. Jervois, as patron, the chief justice (Samuel Way) as president, four vice presidents, and a committee of twenty, all eminent local men. The society was a colonial offshoot of its parent in England, itself having counterparts in, for example, the Société Zoologique d'Acclimatation of France.[34] The South Australian society engaged in various activities, but primary among its objectives was to "introduce, acclimatise, domesticate and then liberate select animal, insect, and bird species from England" that would transform the colony in the home country's image and diversify the economic base of South Australian agriculture. Adapting the colony's alien landscape became the moral (and masculinized) responsibility of the likes of Chief Justice Way, who, in 1881, reported with pride "the work . . . of introducing the songsters and insect-destroying birds of the mother country, so well known to many members in their younger days, with the hope that they may be permanently established here, and impart to our somewhat unmelodious hills the music and harmony of English country life."[35] Domesticating the South Australian landscape facilitated British ascendancy in this colony, while drawing for its justification on much older rationalist conceptions of "human" identity.

Within a year, and following the examples set by the colonies of New South Wales, Victoria, and Tasmania, the RZSSA decided to extend its activities by building a zoo. A tract of land next to the centrally located Botanic Gardens was granted by the state parliament after some dispute about the aesthetic wisdom of juxtaposing botanical concerns and animals, and an aborted attempt to situate the zoo next to Adelaide's Lunatic Asylum.[36] By 1883, construction was underway to erect a place "of recreation and education for the public where they may be familiar with the living specimens of natural history."[37] This would be Adelaide's "Pleasure Garden," in the words of society president Sir Thomas Elder in 1883. Its buildings soon included a keeper's lodge, a ladies' waiting room, separate "paddocks" for ostrich, llama, brahman cattle, and emu, a "most commodious" carnivora house, "sheds" for baboons, monkeys, wombats, and parrots, various aviaries, a bear pit "that will enable visitors to witness to great advantage the antics of the Java sun bear," and a "house" for "Miss Siam," the elephant who in 1883 was donated "so graciously" by President Elder. Miss Siam was the first animal to be named at the zoo, so beginning a tradition at Adelaide (and widely used in other Western zoos) of bestowing titles on animals who were either trainable or charismatic in the eyes of human audiences.[38]

In preparation for the jubilee of the colony of South Australia, zoo director Ronald Minchin traveled to Europe to seek to enlarge the number of exhibits at the Adelaide Zoo. By 1887, it was noted with pride that "the collection of animals exceeds by some hundreds that of any similar institution in the southern hemisphere."[39] The quantity of exhibits was foremost in the minds of zoo officials, who specified "additions to stock" as a badge of honor in each year's annual report well into the twentieth century. The more stock the more striking the image of acquisition and possession, and the more exotic the better, since it was peculiarity that was believed to afford the greatest recreational and educational advantage.[40]

Or so the conventions of colonial imaginations had it. These expectations appear to have made compulsory the display of staple creatures such as elephants, lions, cobra snakes, rhinoceroses, zebras, and bears. These exhibits served as emblems of colonial mastery over the animal world, the former becoming more impressive, it seems, the more exotic the subdued creature. Some of the exhibits, including Miss Siam (the elephant), the hippopotamus, and the polar bears were even given special "houses" to reflect their elite status in the zoo universe. Indeed, so prestigious was Miss Siam that her house was exclusively "furnished" with what was described as a "spacious bath." The ornamentation of her fanciful "Indian-style temple" (Figure 2.3), as well as the Japanese archway constructed in 1892 and the "Egyptian" Hippopotamus House added in 1901, went further to inscribe prevailing stereotypes of racialized difference. The elephant house, in particular, met cultural expectations about the distant lands and peoples over which Imperial influence was extending. This included India, the as yet untamed home of Europe's generic Jumbo. Moreover, it seems significant to note that Miss Siam (and her descendants), as well as Victoria the Hippo, were female. Miss Siam's accommodation was "decorated"—the exotic became wrapped in domesticity—and in time she became the mother figure for children at the zoo. "Othered" people were thus brought closer to nature at the Adelaide Zoo, legitimizing not only the colonial desire to reconstitute the world in an enclosure, but more fundamentally a racialized and gendered conception of normative "human" identity.

The core exotic creatures were also the most valuable in the animal economy that supported this colonial institution. By the late nineteenth century, the Adelaide Zoo was incorporated into a well-co-ordinated Imperial network of animal trading, itself an arm of the regime of extractive capitalism that was beginning to straddle the globe. The RZSSA paid a

Figure 2.3   "Indian-style Temple" that housed Adelaide Zoo's elephant from 1900
Photograph by Kay Anderson

"handsome fee" to a South African firm in 1895 for two zebras (in the hope that it would be possible to acclimatize them at the zoo and ultimately release them to the Adelaide Hills for "draught purposes"). The governor was called on to use his colonial influence in 1896 to help the Adelaide Zoo acquire a hippopotamus, and by 1901 the local newspaper announced this "expensive novelty . . . costing 800 pounds will soon pay for itself in gate receipts."[41] ("Hippo"'s sudden death the following year became a boon for the South Australian Museum, which acquired the skins of dead or destroyed zoo animals for its own displays.)[42] Several other valuable animals were purchased from the local circus operators, Wirth Brothers, who were themselves engaged in purchasing and poaching exotic animals from colonial traders and naturalists abroad. There was also a system of exchange between zoos, with Adelaide looking to negotiate from the Victorian Zoological Society a "reasonable" price for a giraffe and to sell its "surplus zoo-born assets" early in the new century.

Still other animals came from public donations. The most coveted were the wild boar and peafowl given by the king of England and the bull eland and Przewalski's horse granted by the duke of Bedford in exchange for Australian fauna to display at his Woburn estate (later to become England's earliest safari park). His Excellency the Governor reminded the members of the RZSSA of their colonial debts in 1909 when he told the annual meeting that "it is a great compliment to the Society that his Grace [the duke] takes such an interest."[43] Australian fauna and birds were certainly popular with zoos and traders abroad. Many annual reports boasted sales or exchanges that were a profit to the society. Despite the zoo president's observation in 1916 that "foreign animals are well represented at the Gardens, but there are insufficient local fauna,"[44] the overwhelming focus of the collection was on difference, peculiarity, strangeness, and that which was "far away." (An "Australiana exhibit" was not added until 1972 during a period of growing nationalism and urbanization.) One of the appeals of the zoo, it seems, was to provide a vicarious journey abroad. Like other vehicles of mass communication, including the *National Geographic* magazine and nature programs later on in the century,[45] the zoo afforded a local experience of global nature.

Not only did the RZSSA seek in its animal composition practices to cater to colonial imaginations, it also sought to fulfill the desires of an urban audience. The selectivity of animals at the Adelaide Zoo was in part a function, therefore, of their variable capacity to titillate metropolitan expectations. Familiar animals such as sheep, goats, rabbits, donkeys, and ponies, for example—not to mention the thoroughly domesticated companion animals dogs and cats—had no place at the zoo, except much later on in the Children's Zoo (1965), which sought to bring the "farm" to the Adelaide child. By contrast, animals were well qualified for display if they could be made to communicate images of the so-called wild. This (wildly) unspecified term seems to have been used, out of fear and nostalgia, to refer to the jungles, mountains, savannas, swamps, and polar reaches so wondrously removed from human, and especially urban, habitation.[46] Moreover, the more wild and savage the animal, the greater the triumph of domestication and the more glorious the emblem of not only Imperial but also, I have been arguing, a normatively defined "human" control.

Consideration was also given by the society in its animal-recruitment practices to the "educational" potential of its exhibits. Officers of the RZSSA were ever alert to their educational role in the South Australian community, and from the time the zoo opened, the society was never comfortable with the suggestion that the zoo had a solely recreational function. Again, it was the more exotic and (apparently) geographically distant animals that were thought to hold the strongest educational promise. After all, as we saw in the 1909 news report quoted at the beginning of this chapter, education was believed to inhere in nothing more specific than the act of observation, and it was the exotic animals who arrested audience attention. Certainly the members of the South Australian parliament needed little convincing of the zoo's educational function and supported the society's activities with an annual grant.

A range of considerations, then, affected decisions about animal composition at the Adelaide Zoo in its early decades. Officers of the RZSSA—the citizens who enjoyed the power to represent the zoological world to the Adelaide public—drew upon master norms of animal difference to order nature in their image. The officers did not include women, though clearly women were interested in the zoo's operation. In 1887 the governor told the annual meeting of the society that he "was greatly pleased to see so many ladies present . . . [and] hoped on a future occasion that ladies would take part in the proceedings."[47] (This did not transpire until much later in the century.) The officers seemed to have perceived the zoo as a space through which to confer human structure on the chaos of nature. Here they drew upon prevailing rationalist conceptions of "human" identity— ones through which they forged their own moral sense of themselves as masculine and ipso facto authoritative. The zoo grounds were their mandate. The collection was their charge. They were chief crafters of a way of seeing and master authors of a public landscape. In turn, the men presided over a range of male and female staff, who were employed to feed and tend the animals, build and repair enclosures, install heat and light systems, maintain the drainage, fencing, and effluent disposal systems, plant and irrigate the grounds, lay out the paths in an orderly way (in contrast to the "wild"), and more generally create an illusion that only seemed natural in contrast to the surrounding environs of the city of Adelaide.

Just as there were cultural norms governing the composition of animals at the Adelaide Zoo, so were there prevailing conventions for displaying the stock. The exhibition philosophy at the Adelaide Zoo also reflected mutable scientific and popular understandings of the animal universe. Thus in 1883, when the zoo opened, the exhibits were set out in conformity with prevailing Linnaean taxonomies based on visible characteristics—namely reptiles, birds, mammals, and fish—and each exhibit was made to stand as a taxonomic specimen of a broader category. That is, the individual was made to stand as a signifier of species, displaying apparently species-typical behavior. The animals were not, for example, organized in terms of their social patterns or cultural life. Consistent with scientific classification systems, there appeared to exist the presumption that the Animal could be known *as body*, so inscribing and preserving the deeper mind/body, human/ animal dualisms of Cartesian thought.[48]

Exhibit labels, complete with maps of the global distribution of specimens, served to translate scientific thought into a form amenable for popular consumption and instruction. The exhibits showed nature not only confined and subdued, therefore, but also interpreted and classified.[49] To that end, zoo space occupied a critical nexus in the traffic of ideas between the scientific and the popular. There, the boundaries were fused between formal and informal knowledge regimes about animals and nature. A purified Nature—separated out as discrete from Culture—was being translated back into society through acts of scientific mediation.[50] The exhibit maps, for example, drew on the expanding science of zoogeography to imply there was some fixed provenance of the animals in nature;[51] yet, as we have seen, it was other zoos that in reality supplied a large part of the stock. A year after the labels were added, for example, the annual report noted the "stock of animals and birds has been increased greatly by purchases from the Zoological Gardens in Sydney, Melbourne, Auckland, Hamburg, Rotterdam and Toronto."[52]

A rationalist image of discrete types was also communicated through the construction of cellular confinements. There were separate enclosures and partitioned paddocks in front of which visitors would pass as in a museum or prison.[53] While certain (feminized and racialized) celebrities of the zoo community were given distinctive "homes," the accommodation of the zoo's rank and file members consisted mainly of concrete pits and iron cages. Not only did variable "housing" serve to order class distinctions among zoo inhabitants, the bars and cages encoded a bold sense of separation between the penultimate categories of keeper and kept. Indeed, the cages had the multiple functions of maximizing the thrill of proximity to, gaze over, and security from, the exhibits they restrained. An editorial in the *Register* newspaper in November 1892 observed that a "desirable mix of fascination and repulsion is afforded the human being who can look on brute creatures from the right side of the cage." The abiding purpose of all the enclosures was to display "to the greatest advantage the carriage, curiosities, and habits of their inmates."[54] For example, the carnivora house (in which lived seven lions and two tigers) was lined with white tiles in 1896 to "furnish an excellent background for visitors, including natural history students and artists."[55] The enclosure of the black bear was similarly fitted with tiles in 1914. The cages were stark and unfurnished for the most part, although in 1894 director A. Minchin noted that in a "unique performance" the boa-constrictor had "swallowed and disgorged its blanket . . . los[ing] twelve of its teeth."[56]

The comfort of animals was not entirely without consideration during this period, as witness the lengthy discussions in the annual reports between 1912 and 1918 about the cramped conditions of the stock. The susceptibility of the monkeys and birds to "the cold" was also often noted until their timber structures were replaced by brick buildings in the late 1920s. So too was the RZSSA sensitive, during the early decades of the zoo's operation, to periodic critiques on the part of Adelaide men and women about practices of animal cruelty at the zoo.[57] The society's concerns for animal welfare, however, appear to have stemmed largely from anxieties about the mortality of the stock. Annual deaths were reviewed in each report; some, such as Hippo's in 1929, being considered "very serious." As for the problem of congestion, no consideration was ever given to reducing the number of exhibits as a solution until much later in the century.

While most of the zoo enclosures dramatized the distance between the keepers and the kept, there were so-called charismatic exhibits such as Miss Siam, who offered immediate encounters with nature for children who took rides on her back. Images and experiences of both distance from and proximity to animals were thus available at the Adelaide Zoo. Animals, such as the elephant, which could be so tamed as to be playthings, appear to have acquired a special feminized status in the eyes of the public. By 1885, Miss Siam was carrying "thousands of children on her back" and "generally proving an immense attraction", having "so improved in docility as well as size."[58] When she contracted an ulcer in 1894 there was talk of not healing but replacing her. The fate of Miss Siam seems to have been that of being "re-invested with agency and purpose only after being brought captive . . . within the master's sphere of ends."[59]

More generally, the period to 1930 marked the historical stage in nature's colonization at the Adelaide Zoo, where, through an elaborate cultural process, certain leading characters of the animal universe had been gathered, annexed, and re-presented to the Adelaide public as for a set of postage stamps. So crafted was a complex figuration of the colonial enterprise that itself drew on a deeper social project of "human" identity construction and domination. The artifice of distance between human and nonhuman animals, and nonhuman and nonhuman animals, was dramatized in a gallery of images through which the zoological universe was rationally ordered and delivered to the South Australian public. The dominant culture's capacity for domesticating nature was thus gloriously sublimated in the social creation that was the Adelaide Zoo.

"Laugh and Learn": The Zoo as Fairground, 1930–63

At the annual meeting of the RZSSA in 1931, it was noted that "nobody can walk around the Gardens without being impressed by the fact that the animals and birds are comfortably housed, and in better circumstances than those in which they find themselves in their natural habitat."[60] Whatever confidence the society possessed in its efforts to ensure the stock's comfort in the years to 1930 appears to have faded come the Great Depression, when adverse economic conditions were experienced by all Australian zoos. The government grant to the RZSSA was drastically cut and only the "most urgent" works were carried out. Matters grew "especially serious" in 1934 when Mary Ann, Miss Siam's replacement, died. There were no surplus funds to buy another elephant, yet "it was considered imperative to obtain one without delay, as its great popularity and money-earning powers make it a most necessary exhibit."[61] Lilian arrived from the Singapore Zoo later in the year to take up work drawing a passenger trolley (until she was large enough to carry children on her back).

Many other instances of planning and management reveal the complex cultural production that was the Adelaide Zoo during the 1930s and 1940s. The society went to lengths to replace the hippopotamus that died in 1929 when it choked on a rubber ball thrown into its cage by a visitor. Purchase in this case was assisted by *The News* newspaper, which from the 1930s was more generally involved in promoting the zoo's activities. In the local press, the Adelaide Zoo had acquired an authoritative publicity agent. The city's two daily newspapers fastened upon the zoo's entertainment potential and helped fashion an evolving definition of the zoo as an amusement park. The use of newspaper photographs was especially effective in shaping the showcase image; and many a front page from the 1930s featured the latest "important additions" to stock, including Michael and Mary Chimp in 1935. In turn, the society was often moved to gratefully acknowledge the "press and broadcasting stations who have done much to promote the zoo, thereby assisting its finances."[62]

In conjunction with the activities of the press, the council of the RZSSA developed its own strategies to popularize the zoo. During the period of the Great Depression and beyond to the Second World War—when bird and animal imports and exchanges were prohibited and "more common, Australian animals added"[63]—efforts were made to accent the zoo's function as a spectacle. The earlier recreational function of the zoo became transformed into a more specific amusement and entertainment concept. Maximum display value was on the lips of society members as they sought to boost the zoo's coffers with a range of initiatives.

In 1935, it was decided to introduce the ritual of "public feeding time," including Sundays. The eating habits of the carnivora could be featured "for the amusement of the public," it was argued, so commencing a spectacle that afforded a titillating juxtaposition of savagery and captivity within a reassuring wider image of human control. Later in the year it was noted that visitors "evince much interest in the proceedings" at the feeding events of lions and penguins.[64] A monkey yard was donated by Director A. Minchin in the same year, complete with fifty rhesus monkeys and "various gymnastic appliances." The "amusement value" of primates drew increasing notice in the annual reports, and in 1938 it was decided orang-outangs would make a "profitable investment."[65] Gibbons were introduced to the zoo in 1943, and ten years later it was noted that "when they are in the mood to indulge their amazingly agile acrobatics, their cage is one of the most popular attractions in the zoo."[66] With their athletic ability, the primates seem to have been regarded as closer to people, so, perhaps, titillating audience anxieties about the cultural distinction of "human" and "animal."[67] Whatever the basis of audience fascination with the primates, humour seems to have been an especially useful device through which to register (and release) public tensions surrounding ambiguous boundaries.[68]

Another initiative was to mark Lilian's "growing popularity." In 1939, the year admission charges were increased for the first time since the zoo's opening, 1500 pieces of cake were distributed to visitors on the occasion of her (much publicized) ninth birthday. It was also decided to take her on annual walks through Adelaide's central business district to the East End market where she was weighed and photographed. The practice was continued for seven years to 1942, when Lilian began to show "signs of nervousness at the prospect of leaving the zoo grounds."[69] Not discontinued, however, were her rides through the zoo grounds, despite the fact that in 1946 it was reported she had developed blisters on the soles of her feet.

Other "attractions" were added to the zoo grounds in this period that reveal the sense in which the zoo was a site through which to register human mastery over the animal kingdom. Notable was the introduction in 1939 of a circus "for the amusement of children."[70] A trainer was employed,

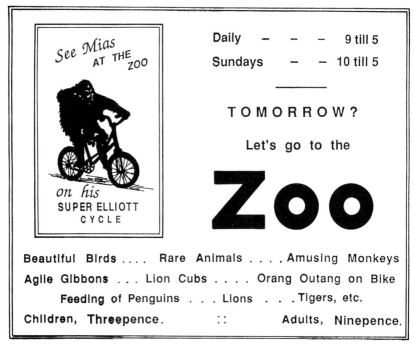

Figure 2.4   Advertisement for the Adelaide Zoo from the *Advertiser*, 5 May 1948

chimpanzees were trained to hold tea parties, and when the circus com-
menced "shows," members of the society observed that "the animals enjoy
the performance as do the many adult spectators who usually outnumber
the children."[71] When, in 1942, the circus attracted criticism from some
visitors for "possible cruel treatment to performing animals," it was stopped,
reminding us of the ambiguity of responses that are prompted in people
by their own separation from animals, as well as the fragility of rationally-
based power relations between people and animals. In this case, a minor
amendment transpired. The circus was substituted with a bicycle-riding act
by Mias the Orang-outang (Figure 2.4). According to the zoo director
there could be "no suggestion of cruelty in getting the orang-outang to
perform as he is very tractable with no vice and appears to enjoy his
performance."[72]

   The importance of product presentation in this era, during which mass
capitalism was being elaborated in South Australia, began to register with
society members from the late 1930s. Many enclosures were fitted with
ornamental fixtures (such as rock pools) to improve their appearance, and
in general the society developed the view "that the manner of exhibition is
as important as the exhibit itself."[73] Display methods were discussed at
length in many of the annual reports between 1930 and the Second World

War, again revealing the sense in which the zoo was a realm conceived *by* human imagination *for* human consumption. The interest of the society in the animals' welfare seems to have eroded almost entirely, no less after the appointment in 1936 of a pathologist to treat disease among the animals. Mr G. McLennan (the pathologist) believed the purpose of the zoo was entirely instrumental: to "use beasts and birds not only to view, but . . . for dissections that thereby we may take light what may be wrought upon the body of man."[74] The use of zoo animals for experiments again suggests the debt of the institution to scientific languages and agendas.

By 1947, when animal quarantine regulations were beginning to seriously constrain the society's ability to import "attractive animals," the attention of the RZSSA (and other societies) turned to breeding programs at the zoo. Inducing animals to breed in captivity seems to have been part of the larger process of domesticating that which was considered "wild." The first annual gathering of Australian zoo directors was held, at which it was decided to list the nation's zoo stock so that temporary exchanges for breeding purposes could be co-ordinated. A female tiger was loaned to the Adelaide Zoo from Bullen Brothers' circus in 1957 for precisely that purpose, but, as transpired in the case of Sally the polar bear, the mother did not show any interest in her cubs and they died. (Such efforts at reproducing what were called "rare species" prefigured the more orchestrated breeding programs of "endangered species" during the 1990s.) A number of attempts were made to induce Sally to breed, using increasingly "natural conditions and diet,"[75] but none were successful. The failures were read (probably correctly) as a measure of the stress suffered by the animals concerned.

Perhaps it is the slogan that was used as a postmark on South Australia's mail during the years 1961 to 1963—"Visit the Zoo: Laugh and Learn"— that captures the blunt turn to commodification adopted by the RZSSA during the period after the Second World War. The phase was marked by a foregrounding of the entertainment value of animals, who were drawn into the vortex of mass capitalism's cultural and economic expansion in the state of South Australia. As the zoo entered the arena of capitalist competition for the recreation dollar, so did it draw more transparently on hierarchical oppositions of human and animal in its appeal strategies. Having crafted a model of nature's domestication earlier in the century, the RZSSA saw fit in this period to appropriate animal difference for wholly rationalist-instrumentalist purposes. By the early 1960s, when the animals' living conditions were in a state of serious neglect and attendance figures peaked, Adelaide Zoo bore supreme witness to the master disciplinary practices that were progressively alienating "humans" from nature.

Conclusion

This chapter has sought to develop a critique of the metropolitan zoo as a space where nature is abstracted from its contexts and shaped into an image and experience by, and for, humans. The zoo is a complexly constructed scene, I have argued, crowding into one space a variety of creatures that would be unlikely to pose so obligingly in real life. Using the case of Adelaide, I have sought to historicize the zoo, nesting it within a range of framing contexts that reveal its insistently cultural foundation. The zoo's debts to colonial and metropolitan regimes of identity formation have been some of the contexts that have been signaled. The chapter has sought to go further, however, to unsettle the binding grip of constructs of "human" identity in the making of the zoo. The rationalist underpinnings of the Adelaide Zoo that are described in the chapter begin to shed light on the arbitrary nature of the boundaries that humans have long erected to distinguish themselves from nonhuman animals. As we have seen, a distinctive way of seeing and relating to animals was crafted out of faith in the capacity of humans for reason—a capacity that was believed to be the defining characteristic of "human" and to set humans apart from other spheres of life.

Yet this conception of "human" had, for centuries past, been itself deeply gendered and racialized. As far back as classical times, reason had been assumed to be the exclusive preserve of European, adult, free men. Women, children, and racialized others were identified with the inferiorized realms of nature and the body. These were the spaces that were progressively domesticated to (rationally based) projects in the course of the long rise to power of history's master subject. The moral order underpinning this power cohered around a set of Cartesian binaries of mind/body and reason/ nature, the critical deconstruction of which begins to disrupt the very basis on which nonhuman animals were set apart from "humans." Attention in turn is directed to the specific disciplinary practices that advanced our accepted conceptions and positionings of "human" and "animal." Also spotlighted are the specific sites at which those practices were articulated and renegotiated.

From the late nineteenth century, the Adelaide Zoo became a site where the muscle of not only human but colonial and metropolitan mastery over nature was flexed and popularized. A most domesticated social product was created, one whose partial perspective lay unseen, off-exhibit, and which was to find novel forms of expression in the decades to the present.

Acknowledgements

I would like to thank David Langdon, assistant director of the Adelaide Zoo, for time spent in interview and for allowing me access to the annual reports of the Royal Zoological Society of South Australia. Thanks to Bruce Matthews, curator of mammals, for an informative guided tour of the Adelaide Zoo. Thanks also to the Royal Geographical Society for permission to reprint, in modified form, the article that was first published in *Transactions of the Institute of British Geographers*. Julie Kesby assisted with proofreading, referencing, and compiling a summary of my notes from the annual reports. The comments of the following faculty and students at the Department of Geography at Syracuse University are gratefully acknowledged: John Agnew, Jim Duncan, Nancy Duncan, Don Meinig, John Mercer, David Robinson, Joanne Sharpe, John Short, and

Judy Walton. I would also like to thank Jacquie Burgess, Cindi Katz, and Val Plumwood for their very helpful and affirming comments.

Notes

1. *Advertiser*, 28 July 1909.
2. B. Mullan and G. Marvin, *Zoo Culture*, London: Weidenfeld and Nicolson, 1987; Y–F. Tuan, *Dominance and Affection: The Making of Pets*, New Haven, Conn. and London: Yale University Press, 1984.
3. C. Tarpy, "New Zoos—Taking Down the Bars," *National Geographic*, July 1993, pp. 2–38.
4. G. Adams, L. Fisher, D. Le Blond, N. Mazur, C. McMahon, T. Peckover, J. Schmiechen, and N. Sharrad, "The Role of the Adelaide Zoo in Conservation," Report prepared for the Royal Zoological Study of South Australia, Adelaide: Mawson Graduate Centre for Environmental Studies, University of Adelaide, 1991; A. Townsend, "Attitudes, Perception and Behaviour among Visitors at the Adelaide Zoo," Honours thesis, University of Adelaide, 1988.
5. Only a small amount of research has been published on the way socially constructed nature has been received and interpreted by human users. See, for example, J. Burgess, "The Production and Consumption of Environmental Meanings in the Mass Media: A Research Agenda for the 1990s," *Transactions of the Institute of British Geographers*, vol. 15, no. 2, 1990, pp. 139–61; Burgess, "The Cultural Politics of Nature Conservation and Economic Development," in K. Anderson and F. Gale, eds, *Inventing Places: Studies in Cultural Geography*, Melbourne: Longman Cheshire, 1992, pp. 235–54.
6. See, for example, N. Evernden, *The Social Creation of Nature*, Baltimore, Md.: Johns Hopkins Press, 1992; M. Fitzsimmons, "The Matter of Nature," *Antipode* 21, 1989, pp. 106–20; C. Katz and A. Kirby, "In the Nature of Things: The Environment and Everyday Life," *Transactions of the Institute of British Geographers* 16, 1991, pp. 259–71; C. Nesmith and S. Radcliffe, "(Re)mapping Mother Earth: A Geographical Perspective on Environmental Feminisms," *Environment and Planning D: Society and Space*, vol. 11, no. 4, 1993, pp. 379–94; N. Smith, *Uneven Development: Nature, Capital and the Production of Space*, Oxford: Blackwell, 1984; S. Whatmore and S. Boucher, "Bargaining with Nature: The Discourse and Practice of 'Environmental Planning Gain,'" *Transactions of the Institute of British Geographers*, new series, vol. 18, no. 2, 1993, pp. 166–79; A. Wilson, *The Culture of Nature: North American Landscape from Disney to the Exxon Valdez*, Oxford and Cambridge, Mass.: Blackwell Publishers, 1992; also see many recent contributions to the journal *Ecumene*.
7. N. Jardine, J. Secord, and E. Spary, eds, *Cultures of Natural History*, Cambridge: Cambridge University Press, 1996.
8. M. Foucault, "Disciplinary Power and Subjection," in C. Gordon, ed., *Power/Knowledge: Selected Interviews and Other Writings of Michel Foucault 1972–1977*, Brighton: Harvester Press, 1980, Sx 20–40.
9. B. Noske, *Human and Other Animals*, London: Pluto Press, 1989; A. Manning and J. Serpell, *Animals & Human Society*, London and New York: Routledge, 1994.
10. L. Zuckerman, ed., *Great Zoos of the World: Their Origins and Significance*, London: Weidenfeld and Nicolson, 1979.
11. Mullan and Marvin, *Zoo Culture*.
12. Erased in the rise to dominance of this Linnaean scientific model of plant and animal classification were distinctive regional systems of plants' names. See K. Thomas, *Man and the Natural World: Changing Attitudes in England 1500–1800*, London: Allen Lane, 1983.
13. H. Ritvo, *The Animal Estate: The English and Other Creatures of the Victorian Age*, Harmondsworth: Penguin, 1987.
14. G. Lloyd, *The Man of Reason*, London: Methuen, 1984; V. Plumwood, *Feminism and the Mastery of Nature*, London: Routledge, 1993.

15. Perhaps there are legacies of this thinking in the inclusion of certain marginalized humans, such as pygmies, at some nineteenth-century zoos such as the Bronx Zoo.

16. Plumwood, *Feminism and the Mastery of Nature*.

17. Plumwood, *Feminism and the Mastery of Nature*, p. 5.

18. D. Haraway, "Situated Knowledges: The Science Question in Feminism and the Privilege of Partial Perspective," *Feminist Studies*, vol. 14, no. 3, 1988, pp. 575–99; S. Harding, ed., *The "Racial" Economy of Science: Toward a Democratic Future*, Bloomington, Ind.: Indiana University Press, 1993; Lloyd, *The Man of Reason*. In geography, see A. Blunt and G. Rose, eds, *Writing Women and Space: Colonial and Post-colonial Geographies*, New York: Guilford, 1994.

19. N. Duncan and J. Duncan, "(Re)reading the landscape," *Environment and Planning D: Society and Space* 6, 1988, pp. 117–26.

20. Haraway, "Situated Knowledges," p. 595.

21. M. Jay, "Scopic Regimes of Modernity," in S. Lash and J. Friedman, eds, *Modernity and Identity*, Oxford: Blackwell, 1992, pp. 178–95.

22. D. Haraway, "Teddy Bear Patriarchy: Taxidermy in the Garden of Eden, New York City, 1908–1936," *Social Text*, 1983, pp. 20–64.

23. T. Bennett, "The Exhibitionary Complex," *New Formations* 4, 1988, pp. 73–83.

24. Note the relevance here of the important critique of forms and practices of visualization in cultural geography, beginning with D. Cosgrove, "Prospect, Perspective and the Evolution of the Landscape Idea," *Transactions of the Institute of British Geographers* 10, 1985, pp. 45–62, and continuing in such works as T. Barnes and J. Duncan, eds, *Writing Worlds: Discourse, Text and Metaphor in the Representation of Landscape*, London: Routledge, 1992; D. Cosgrove and S. Daniel, eds, *The Iconography of Landscape*, Cambridge: Cambridge University Press, 1988; S. Daniels, *Fields of Vision: Landscape Imagery and National Identity in England and the United States*, Oxford: Oxford University Press, 1993; J. Ryan, "Visualizing Imperial Geography: Sir Harold Mackinder and the Colonial Office Visual Instruction Committee," *Ecumene*, vol. 1, no. 2, pp. 157–76; F. Driver, "Visualizing Geography: A Journey to the Heart of the Discipline," *Progress in Human Geography*, vol. 19, no. 1, 1995, pp. 123–34.

25. In geography, see E. Issac, *Geography of Domestication*, Englewood Cliffs, N.J.: Prentice Hall, 1970.

26. I am grateful to John Agnew for highlighting this connection. See also M. Warner, "Monstrous Mothers," in *Managing Monsters: Six Myths of Our Time*, London: Vintage, 1994, regarding the idea of female, untamed nature whose ungoverned energy must be leashed.

27. See K. Anderson, "A Walk on the Wild Side: Toward a Critical Geography of Domestication," *Progress in Human Geography* 21, 1997, pp. 463–85.

28. L. Birke, *Feminism, Animals and Science: The Naming of the Shrew*, Buckingham and Philadelphia, Pa.: Open University Press, 1994, p. 34.

29. The ambivalences and complications in the human relationship with plants and animals are highlighted in J. Goody, *The Culture of Flowers*, Cambridge: Cambridge University Press, 1993, a richly detailed study of different cultural practices surrounding flowers in Western and non-Western societies.

30. On the nostalgia for lost natures, see R. Williams, *The Country and the City*, New York: Oxford University Press, 1973; for the animals who were progressively removed from everyday life, see J. Berger, *About Looking*, New York: Pantheon Books, 1980.

31. *Daily Telegraph*, 13 June 1991.

32. K. Anderson, "In the Place of Trans-disciplinary Space," *Australian Geographical Studies*, no. 34, vol. 1, 1996, pp. 121–5.

33. See K. Anderson, "Culture and Nature at the Adelaide Zoo: At the Frontiers of 'Human' Geography," *Transactions of the Institute of British Geographers* 20, 1995, pp. 275–94.

34. See H. Ritvo, *The Animal Estate*; M. Osborne, *Nature, the Exotic, and the Science of French Colonialism*, Bloomington, Ind.: Indiana University Press, 1994.

35. South Australian Acclimatization Society (SAAS), 3rd Annual Report, 1881, p. 6.

36. T. Bennet, "The Exhibitionary Complex," *New Formations* 4, 1988, pp. 73–83, argues

in his paper on the history of museums and exhibitions that the institutional mechanisms of "confinement" (emphasized by Foucault in his study of prisons) were complemented in the nineteenth century by those of "exhibition." This makes all the more noteworthy the consideration given by the Royal Zoological Society of South Australia to juxtaposing the asylum and the zoo.

37. SAAS, Third Annual Report, 1881, p. 5.
38. See Mullan and Marvin, *Zoo Culture.*
39. Royal Zoological Society of South Australia (RZSSA), Ninth Annual Report, 1887, Adelaide Zoo Library, p. 6.
40. See T. Y. Rothenburg, "Voyeurs of Imperialism: the National Geographic Magazine before WWII," in A. Godlewska and N. Smith, eds, *Geography and Empire*, Oxford: Blackwell, 1994, pp. 155–72, concerning the confluence of ideas surrounding the exotic, race, and gender in the appropriation of resources under imperialism.
41. *The Advertiser*, 3 August 1901.
42. Apparently, there were other uses for dead zoo animals: at their functions, other societies, such as London's Royal Zoological Society, treated members to the remains of whatever had died in the Gardens! See D. Stoddart, *On Geography and History*, Oxford: Basil Blackwell, 1986, p. 22.
43. RZSSA, Thirty-First Annual Report, 1909, Adelaide Zoo Library, p. 31.
44. RZSSA, Thirty-Eighth Annual Report, 1916, Adelaide Zoo Library, p. 10.
45. See Wilson, *The Culture of Nature.*
46. These classificatory codings also crept into the tradition of world regional geography, which was itself yoked to British expansionism of the nineteenth and twentieth centuries: see D. Livingstone, *The Geographical Tradition: Episodes in the History of a Contested Enterprise*, Oxford and Cambridge, Mass.: Blackwell, 1992. The notion of "wild" (as distinct from "civilized," that is, domesticated) environments coursed its way through geography's own mappings.
47. RZSSA, Ninth Annual Report, 1887, p. 7.
48. I am grateful to Val Plumwood for this point.
49. Ritvo, *The Animal Estate.*
50. B. Latour, *We Have Never Been Modern*, London: Harvester Wheatsheaf, 1993.
51. P. Darlington, *Zoogeography: The Geographic Distribution of Animals*, New York: John Wiley and Sons, 1957.
52. RZSSA, Fifty-Third Annual Report, 1931, Adelaide Zoo Library, p. 3
53. Berger, *About Looking.*
54. RZSSA, Thirty-Third Annual Report, 1911, Adelaide Zoo Library, p. 4.
55. RZSSA, Eighteenth Annual Report, 1896, Adelaide Zoo Library, p. 8.
56. RZSSA, Sixteenth Annual Report, 1894, Adelaide Zoo Library, p. 4.
57. See, for example, the editorials in the Adelaide newspapers concerning the ethics of keeping animals: *Register*, 31 January 1896; *Advertiser*, 13 May 1910.
58. RZSSA, Seventh Annual Report, 1885, Adelaide Zoo Library, p. 6.
59. V. Plumwood, *Feminism and the Mastery of Nature*, p. 192.
60. RZSSA, Fifty-Third Annual Report, 1931, p. 1.
61. RZSSA, Fifty-Sixth Annual Report, 1934, Adelaide Zoo Library, p. 8.
62. RZSSA, Seventieth Annual Report, 1948, Adelaide Zoo Library, p. 3.
63. RZSSA, Sixty-Second Annual Report, 1940, Adelaide Zoo Library, p. 9.
64. RZSSA, Fifty-Seventh Annual Report, 1935, Adelaide Zoo Library, p. 6.
65. RZSSA, Sixtieth Annual Report, 1938, Adelaide Zoo Library, p. 8.
66. RZSSA, Seventy-Sixth Annual Report, 1953–54, Adelaide Zoo Library, p. 3.
67. D. Haraway, *Primate Vision: Gender, Race and Nature in the World of Modern Science*, New York: Routledge, 1989.
68. I am grateful to Felix Driver for this point.
69. RZSSA, Sixty-Fourth Annual Report, 1942, Adelaide Zoo Library, p. 4.
70. RZSSA, Sixty-First Annual Report, 1939, Adelaide Zoo Library, p. 11.
71. RZSSA, Sixty-First Annual Report, 1939, p. 13.

72. RZSSA, Minutes of Council, 23 November 1942, Mortlake Library of South Australia.

73. RZSSA, Sixtieth Annual Report, 1938, p. 7.

74. RZSSA, Fifty-Eighth Annual Report, 1936, Adelaide Zoo Library, p. 5.

75. RZSSA, Eightieth Annual Report, 1957–59, Adelaide Zoo Library, p. 7.

3

Animals, Geography, and the City:
Notes on Inclusions and Exclusions
Chris Philo

Introductory Comments

Tuan's text *Dominance and Affection: The Making of Pets* has long intrigued me as a possible starting point for rethinking the way in which animals might be researched by human geographers, and in this paper I offer notes on such a project.[1] Tuan's objective in *Dominance and Affection* is to recover hidden dimensions to the workings of power in "human reality," and as such he glances beyond the obvious exercise of economic and political power (by those with resources over those without) to the more subtle social and cultural processes through which an interweaving of dominance and affection fixes "lesser" beings in the almost playful grasp of "superior" ones. This is why the focus of the text turns to the unequal power relations which run between certain human beings—notably older male ones with a measure of social status—and a range of other beings in the world, including women, children, slaves, "dwarfs" and "fools," but also including animals in zoos and even plants in gardens. Tuan's approach opens up the possibility of thinking about animals using concepts more commonly employed by human geographers when studying minority or "outsider" human groups and when describing something of the lives and the places that these groups strive to create "from within,"[2] at the same time as explaining how such groups frequently become caught up in webs of power structured very largely "from without" (by other groups based elsewhere and possessing quite different interests). In the essays on animals in *Dominance and Affection* Tuan thereby begins to regard animals very much as a "social" group ensnared in a struggle with humans, and he also directs attention to how animals are conceptualized by humans on scales oscillating between reverence and revulsion, compassion and control, utilitarianism and disinterest. Precisely how different human communities think, feel, and talk (or "discourse") about the animals nearby will obviously shape their socio-spatial practices towards these beings on an everyday basis, with important consequences for the extent to which the different animal species

51

present are either included in or excluded from common sites of human activity.

Also, the possibility is raised here of thinking about animals as a social group with at least some potential for what might be termed "transgression" or even "resistance" when wriggling out of the cages, fields, and wildernesses allotted to them by their human neighbors (and this will most notably be true of wild animals). There are many difficult theoretical issues bound up in making such a statement, chiefly because it borders on attributing "agency" and "intentionality" to animals in a manner normally only reserved for human beings,[3] and also because it raises questions as to whether it is appropriate to conceive of transgression or resistance in a situation where the parties involved—in this case animals and humans—seemingly cannot even begin to share the same systems of (political) meaning. Furthermore, it may be that applying such terms to animals is to risk being anthropocentric, interpreting nonhuman beings through the lens of the human world and sliding into precisely the traps critiqued below. For the purposes of what follows, and despite wishing to retain some sense of animals as having their own "will" to behave contrary to human expectations of them, I have decided to simplify my argument by writing of animals trangressing but not of them resisting. Here, I borrow a distinction drawn by Cresswell:

> Resistance seems to imply intention—purposeful action directed against some disliked entity with the intention of changing it or lessening its effect. . . . Transgression, in distinction to resistance, does not . . . rest on the intentions of actors but on the *results*—on the "being noticed" of a particular action. . . . Transgression is judged by those who react to it, while resistance rests on the intentions of the actor(s).[4]

It is also helpful that Cresswell supposes transgression to involve "crossing lines," by which he means both the metaphorical crossing of social boundaries (norms, conventions, and expectations) and the more concrete crossings leading to what he terms "out of place" phenomena (matter, actions, and happenings that are not in the worldly location expected of them). It seems to me that in these terms many animals (domesticated and wild) are on occasion trangressive of the socio-spatial order which is created and policed around them by human beings, becoming "matter out of place" in the process; and it is in this respect that animals often squeeze out of the places—or out of the roles that they are supposed to play in certain places—which human beings envisage for them.

My purpose in this chapter is to follow Tuan's lead in advocating a social-cultural perspective in geographical work on animals, and I propose to develop my argument in two ways. In the first section I explore aspects of how academic geography has treated animals to date, particularly in terms of their relation to humans, and I suggest that this treatment has tended to be at best highly partial and at worst wholly exclusionary. The tendency has been to consider animals as marginal "thing-like" beings devoid of inner lives, apprehensions, or sensibilities, with little attempt made to probe the often taken-for-granted assumptions underlying the different uses to which human communities have put animals in different times and places. This critique is supported by a brief survey of the neglected subdisciplinary field known as "animal geography," although here I also retrieve indications of a more sensitive approach that examines in detail animal-human relations (rather than simply taking them as given), as well as taking seriously non-utilitarian aspects of how animals become embedded in broader societal orderings of respect and disgust. The suggestions here about the ways in which such orderings connect to spatial practices of including or excluding animals, with different practices being directed at different animals, both lead to the next section of the paper (dealing with empirical materials) and more generally signpost the possibilities for a "new" animal geography. In the second section of the chapter I offer a description of how such a revived animal geography might substantively tackle animal-human relations in the specific case of live meat markets in nineteenth-century cities, and in so doing I tease out the claims of a decidedly anti-animal agenda that wished to exclude livestock animals from the city on a number of grounds ranging from the medical to the moral.

I must underline the specific focus of what follows, though, in that I regard it as constituting a thought-experiment on the fringes of "social geography"—an attempt to see what happens when animals are indeed treated as another social group (a very "other" social group) present in the human world—and I am not casting it as a contribution to the larger and more tangled realm of inquiries into the themes of nature and culture in the city.[5] Inevitably, my chapter will bump up against a range of literatures that could have been consulted more extensively, such as texts dealing with the nature-culture axis or laying out the conceptual bases for critiquing conventional intellectual responses to animals, but I have deliberately sought to confine myself to asking about the possible responses of a social geographer when confronted by the intrusive reality of (say) cows, sheep, and pigs mingling with people in the spaces of a large urban area.

Animals and Geography

Just as the sensitizing of human-geographical inquiry to all manner of "other" human social groups (defined in terms such as gender, sexuality, class, ethnicity, and health) has commonly involved a critical reflection on the failings of earlier geographical treatments (noting their negligent or prejudicial responses to, for instance, women, homosexuals, workers, people of colour, and people with illnesses), so I will begin by critiquing past treatments of animals within academic geography. The geographical litera-ture as a whole has largely overlooked animals as distinctive objects of study, often subsuming them within broader discussions of nature and environ-ment, and rarely making them into a special issue deserving of special consideration. It is true that they have received attention from physical geographers, notably in the field of biogeography when examining the spatial distributions of animals and plants relative to natural-environmental contexts, but they have tended only to be referenced by human geographers when considering how they are raised and their products (meat, milk, fur, hides) utilized and sold by human communities. This is why animals such as cows, sheep, and pigs have figured quite prominently in agricultural geography, as well as being present in the background of studies in rural geography. Yet I cannot help feeling that, even in the texts in which animals do make an appearance, there is still something missing: a sense of animals *as animals*; as beings with their own lives, needs, and (perhaps) self-awarenesses, rather than merely as entities to be trapped, counted, mapped, and analyzed; as beings whose lives are indelibly shaped by the uses that humans formulate for them, but whose fates resulting from these taken-for-granted uses (along with the human rationales behind these uses) are almost never subjected to critical scrutiny.

The issues in this respect became clearer when I encountered a book by Anderson called *The Geography of Living Things*, which began promisingly with this statement:

> Any living beings present in the environment, whether plants or animals, are all too often looked upon in much the same way as the dead physical features of the landscape, as material ready to man's [*sic*] hand to use as he sees fit. Similarly, in much political writing, the earth with all its living inhabitants is tacitly regarded as if it were a lifeless, passive medium, clay in the hands of man the potter, submissively waiting for him to mould it to his own ends exactly as seems best to him.[6]

Given such a rejection of anthropocentrism, it is perplexing that the emphasis in the rest of the book continued to be so squarely on humanity in its dominion over animate nature, and on virtually every page Anderson concentrated so much on the well-being of humans that animals and plants appeared to have been placed on earth solely to service these "superior" beings. The river thus became "the home of fish and other shy wild creatures that feed man or pester him, and in open or hidden ways affect his life and activities,"[7] rather than the home for countless life-forms whose lives in the river might possess their own "stories" quite apart from any relationship to humans, their needs and illnesses. Humanity remained the chief point of reference for Anderson after all, the site of questions to be asked and of problems to be solved, and nonhuman "living things" were largely ignored as beings with their own lives and geographies beyond the charmed circle of human existence. Anderson's book has not been influential and is now largely forgotten, but it does starkly illuminate what is arguably a broader malaise within academic geography: the tendency either to overlook nonhuman "living things" altogether or to investigate them only insofar as they have an impact upon the lives of human beings. This observation does not apply to most studies in biogeography, where nonhumans are routinely considered without reference to humans, but the real issue that arises to haunt me from the pages of Anderson's book is the fate of nonhuman animals in *human*-geographical texts that *do* talk about animals as participants in the world alongside human communities.

The issues signaled here can also be approached with reference to the minor subdiscipline of animal geography, which has featured in the literature of academic geography, and it is instructive to review the sporadic claims that have been made on its behalf. Growing out of papers such as Guppy's "The Distribution of Aquatic Plants and Animals," something called animal geography gradually became recognized as the study of animal populations in terms of their spatial distributions and environmental associations.[8] Thomson, a well-known professor of natural history, contributed a short paper to the *Geographical Teacher* on "the geographical distribution of animals," for instance, and in so doing outlined how such distributions are inevitably bound up with spatial and temporal patterns in climate, earth surface configuration, plant life, and even "human evolution."[9] He quoted from earlier texts by authors such as Lydekker and Wallace, but it is revealing that they (like Thomson) would use the term geography without necessarily having a disciplinary affiliation with an emerging academic geography. This means that it was probably only with the posthumous publication of the geographer Newbigin's *Plant and Animal Geography* that something called animal geography became accepted as part of this new disciplinary structure.[10] The precise emphasis may have differed from author to author, with some leaning more towards describing spatial ("regional") patterns whereas others sought more to explain environmental ("ecological") influences, but by mid century animal geography had gained at least a measure of independent identity, attested to in specific texts and papers.[11] Yet its distinctiveness remained a matter of dispute, as was evident from the readiness with which it was renamed "zoogeography," itself "a discipline cultivated by zoologists even before the days of Humboldt and Ritter,"[12] and one that was more self-consciously dependent upon the "scientific" realms of zoology, ecology, paleontology, and (ultimately) biology than upon any geographical tool-box of ideas and methods. Several of the authors referenced here did endeavor to reclaim animal geography from zoogeography and its grounding in the systematic sciences, but they were unsuccessful, with the result that animal geography has disappeared from the discipline today as anything but a trace within the burgeoning study of biogeography.

This short story about the patchy career of animal geography could be taken as a parable about the general neglect of animals in academic geography, but it is also salient to consider the specific position that was occasionally claimed for animal geography in relation to human geography. Indeed, when noting that it was "the branch of geography least practised by geographers," one observer went on to speculate that this was because the concerns of animal geography were "too remote from the central problems of human geography" and hence unlikely to be taken seriously by geographers interested in human communities, their products, activities, and landscapes.[13] On occasion it has been suggested that distributional-environmental principles pertaining to animal life may not be so different from those controlling human life, but of greater interest are those rare statements that have sought to reconstitute or to enlarge animal geography by exploring the relations between animals and humans. Thomson specu-lated about the need to consider "what all this complex distribution of animal life has meant and still means" to human beings,[14] and over forty years later Cansdale contributed a series of essays to the *Geographical Magazine* that introduced animal geography to the long-running struggle whereby humans have domesticated animals as ready-to-hand sources of food, clothing, locomotion, pets, allies, "guinea pigs" and "biological controls."[15] In a similar vein, Bennett floated the possibility of what he called a "cultural animal geography": "The proposed field would encompass those aspects of animal geography which accumulate, analyse and systema-tise data relevant to the interactions of animals and human cultures."[16] Bennett envisaged a deep historical perspective for this field, one examining how from earliest times humans in their role as "ecological dominant" have been affecting animals, moving them, domesticating them, and destroying them; and he also urged a sensitivity to how human beings have often endured the presence of animals as robbers of crops, predators on domestic livestock, and carriers of disease.

This version of how geographers could deal with animals is one that has affinities with what might be termed, in the wake of Sauer's pioneering work on *Seeds, Spades, Hearths and Herds,*[17] a "Sauerian" approach to the role of animal domestication in the process whereby "natural landscapes" are converted into "cultural landscapes." Several geographers have clearly been inspired by Sauer's example, notably Issac in his survey of the world geographies underlying both plant and animal domestication, as well as Donkin and Gade in their studies of specific domesticated animals such as the peccary, the muscovy duck, the guinea fowl, the llama, and the guinea pig.[18] The critique of anthropocentrism can probably be turned on Sauer and his followers for some of their statements, but my view is that their work began to negotiate this critique by remaining alert to the modes of being of the animals themselves—to their own physiologies and behaviors indepen-dent of the human imprint upon them—and by acknowledging how animals can become bound into relationships with human communities that have more to do with their centrality to myth and legend than with the satisfaction of more utilitarian human needs.

The Sauerian approach appreciated that animal lives on this planet are often greatly influenced by the humans who share it with them, but at the same time it kept the human factor "in suspension" so that animals continued to be central to the narrative, with their experiences, activities, and sufferings being granted a certain space (and even dignity) in the presentation. Attention was thereby drawn to the unequal power relations between animals and humans, and this one-sided struggle was opened up to scrutiny (rather than simply being assumed), while claims were made about the "agency" of animals giving them at least some capacity to rebel against the imposition of human requirements. The "wildness" of animals was explored at some length in this respect, notably when stressing how so often animals have resisted the domestication process simply by running away. Donkin became particularly interested in "transitional species" of animals that "lie close to the frontier between the wild and the domestic,"[19] implying a fascination with those qualities of animals themselves that foster a lack of "co-operation" with their human neighbors. In a general statement about domestication, he observed that animals "possess a variety of psychological characteristics which, to a greater or lesser extent, impede the overlap of social media,"[20] and he went on to argue that the intrinsic properties of the animals themselves play a far greater role than humans in directing which species, when, and where, have become domesticated. He indicated that a certain "psychological fitness" to co-habit with human beings marked out some animals as better candidates for domestication than others; and in his earlier work on the peccary he discussed the "pioneer qualities" of the pig which rendered it "admirably suited to an age of exploration and conquest," in part because pigs "were omnivorous and, like the *conquistadores* themselves, hardy, mobile and physically adaptable."[21] This attention to the "psycho-biological make-up" of animals strikes me as very different from anything claimed about animals in the literatures of, say, biogeography or agricultural and rural geography.

A further noteworthy aspect of this Sauerian animal geography was a skepticism about seeing animal-human relations in solely or even principally economic terms. Crucial to Sauer and his followers was the idea that the first stage of animal domestication would have involved a *non*-economic contact; they rejected a reductively economic explanation—one regarding domestication as a kneejerk response to human needs for securing more plentiful and reliable sources of meat, milk, fur, and hides—and stressed instead the importance of myth, legend, and ceremony in the domestication process. Donkin argued that "[a]n originally economic motive is improbable," and added that "[e]conomic benefits would only become apparent after domestication (regular breeding in captivity) had been achieved,"[22] something probably occurring in the first instance to fulfill a sacrificial role integral to a certain strain of "religious" impulse.[23] This non-economic explanation for the origins of domestication has not gone without dispute, and Rodrigue recently complained about what she saw as a Sauerian orthodoxy in this connection,[24] but one implication of such non-economic thinking is the possibility of conceptualizing animals as more than just things for either the use value of some human communities or the exchange value of others. This possibility calls for researchers to reflect on the insertion of animals into folklore, religious belief, and cosmology, those symbolic sites through which societies think their own existence and formulate codes of practice accordingly. Indeed, exploring a given human community's discourses on such matters may offer vital clues about the everyday practices it pursues towards animals, as well as helping to account both for why different animals may be treated differently in the same community (perhaps the horse being venerated while the donkey is despised) and why the same animal may be treated differently by different communities. In this latter respect Sauer himself wrote about the position of the dog in different parts of Ancient Asia:

> In the Old Planting [Southeast Asian] culture the dog, as a prized and respected creature, came to be an object of sacrifice, and of ceremonial consumption by the participants. As familial and religious connotations became blurred, the dog became a food item, especially at feasts. Because the dog held a position of high respect in this culture, it readily became an object of antipathy in neighbouring cultures of other ceremonial and religious orientations. These latter rejected the animal as despised and unclean.[25]

There is already a substantial anthropological literature on how animals become positioned within conceptual frameworks of good and evil, sacred and profane, pure and polluted,[26] but I remain excited by how being alert to such materials might assist an animal geography in which the spotlight is turned on how animals—like so many outsider human groups—become constructed by human communities as one thing rather than another (as "this" kind of phenomenon with "these" characteristics) and then subjected to related socio-spatial practices of inclusion or exclusion.

Animals and the City

This brief historiographic survey of how geographers—and particularly human geographers—have treated animals in their studies raises a number of issues, notably about how a reshaped animal geography might work through the twin maneuvers of paying attention to both the "psycho-biological make-up" and the discursive constitution of animals. The first of these strategies urgently needs more sustained consideration, perhaps by consulting emerging literatures on "animal consciousness,"[27] but it is only the second maneuver that I intend to pursue in the remainder of this paper. I propose to do this by looking in more detail at animals in the city as a flash point in the struggle between animals and humans, and one in which the discourses that humans utter and script about animals in a range of registers—from the medical to the moral—often embrace a resolutely anti-animal agenda hostile to the presence of live animals and therefore urging their socio-spatial exclusion. At the same time, questions about animals transgressing human expectations of what animals should be like, how they should behave, and where they should or should not be present will become more prominent, hence exploring the implications of regarding animals as a social group caught up in the maelstrom of city life (Figure 3.1).

There are many dimensions to how animal-human relations occur in,

Figure 3.1 "Cardiff," a watercolor by T. Rowlandson, plate number 67 in
A. P. Oppé's 1923 *Thomas Rowlandson: His Drawings and Watercolours*
(The Studio, London); the picture shows the center of a provincial
city, Cardiff in Wales, probably circa 1800–1810, and it conveys an
impression of humans and animals (cows, pigs, horses, and dogs) all sharing
the same small expanse of urban space

through, and across the city, and it would be possible to emulate Tuan by
studying pets as both an urban occurrence and (on occasion) an urban
problem, notably in the case of dogs fouling public spaces or becoming
strays frequenting city "back regions."[28] Alternatively, attention might be
given to the relationship between cities and zoos,[29] particularly given the
argument that a properly constituted zoo demands more land than is
available in most built-up areas. Another possibility would be to investigate
the relationship between cities and circuses. But it is the experience of
livestock animals in the city that fascinates me, and a useful place for me to
begin is with this observation:

> The idea of finding animal husbandry in an English city in the present or the
> recent past might appear strange in view of the current pressure of
> urbanisation upon agricultural land use. The built-up area somehow seems
> an alien environment in which to keep horses, cows, pigs and sheep, but in
> mid-nineteenth century London the idea of a clear-cut distinction between
> urban and rural life had yet to develop.[30]

This observation derives from Atkins's discussion of nineteenth-century
London's intra-urban milk supply, in the course of which he identified and
mapped the remnants of a "back-yard agriculture" with dairy cows being
kept in and around urban dwellings (and sometimes in sheds below ground)

in much of the metropolitan area. His paper traded on the surprise that readers might feel about his findings given the prevalent assumption that the city is an environment where livestock animals have no place, and a similar note of surprise is sounded in geographical papers dealing with Southern California's "dairy cities" and with the "industrialised drylot dairying" of herds kept on the fringes of large metropolitan centers in the United States.[31]

These accounts also recognize that the presence of livestock animals in cities is often contested, since city dwellers commonly object to such animals being part of the urban scene because of, amongst other problems, the "odours, flies and unseemly sights associated with animal husbandry."[32] Atkins made it clear that by the nineteenth century cow-keeping in London was a practice very much under threat, chiefly as a result of an emerging sanitary discourse critical of the scant interest shown by London cow-keepers "in cleanliness for the sake of any neighbouring human population."[33] The well-known ambitions of Chadwick, Simon, and other mid-century public health reformers looking to cleanse a filthy urban environment included a fierce opposition to livestock animals in the city, and they were "especially enthusiastic for the exclusion of cattle slaughtering, milk production and other so-called 'noxious trades' which created both dirt and smell as a byproduct of their operation."[34] There are two larger and intertwining stories here: one involving the long-term process whereby all sorts of phenomena have become categorized in certain ways and allotted to certain spatio-temporal containers, thus raising the difficulties of what to do with "matter out of place"[35]; the other involving the equally long-term splitting apart of the urban and the rural as distinctive entities conceptually associated with particular human activities and attributes (the industrial and civilized city, the agricultural and barbarian countryside.[36] My argument is simply that animals as a social group have become inextricably bound up in these stories, much as have certain outsider human groups, and as a result animals have become envisioned in particular ways with particular practical consequences: one of which is that some animals (cats and dogs) have been turned into pets valued as an element of the urban world whereas other animals (cows, sheep, and pigs) have become matter that should be expelled to the rural world.

The background to this quite specific contestation of city spaces by livestock animals and humans lies partly in the historical geography of the meat industry, or to be more precise in the changing spatial structure of this industry, which led it into large urban areas such as London and Chicago. It appears that by the nineteenth century most of the urban centers of Britain and the United States supported some kind of meat industry, but it is also evident that several cities in particular were becoming known for their meat industry, live meat markets, and concentration of slaughterhouses and related activities. In Britain, London was obviously one such city, and from AD 950 onwards a sizeable network of traders and slaughterers had grown up around the Smithfield live meat market at the heart of the capital,[37] but other cities such as Leeds (with its six or seven large slaughterhouses, leather industry, neighboring woollen industry and substantial working-class population prepared to consume edible offal) also rose to prominence on this count.[38] In the United States there was a centralizing trend in the meat-packing industry during the nineteenth century, such that an earlier period based upon the river transport of animals and the prominence of smaller river towns was followed by a period based upon rail transport and the ascendancy of cities such as Cincinnati and Louisville;[39] but by the 1860s it was Chicago that had become the best-known meat city.[40] I will concentrate particularly on the London experience in what follows, drawing in part on primary research, but I will also comment briefly on similarities in the Chicago experience. It should be noted that elsewhere slaughterhouses in London and Chicago are also discussed,[41] but I can only focus here upon live meat markets as sites of human-animal encounter in these cities.

Concern about Smithfield in London hinged around the spatial constrictions of the market itself, which meant that market days on Mondays and Fridays gave rise to "scenes of indescribable confusion,"[42] and it can be seen from nineteenth-century reports that many parties regarded this volatile mixing of beasts and people as undesirable for all concerned. When the market had first been instituted in the tenth century the situation had been different,[43] but by the nineteenth century the growth of the city had completely hemmed it in on all sides by houses and shops (Figure 3.2):

> I think it must be evident to anybody who goes into Smithfield market that, however suitable it may have been for the purpose originally, it has grown beyond it, both geographically and also in the size of it; and that which was the most suitable place originally has become exceedingly unsuitable and a great public nuisance. I am clearly of the opinion that the space of Smithfield is exceedingly inadequate to the purpose, and that the evils are very much aggravated by the want of space, both as to cruelty to the beasts and confusion in the arrangement.[44]

A shop assistant from a furnishing ironmonger near the market neatly captured the prevailing view that there was something deeply wrong, both distasteful and ludicrous, in allowing livestock animals to violate human space:

> On Monday last we had one beast put his head through the window; we are obliged to have a person at the door to keep them off; and last Monday week we had a sheep got into the shop and fell down the cellar steps into the cellar amongst the workmen: I think that fewer customers come to the shop on Monday; the ladies would not come to the shop if there was a crowd of bullocks.[45]

Figure 3.2 "Smithfield Market," an aquatinted watercolor by A. Pugin and
T. Rowlandson, plate number 8 in R. Ackerman's 1808–1810 *The Microcosm of
London* (R. Ackermann's Repository of Arts, London); the picture gives a good
impression of the market hemmed in on all sides by city buildings

Reproduced by kind permission of the Guildhall Library, Corporation of London

The owner of a shoe warehouse in Smithfield continued the theme of
insufficient space, but added the revealing remark that he considered the
whole situation to be "highly inconsistent with the rights of humanity and
the interests of the public,"[46] and in so doing he voiced the view that animals
such as sheep and cattle should not be permitted into London (at least not
in such an uncontrolled fashion).

The above materials are taken from an 1828 report, and similar arguments can be found in a report of 1849, in which a strong recommendation
was made to combine all of London's meat markets on a new site adjacent
to railway termini (animals from all over the country were increasingly
being transported to London by rail rather than by road) and away from
dense concentrations of human population.[47] Perren outlined the
entrenched opposition that faced the relocation proposals,[48] as well as
describing the eventual victory of the relocationists following an act of 1851
and the opening of the Metropolitan Cattle Market at Copenhagen Fields
in 1855, at which point a new measure of spatial ordering (and to some
extent exclusion) was imposed upon the animals by their human masters.
What is more interesting in this context, though, is the further gloss that
witnesses before the 1849 committee gave to the issue of animal-human

mixing in and around Smithfield. Mention was made of the sheer physical danger to which people were subjected by the driving of large animals at speed through congested streets adjacent to the market, and one physician stressed that "the driving of cattle" associated with Smithfield was "very dangerous to the inhabitants."[49] In addition, he recounted one particular incident when cattle from a drove bolted and ended up blocking a pavement outside of his house in Brook Street, and added that "I am quite sure if women or children had been there they would have been frightened."[50] In a later report these fears were echoed by representatives of the police, and a Mr Kittle (superintendent of the Metropolitan Police) denounced the driving of cattle and sheep through the streets as "a most substantial evil," and then declared that "in the densely populated neighbourhoods, such as Seven Dials, or in a close neighbourhood of that kind, driving cattle through them must necessarily be dangerous."[51] Kittle favored a new system in which railways were used for bringing cattle into London, the implication being that highways ought to be left to humans and possibly to horses, and the authors of the report indicated that the police were even unhappy with a system that restricted the driving of cattle through the streets to the hours between 7 P.M.. and 10 A.M.[52]

The materials taken from the primary documents here are very much to do with health and safety, as well as questioning the disorderliness endemic to having livestock animals charging through urban streets, but perhaps even more intriguing is the grid of moral assumptions informing the views of some contemporaries about "beasts" in the center of London. The evidence of a Mr J.T. Norris, the owner of a printing establishment located close to Smithfield, is most revealing in this respect, not least because of the form of certain questions addressed to him by the 1849 committee. An initial observation to the effect that the chief trades "encouraged by the existence of Smithfield" were "gin shops and public houses" led into a discussion of the locality's "degeneracy" and a claim about its "respectable" citizens wanting to see the market's removal.[53] Then the following question and answer were heard:

> Can you speak as to the morality of the neighbourhood of Smithfield, whether it is as pure as the rest of the city of London?—I am afraid it is not; the language I hear used as I pass, and the frequent visits of a good many concerned in the business of the market to the gin shops, the fighting and the disturbances that occur thereabout, make me say that it exhibits the lowest state of morals to be found anywhere in the metropolis, excepting perhaps St. Giles.[54]

Although not explicit, a transference of associations was clearly occurring here, in that a locality regularly inundated by large beasts devoid of human qualities was automatically being cast as one in which the "higher" processes of human thought and sensibility were impaired in the people living and working there. The anticipation was that these people would be debased, bestial in their habits and strangely similar in disposition to the animals with which they shared their spaces. That such a transference was occurring can be seen more starkly in a further exchange between the committee and Norris:

> Have you seen any particular cases of immorality in that neighbourhood? [Norris is initially reluctant to answer but eventually explains.] I . . . know that I have seen that which to me in a refined city is very unbecoming: I have seen bullocks driven from the market, which have been imperfectly operated upon, jumping on the backs of the cows in the public streets, in the presence of passengers of both sexes. I think that it is an offence to decency: it is unbecoming in a great city, and forms a reason in my mind why such scenes should be at a distance, and out of the way of the observation of females and children.[55]

The untamed sexuality of the animals being freely expressed in the "public streets" was regarded by Norris as thoroughly degrading; he evidently supposed it not only to disturb the vulnerable minds of women and children, but also to act as a likely stimulus to improper sexual practices on the part of the impressionable people living and working in Smithfield. In this latter connection Norris was even asked if he had learned of "any improper practice between the sexes occurring there," to which he replied that he had "heard of it frequently" but that he would rather not specify details.[56] There can be no doubting the chain of thought here leading to the conclusion that livestock animals should be kept "at a distance" from the normal spaces of the "refined city" for the good of the "public morals."[57]

An important point to make here—and one that squares with my insistence above on an animal geography that allows an enlarged space for animals *as animals*—is that the unrestrained sexuality of the bullocks witnessed by Norris, along with the transgressions of shop space by sheep and cattle, ran very much against the wishes of London's nineteenth-century human community. Similarly, from the perspective of the animals themselves it is likely that being driven to and from Smithfield (as well as being exhibited there) would have involved much pain and possibly terror, and I should acknowledge that within the human discussion of this process many people did express concern at what the sheep and cattle might actually suffer as a result of the congestion associated with the market. Mr Gurney's worries about "cruelty to the beasts" have already been noted,[58] and Mr Hickson commented before the same committee that "the cruelty practised on the cattle in beating them into rings no person can believe unless they have seen it."[59] The 1849 committee heard similar objections and referenced a report from a Mr Stafford in which—in the course of noting the "great inconvenience" felt by many people from having cattle driven past their houses—he stated that "a large portion of the public are impressed with a belief that unnecessary pain is inflicted upon animals as a consequence of the confined space of the market."[60] Moreover, Norris asserted that the commonplace violence directed at the animals "in consequence of the insufficiency of space" actually "educate[d] the men in the practice of violence and cruelty, so that they seem to have no restraint on the use of it,"[61] and this line of reasoning implied a further link here between localities hosting large numbers of livestock animals and the debased behavior of its human occupants and workers.

The evidence from the secondary histories is that a city such as Chicago—with its markets, stockyards, and driving of pigs and cattle—was host to much the same set of debates as those arising in nineteenth-century London over the limited spaces being contested by animals and humans. One historian of Chicago duly identified the "chronic traffic jams between bipeds and quadrupeds fighting for the right-of-way on city streets,"[62] and a contemporary commentator spoke of the spatially dispersed livestock pens and packing plants, which obliged drovers "to drive their animals through the crowded streets of the city, from yard to yard, thereby undergoing the greatest inconvenience."[63] One particular incident illustrated the chaos and threat endemic to this state of affairs:

> In November 1863 an impatient drover ignored the warning bell and took his cattle on to the Rush Street bridge when it was about to open. As the span began moving, cattle stampeded to one end, blocking the only escape route for pedestrians and causing the iron bridge to twist and break. Fifty cattle and a dozen people went into the Chicago River, and taxpayers were set back $10,000.[64]

Partly in response to this situation and partly in an attempt to escape from "nuisance ordinances" prompted by the slaughterhouses, a specific solution was proposed which introduced a very new ingredient to the geography of the city's meat industry: "a novel centralised stockyard beyond the urban boundaries."[65] The Union Stock Yard duly opened in 1865, situated in the township of Lake where it was accessible to the railroads and the local packers, and it was the equivalent of the new Copenhagen Fields market opened in London in 1855, albeit rather more grandly conceived. In essence both of these spatial solutions entailed removing livestock animals from the throng of the city (which was increasingly identified as a place for people rather than for beasts) and taking them to more secluded country-side surroundings (which were deemed appropriate for beasts who proved so difficult to manage in congested marketplaces and urban streets).[66]

Concluding Thoughts

The purpose of this chapter has been to consider the possibility of a revived animal geography—a social-cultural animal geography[67]—which counters the anthropocentrism of much existing geographical work on animal-human relations, as found in Anderson's highly partial approach to "living things" and as symbolized by the effective exclusion of animal geography

from the fold of practised human-geographical subdisciplines.[68] In seeking
to forge an approach more sensitive to animals as "living things" encounter-
ing, being constructed by, and sometimes trangressing what is expected of
them from the human world, I have drawn inspiration from sources as
diverse as Tuan and Sauerian inquiries into animal domestication. The
result should certainly be regarded as provisional, as a small experiment in
which the idea of regarding animals as a social group possessing both inner
experiences and outer determinations (much like many "outsider" or
minority human groups) has been shaken to see what it will reveal.
Threading through my paper is also a more particular argument about how
animals become inserted into human discourses: a recognition of the
cosmo-religious conceptual frameworks that may often have encompassed
animals in other human societies in other times and spaces, but also a
suggestion that in societies nearer in time and space animals have inevitably
been defined, categorized, interpreted, praised, criticized, hated, and loved
in a diversity of ways that have commonly had spatial implications. It might
be appropriate here to think of a continuum between inclusion and
exclusion, with animals such as dogs and cats tending to be at the
inclusionary extreme (as "companion animals" or pets readily accepted into
the everyday spaces of much human life) and with animals such as lions and
bears tending to be at the exclusionary extreme (as wild and even dangerous
animals more normally kept at a physical distance from these everyday
spaces). The experiences of many animals do not fit easily onto this
continuum. Are we to say that a fish in its bowl is included or excluded,
since it occupies "its own world, one that does not impinge on ordinary
human living space"?[69] Are we to say that an urban fox is included or
excluded, since it deliberately utilizes city spaces even if humans do not
want it to? And yet I would maintain that to ask questions about animal-
human relations using simple notions of inclusion and exclusion (however
scrambled they may have to become in the process) will remain a useful
tool for any refashioned practice of animal geography.

   I have sought to secure my claims about the value of such an animal
geography in the substantive vignette sketched out, which recovered something of the discourses (medical, hygienic, organizational, and moral) that
have gradually coded animals such as cows, sheep, and pigs as impure,
polluting, disruptive, and discomforting occupants of city spaces where
humans are supposed to live and work. The hope is that the assumptions
coded into the animal-human relations involved here have been exposed
for scholarly attention, rather than being left in some taken-for-granted
shadowland, and that as far as is possible a glimpse has been caught of the
animals themselves—of their experiences, sufferings, and unwitting acts of
transgression—albeit through the distorting lenses of historical documents
written by humans. At the same time what should have become clear is
the definite and growing will to expel certain categories of animal, along
with the meat markets where they were traded and the slaughterhouses
where they were killed, all of which duly ended up being excluded from

nineteenth-century cities such as London and Chicago. My final thought is that the possibilities opened up in such studies for understanding the socio-spatial processes of exclusion to which particular animals are so often subjected, when in effect being chased from the streets and consigned to the fields, depend precisely on achieving a (re)inclusion of these animals in contemporary human-geographical theorizing and research.

Acknowledgements

I would like to thank Jody Emel and Jennifer Wolch for their encouragement and advice in the preparation of this chapter. It is a modified version of an essay published originally in *Environment and Planning D, Society and Space*, vol. 13, no. 6, December 1995, pp. 655–81, and thanks are due also to its publishers, Pion Limited, for permission to reprint. Versions were given to seminars in both the Department of Geography, Royal Holloway, and the Department of Archaeology, Southampton, and thanks are due to participants in these seminars for their insights and support; but I should mention in particular Felix Driver, Dave Gilbert, Rob Potter, and Julian Thomas. Various colleagues in Lampeter past and present have given me help and suggested materials to look at, and here I should mention in particular Tim Cresswell, Clare Fisher, Jane Norris-Hill, Miles Ogborn, and Hester Parr. Additional thanks go to Denis Cosgrove for his valuable comments on an earlier draft of the paper, to two anonymous referees of that version for their constructive criticisms, and to Kay Anderson for letting me see her excellent piece on the Adelaide Zoo prior to publication (see chapter 2). Finally, I would like to thank Robin Donkin for first stimulating my interest in animal geographies, and for his generous response to the earlier incarnation of this chapter.

Notes

1. See Y-F. Tuan, *Dominance and Affection: The Making of Pets*, New Haven, Conn. and London: Yale University Press, 1984. In one sense, human beings are animals, of course, but in this paper I reserve the term "animals" for nonhuman animals, which are animate (unlike plants) and which can be taken to include everything from the elephant to the ant, the whale to the starfish, the eagle to the hummingbird. I acknowledge that such an attempt to delineate nonhuman animals is itself open to criticism.

2. C. Philo, "'The Same and Other': On Geographies, Madness and Outsiders," Loughborough University of Technology, Department of Geography, Occasional paper no. 11; D. Sibley, *Outsiders in Urban Societies*, Oxford: Basil Blackwell, 1981; Sibley, "Outsiders in Society and Space," in K. Anderson and F. Gale, eds, *Inventiong Places: Studies in Cultural Geography*, Melbourne: Longman Cheshire, 1992, pp. 107–22; and Sibley, *Geographies of Exclusion: Society and Difference in the West*, London: Routledge, 1995.

3. D. Haraway, *Simians, Cyborgs, and Women: The Reinvention of Nature*, London: Routledge, 1991.

4. T. Cresswell, *In Place/Out of Place: Geography, Ideology and Transgression*, London: University of Minnesota Press, 1996, pp. 22–3.

5. As recently surveyed in W. Cronon, *Nature's Metropolis: Chicago and the Great West*, New York: Norton, 1992.

6. M. S. Anderson, *Geography of Living Things*, London: English Universities Press, 1951, p. 1.

7. Anderson, *Geography of Living Things*, p. 7.

8. For further brief notes on the earliest "geographical" writing on the spatial patterns and ecologies of mammals and other fauna, see T. W. Freeman, "The Royal Geographical Society and the Development of Geography," in E. H. Brown, ed, *Geography, Yesterday and Tomorrow*, Oxford: Oxford University Press, 1980, pp. 21–2.

9. J. A. Thomson, "How to Teach the Geographical Distribution of Animals," *Geographical Teacher* 3, 1905, pp. 116–19.

10. M. I. Newbigin, *Plant and Animal Geography*, 1936; reprint, London: Methuen & Co., 1968.

11. W. George, *Animal Geography*, London: Heinemann, 1962; C. F. Bennett, "Cultural Animal Geography: An Inviting Field of Research," *Professional Geographer*, vol. 12, no. 5, 1960, pp. 12–14; C. F. Bennett, "Animal Geography in Geography Textbooks: A Critical Analysis," *Professional Geographer*, vol. 13, no. 2, 1961, pp. 13–16; G. S. Cansdale, "Some Problems in Animal Geography," *Geographical Magazine* 22, 1949, pp. 108–9; J. L. Davies, "Aim and Method in Zoogeography," *Geographical Review* 51, 1961, pp. 412–17; L. C. Stuart, "Animal Geography," in P. E. James and C. F. Jones, eds, *American Geography: Inventory and Prospect*, New York: Syracuse University Press, 1954, pp. 442–51; C. E. Williams, "A Laboratory Exercise in Animal Geography," *Professional Geographer*, vol. 14, no. 1, 1962, pp. 32–40.

12. Davies, "Aim and Method," p. 412.

13. Davies, "Aim and Method," p. 412.

14. Thomson, "How to Teach," pp. 116–19.

15. Cansdale, "Some Problems in Animal Geography"; G. S. Cansdale, "Animal and Man I: The Results of Competition and Conflict," *Geographical Magazine* 23, 1950, pp. 74–84; Cansdale, "Animals and Man II: Domestication—Providers of Food and Clothing," *Geographical Magazine* 23, 1951, pp. 390–99; Cansdale, "Animals and Man III: Domestication—Transport Animals, Pets and Allies," *Geographical Magazine* 23, 1951, pp. 524–34; Cansdale, "Animals and Man IV: Chance and Deliberate Introductions," *Geographical Magazine* 24, 1951, pp. 155–62; Cansdale, "Animals and Man V: Biological Control," *Geographical Magazine* 24, 1951, pp. 211–17. Cansdale was actually the super-intendent of the London Zoo, and as such wrote in the less specialist languages of zoogeography and more in a populist vein suited to *Geographical Magazine*. He authored a book based on the essays in the journal (G. S. Cansdale, *Animals and Man*, London: Hutchinson, 1952) and subsequently another book discussing the animals of the "Bible lands" in a manner keenly alert to their "geographical background" (G. S. Cansdale, *Animals of Bible Lands*, Exeter: Paternoster Press, 1970).

16. Bennett, "Cultural Animal Geography," p. 13.

17. C. O. Sauer, *Seeds, Spades, Hearths and Herds*, Cambridge, Mass.: MIT Press, 1969.

18. For a survey of world geographies underlying plant and animal domestication, see E. Issac, *Geography of Domestication*, Englewood-Cliffs, N.J.: Prentice-Hall, 1970; for specific examples, see R. A. Donkin, *The Peccary: With Observations on the Introduction of Pigs to the New World*, a book-length study comprising *Transactions of the American Philosophical Society*, vol. 75, part 5, 1985; Donkin, *The Muscovy Duck, Catrina Moschata Domestica: Origins, Dispersal and Associated Aspects of the Geography of Domestication*, Rotterdam: A. A. Balkema, 1989; Donkin, *Meleagrides: An Historical and Ethnographical Study of the Guinea Fowl*, London: Ethnographical Ltd, 1991; D. W. Gade, "The Guinea Pig in Andean Folk Culture," *Geographical Review* 57, 1967, pp. 213–14; Gade, "The Llama, Alpaca, and Vicuna: Fact Versus Fiction," *Journal of Geography* 68, 1969, pp. 339–43. See also R. P. Palmieri, "The Domestication, Exploitation and Social Functions of the Yak in Tibet and Adjoining Areas," *Proceedings of the Association of American Geographers* 4, 1972, pp. 80–83; P. A. D. Stouse, "The Distribution of Llamas in Bolivia," *Proceedings of the Association of American Geographers* 2, 1970, pp. 136–40.

19. Donkin, *Meleagrides*, p. ix.

20. Donkin, *The Muscovy Duck*, p. 1.

21. Donkin, *The Peccary*, p. 41.

22. Donkin, *Meleagrides*, p. ix.

23. The argument about animals being kept for sacrifice during "religious" ceremonies owes much to the late-nineteenth-century thinking of Hahn, a German historian-geographer who has regularly been cited as an influence on academic geography. See Donkin, *Meleagrides*; Issac, *Geography of Domestication*, esp. pp. 6–7; F. L. Kramer, "Eduard

Hahn and the End of the 'Three Stages of Man,'" *Geographical Review* 57, 1967, pp. 73–89; Palmieri, "Domestication, Exploitation."

24. Rodrigue claimed that the prior existence within a society of storage facilities and trade networks would have prompted animal domestication, thereby effectively restating a political-economic over a more social-cultural explanation. The debate here about animal domestication is underlain by a deeper disagreement between what Rodrigue called "the Hahn-Sauer-Issac-Simoonses tradition in geography," which she reckoned to be compromised by its "philosophical idealism," and the more empirically rooted materialist approaches taken by the disciplines of archeology and anthropology: C. M. Rodrigue, "On the Dehiscence of Debate in the Geography of Domestication: A Reply to Mark Blumler," *Professional Geographer* 45, 1993, pp. 363–6. See also Rodrigue, "Can Religion Account for Early Animal Domestications? A Critical Assessment of the Cultural-Geographical Argument, Based on Near Eastern Archaeological Data," *Professional Geographer* 44, 1992, pp. 417–30; and M. A. Blumler, "On the Tension between Cultural Geography and Anthropology: Commentary on Rodrigue's 'Early Animal Domestication,'" *Professional Geographer* 45, 1993, pp. 359–63.

25. Sauer, *Seeds, Spades*, p. 31.

26. See M. Douglas, *Purity and Danger*, London: Routledge, 1966.

27. J. Wolch and J. Emel, "Bringing the Animals Back in," *Environment and Planning D: Society and Space* 13, 1995, pp. 632–6.

28. Anonymous, "Do Cities Really Need Dogs?" *Time Magazine*, 20 July 1970, p. 29.

29. K. Anderson, "Culture and Nature at the Adelaide Zoo: At the Frontiers in 'Human' Geography," *Transactions of the Institute of British Geographers*, new series 20, pp. 275–94.

30. P. J. Atkins, "London's Intra-urban Milk Supply, Circa 1790–1914," *Transactions of the Institute of British Geographers*, new series 2, 1977, p. 383.

31. H. F. Gregor, "Industrialized Drylot Dairying: An Overview," *Economic Geography* 39, 1963, pp. 299–318; G. F. Fielding, "Dairying in Cities Designed to Keep People Out," *Professional Geographer*, vol. 14, no. 1, 1962, pp. 12–17.

32. Fielding, "Dairying in Cities," p. 16.

33. Atkins, "London's Intra-urban Milk Supply," p. 389.

34. Atkins, "London's Intra-urban Milk Supply," p. 389.

35. T. Cresswell, "The Crucial 'Where' of Graffiti: a Geographical Analysis of Reactions to Graffiti in New York," *Environment and Planning D: Society and Space* 10, 1992, pp. 329–344; Cresswell, *In Place/Out of Place*; Douglas, *Purity and Danger*; D. Sibley, "Purification of Space," *Environment and Planning D: Society and Space* 6, 1988, pp. 409–21; and Sibley, *Geographies of Exclusion*.

36. R. Williams, *The Country and the City*, London: Chatto & Windus, 1973.

37. R. Perren, *The Meat Trade in Britain, 1840–1914*, London: Routledge & Kegan Paul, 1978, p. 32.

38. Perren, *The Meat Trade in Britain*, p. 46, highlighted the difference between provincial centers sending live meat to London and others (such as Aberdeen, Edinburgh, and Leeds) sending dead meat, therefore possessing their own slaughterers. For further details on the geography of the nineteenth-century British meat market, see M. P. Phillips, "Market Exchange and Social Relations: The Practices of Food Circulation in and to the Three Towns of Plymouth, Devonport and Stonehouse, 1800–circa 1870," Ph.D. diss., University of Exeter, 1991.

39. J. M. Scaggs, *Prime Cut: Livestock Raising and Meatpacking in the United States, 1607–1983*, College Station, Texas: Texas A and M University Press, 1986; M. Walsh, "The Spatial Evolution of the Mid-Western Pork Industry, 1835–1875," *Journal of Historical Geography* 4, 1978, pp. 1–22; Walsh, *The Rise of the Midwestern Meat Packing Industry*, Lexington, Kentucky: University Press of Kentucky, 1982.

40. W. Cronon, *Nature's Metropolis: Chicago and the Great West*, New York: Norton, 1992; T. Jablonsky, *Pride in the Jungle: Community and Everyday Life in Back of Yards, Chicago and Baltimore*, Baltimore, Md.: Johns Hopkins University Press, 1993; L. C. Wade, *Chicago's*

Pride: The Stockyards, Packingtown and Environs in the Nineteenth Century, Urbana and Chicago, Ill.: University of Illinois Press, 1987.

41. C. Philo, "Animals, Geography and the City: Notes on Inclusions and Exclusions," *Environment and Planning D: Society and Space* 13, 1995, pp. 655–681.

42. Perren, *The Meat Trade in Britain*, p. 32.

43. Smithfield was apparently first organized as a market circa 1615 by the authorities of the City of London, who "reduced the rude, vast place of Smithfield into a faire and comely order" (according to a contemporary observer, in W. J. Passingham, *London's Markets: Their Origin and History*, London: Sampson Low, Marston & Co., 1935, p. 5), but by the early nineteenth century the congestion of people and animals resulting from the city's crowding of the site had rendered it "a public nuisance and a public scandal" (Passingham, *London's Markets*, p. 7). For further details of the market's history, with special reference to the nineteenth-century conflicts leading to its closure as a live-meat market, see A. Forshaw and T. Bergström, *Smithfield, Past and Present*, London: Robert Hale, 1990, pp. 20–21, 34–6, 54–60.

44. Gurney, in *Second Report from the Select Committee on the State of Smithfield Market*, PP 1828 VIII [3], pp. 5, 36–7.

45. Padmore, in *Second Report from the Select Committee*, pp. 6, 147.

46. Hickson, in *Second Report from the Select Committee*, pp. 6–7, 16.

47. *Report from the Select Committee on Smithfield Market*, PP 1849 XIX [242], p. 247.

48. Perren, *The Meat Trade in Britain*, pp. 36–41.

49. Webster, in *Report from the Select Committee*, p. 400.

50. Webster, in *Report from the Select Committee*, p. 400.

51. Kittle, in *Report from the Select Committee on Noxious Businesses*, PP 1873 X [431], pp. 627, 630.

52. Committee, in *Report on Noxious Businesses*, p. 434.

53. Norris, in *Report from the Select Committee*, pp. 367–9.

54. Committee and Norris, in *Report from the Select Committee*, p. 372.

55. Committee and Norris, in *Report from the Select Committee*, p. 372.

56. Committee and Norris, in *Report from the Select Committee*, p. 372. Later on in his evidence, Norris was asked about parallels between the situation in Smithfield and that occurring in Regent Street "at the close of the evening" (the reference was obviously to the activities of prostitutes), and he replied that—although not familiar with Regent Street in the evening—"I can imagine the immorality to which you allude . . . but I should make a distinction between the open gross violation of the public decency which I understand to occur in Smithfield, and the more refined amours of Regent Street: that which offends the eye should be put out of sight." *Report from the Select Committee*, p. 373.

57. The term "public morals" was actually used in this context by Sir Charles Douglas, and more particularly what he urged was that "every consideration for public morals, and especially for the health of the metropolis, demands [Smithfield's] removal at the earliest possible period." Douglas, in *Report from the Select Committee*, p. 253.

58. Gurney, in *Second Report from the Select Committee*, pp. 5, 37.

59. Hickson, in *Second Report from the Select Committee*, pp. 6–7, 16.

60. Stafford, in *Report from the Select Committee*, p. 254.

61. Norris, in *Report from the Select Committee*, p. 367.

62. Wade, *Chicago's Pride*, p. xii.

63. Quoted in Scaggs, *Prime Cut*, pp. 45–6.

64. Wade, *Chicago's Pride*, p. 47.

65. Wade, *Chicago's Pride*, pp. xii–xiii.

66. The new sites have long since been incorporated into the urban fabrics of London and Chicago, and so the "out of town" qualities that were vital to their initial selection are no longer readily apparent. Perren, *The Meat Trade in Britain*, pp. 40–41, noted that in Britain this spatial solution was arrived at in a number of provincial towns and cities, all of which witnessed the resiting of their live-meat markets to "less congested areas" (usually, "quieter suburbs") near to "out of town" railway stations; and he also identified a process

whereby street markets were increasingly replaced by enclosed ones that screened the animals off from the surrounding public spaces.

67. See also Anderson, "Culture and Nature at the Adelaide Zoo"; and D. Matless, "Moral Geography in Broadland," *Ecumene: A Journal of Environment, Culture, Meaning* 2, 1994, esp. pp. 136–44.

68. Anderson, *Geography of Living Things.*

69. Tuan, *Dominance and Affection*, p. 948.

4

Le Pratique Sauvage:
Race, Place, and the Human-Animal Divide

Glen Elder, Jennifer Wolch, and Jody Emel

Introduction

Former Hollywood sex goddess and French animal rights activist Brigitte Bardot was reported to be considering self-imposed exile from her French homeland. Her reasons: the invasion of France by late-twentieth-century Moslem infidels with their ritualistic animal slaughter practices. Bardot stated that "from year to year we see mosques flourish across France, while our church bells fall silent." To further polarize her anti-immigrant stance, Bardot invoked the image of "tens of thousands of poor beasts whose throats are slit . . . with blades that are more or less sharpened by clumsy sacrificers who have to repeat their gesture several times, while kids splashed with blood bathe in this magma of terror, of blood spurting from badly-slit jugulars."[1] Despite its provocative images of idolatrous heathens, messy blood-letting, and animal terror, Bardot's high-camp performance was in fact unoriginal. Her role and lines were snatched from a long string of previous performers, including explorers, colonialists, slave holders, modern-day racists and xenophobes, and right-wing politicians. All of these characters have constructed racial difference by casting the Other as "savage" or uncivilized on the basis of their interactions with animals.

Visceral arguments about animals and race have in fact become increasingly common, particularly in Europe and the United States. Animals and their bodies appear to be one site of struggle over the protection of national identity and the production of cultural difference. Why are animals used (and so useful) in such sociopolitical conflicts? In this chapter we attempt to address this question. Following a series of illustrative case studies about conflicts which involve animals and race in the US, we argue that practices that bring harm to animals are being used to racialize immigrant groups. On the basis of postcolonial theories of racialization and the impacts of postmodern time-space compression in a globalizing economy, we suggest that this process of animal-linked racialization works to sustain power relations between dominant groups and subordinate immigrants, deny their

72

legitimacy as citizen-subjects, and restrict the material benefits that derive from such status.

Animal practices are extraordinarily powerful as a basis for creating difference and hence racialization. This is because they serve as defining moments in the social construction of the human-animal divide. While universally understood in literal terms, the divide is a shifting metaphorical line built up on the basis of human-animal interaction patterns, ideas about hierarchies of living things (both human and nonhuman), and the symbolic roles played by specific animals in society. Certain sorts of animals (such as apes, pets, or revered species) become positioned on the human side of this metaphorical line, rendering some practices unacceptable. But other harmful practices are normalized, to reduce the guilt (or at least the ambivalence) associated with inflicting pain or death, and to justify them as defensible behaviors differentiated from the seemingly wanton violence observed in nonhuman nature.

Norms of legitimate animal practice are neither consistent nor universal. Instead, codes for harmful animal practices are heavily dependent on the immediate context of an event. Here, the critical dimensions of context include the animal species, human actor(s), rationale for and methods of harm, and site of action involved in the practice. And because animal practices emerge over long periods of time as part of highly variable cultural landscapes, place is also implicated in constructing the human-animal divide. When distinct, place-based animal practices are suddenly inserted into new locales by immigrants and are thus decontextualized conflict erupts. Those newcomers who violate or transgress the many-layered cultural boundary between people and animals become branded as "savage," "primitive," or "uncivilized" and risk dehumanization, that is, being symbolically allocated to the far side of the human-animal divide.

Driven by anxiety over declining global hegemony, economic and social polarization, and growing population diversity that threatens the country's image as "white," dominant groups in the US are waging an intense battle to maintain their positions of material and political power. Moreover, they seek to protect a socially constructed national identity built upon some particular (and typically reified) categories of people and places in part defined in contradistinction to others.[2] In this situation, racialization of those immigrants whose darker skin color feeds into entrenched racial ideologies, stereotypes, and discursive practices serves to demarcate the boundaries of national culture and belonging to place, and to exclude those who do not "fit." Conflicts over animal practices, rooted in deep-seated cultural beliefs and social norms, fuel ongoing efforts to racialize and devalue certain groups of immigrants. Animal practices have thus become tools of a cultural imperialism designed to delegitimize subjectivity and citizenship of immigrants under time-space conditions of postmodernity and social relations of postcoloniality.

Our readings of the links between race, place, and animals imply that violence done to animals and the pain inflicted on them are inevitably interpreted in culturally- and place-specific ways. It is therefore both difficult and inappropriate to characterize one type of harm or death as more painful or humane than another. This categorically does not imply, however, that animal suffering, agony, and death are mere social constructs; *they are only too real*. Indeed, our ultimate purpose is to stimulate a profound rethinking of *all* "savage practices" toward animals as well as toward "othered" people. As our title suggests, we promote a "wild practice"(or *pratique sauvage*) in which heterogeneous others use their marginality as a position from which to pursue a radically open, anarchic, and inclusive politics.[3] We conclude by raising the possibility that a truly inclusive *pratique sauvage* could encompass animals, the ultimate other.

Postcolonial Animal Stories

We launch our arguments by telling a series of stories drawn from recent events in the US. Unlike colonial animal stories such as *Babar*, in which the animals are representations of colonists and "natives," these postcolonial stories focus on the treatment of animals by subaltern groups and the ways these practices are used to devalue them. Their practices, interpreted as "out of place" by dominant groups, serve to position them at the very edge of humanity—to racialize and dehumanize them through a complicated set of associations that measure their distance from modernity and civilization and the ideals of white America.

The rescue dog
Late in 1995, a three-month-old German shepherd puppy was beaten to death in a residential neighborhood of Fresno, one of the fastest growing urban regions in California's vast Central Valley.[4] The puppy death created a public furor. Neighbors complained to local authorities, and the man responsible for the dog's death was taken into custody on felony charges of animal cruelty. Later these charges were reduced to misdemeanor cruelty, to which the defendant pleaded guilty. The man charged in the case was Chia Thai Moua, a Hmong immigrant from Laos who had come to the United States in the 1970s. Moua was also what the press reports termed a "shaman." Curiously, his shaman's logic in turning to the puppy was precisely that of so many others who use dogs to serve people: he was trying to rescue another human (in this case, his wife). He explained that he had killed the dog in order to "appease an evil spirit" that had come to plague her in the form of diabetes. The sacrifice could drive out the spirit and effect a cure. According to Hmong beliefs, "a dog's night vision and keen sense of smell can track down more elusive evil spirits and barter for a sick person's lost soul." Other animals, such as chickens and pigs, are sacrificed first, but if the killing of such animals does not solve the problem, then, according to Moua, "If it is a serious case . . . I have no other choice" but to "resort" to a dog. Moua stated that each year he performs a special ceremony to release the souls of all the animals who have helped him, so that they can

be reborn. Thus, according to Moua, Hmong people from the highlands of Laos "are not cruel to animals. . . . We love them. . . . Everything I kill will be reborn again."

Moua's reliance on the Hmong conception of the human-animal border and the appropriate uses for certain animals puts him at odds with mainstream American ideas on the subject. He killed a dog. His reasons for doing so had no resonance or legitimacy for members of the dominant culture, who only sanction a limited number of contexts for dog killing. Dogs can be "laboratory workers" and "give" their lives to science, or they can be "entertainment workers" and be legitimately killed when no longer "employable"—witness the large numbers of "surplus" racing greyhound dogs that are killed each year. (Note that some forms of entertainment such as dog fighting, in which the *purpose* of the event, rather than the *result*, is dog injury and death, are strictly illegal.) But neither canine "lab workers" nor "entertainment workers" can be pets: dogs are usually purpose-bred for both the laboratory and the track.

Because Moua killed the puppy in his home, the dog was automatically a pet (and a pet of a revered breed at that). People are expected to dote on pet puppies in their homes, lavishing on them toys, tidbits, and attention. Barring unfortunate accidents, humans are not supposed to kill pets, except for veterinarians or euthanasia technicians in an animal shelter. Moua was neither. Worse, instead of using medicalized instruments such as the scalpel or syringe, to be wielded in the name of science or "kindness," Moua used a method (bludgeoning) widely seen as "inhuman"— a gross act of physical force that suggests a deeply disturbing animality.

An insightful head investigator for Fresno's Humane Society claimed that he could "count on my hand the actual cases [of Hmong dog sacrifices] I know about. . . . A lot of the false complaining is racism, pure and simple." Nonetheless, the publicity around Moua's deed and arrest did nothing to resolve ethnic tensions between the Anglo population of Fresno and the sizable Hmong population, which continue to fester.

Bambi's brother
One night in 1995, four men drove into the Angeles National Forest in southern California. One of the men proceeded to immobilize a deer with a spotlight, shoot the animal in the throat, and load the struggling animal into the trunk of the car. Back in town, the car was pulled over by a police officer, who, upon hearing thumping noises coming from the trunk, demanded that it be opened. There was the deer. Veterinarians were not able to save the deer, despite an emergency operation. All four men were arrested and charged with crimes; the shooter pleaded guilty to conspiracy for poaching and premeditated cruelty to an animal, and was sentenced to a year in jail, 100 hours of community service, and ordered to pay a $200 fine. The other three were convicted of lesser charges, for which one got a six-month jail sentence and all were ordered to perform community service and pay fines.[5]

The men involved in this case were Latinos: a photograph of all four appeared in the *Los Angeles Times* article covering the incident, and the images portray them as severe men with classically *indio* and *mestizo* features. The photograph caption identifies them as "Gunman" Enrique Chavez and his companions, Alfredo Guerrero, Juan Velasquez, and Ramon Espinoza (no quotation marks in original). The caption labels these Latino men as "killers" instead of a group of "sport hunters" because they look wrong for the part and acted in the wrong time-place. They were clearly not white men, nor were they wearing "hunting attire" or driving a truck. Like many sport hunters, they claimed to have shot the deer because they wanted some venison; but sport hunters supposedly shoot to kill, whereas in this case the deer was shot in the throat and left to bleed and suffer as the men made their way home where they intended to kill and eat the animal. Their rationale for this seemingly barbaric method was that it kept the meat very fresh, a quality highly valued by them but not by the broader Anglo culture (most of whom are happier eating severely anemic and thus "white," crate-raised veal). Lastly, by shooting a deer without a license in a place where hunting was not in season, these four were poaching—killing the wrong animal in the wrong place and time—and thus violating that period of the year when we are supposed to regard the forest as a piece of pristine nature reserved for all the wonderful wild creatures. They were, therefore *illegal.*

Enrique Chavez's Anglo lawyer recommended that his client spend his community service hours in a local animal shelter so that Chavez could "learn how to treat animals better"—a bitterly ironic prescription for anyone aware of how many millions of so-called pets are killed annually in such shelters, all in the name of "kindness."

The bowser bag

Two Long Beach men were charged with cruelty to animals for allegedly killing a German shepherd puppy and eating the dog for dinner on a March evening in 1989. A Los Angeles area judge ruled that there was no law against eating dogs, and that the animal had not been killed in an inhumane fashion. The charges were therefore dropped.

The case did not die, however. Rather, it spurred the introduction of a law, signed by then-Governor George Deukmejian, making pet-eating a criminal misdemeanor, punishable by a six-month jail term and a $1000 fine. Pets are defined in this statute as any animal commonly kept as a pet. Killing and eating wildlife, poultry, livestock, fish, or shellfish remain legal since these sorts of creatures fall beyond accepted definitions of "pet."[6]

But all this is beside the point, which is that Americans eat hot dogs, not dogs. In fact, given the status of most pet dogs and cats as quasi-human members of the family, eating a dog or cat is much too close to cannibalism for comfort. Indeed, the puppy involved was killed in an apartment complex, at home, it was all in the family. But the two men above were not "American," they were refugees from Cambodia. Trying to minimize the backlash against his community, the head of the Cambodia Association of America claimed that "Cambodians don't eat dogs," but it is widely known that many people from various parts of Asia do. (Isn't this how chow-dogs got their name?) And some Asians eat cats as well; civet cats, for example, are eaten in many parts of China and Southeast Asia. But in the Asian context, dogs and cats are "speciality" meats, considered "delicacy" foods. While most people see nothing wrong with eating many animals for food (including baby animals) and even taboo animals under conditions of duress, killing a cute helpless puppy for a luxury meal is another story—an act guided by self-indulgence, not the hand of necessity.

As initially drafted, the pet-protection bill only covered cats and dogs. Protests by Asian civic organizations led to an extension of the killing ban to all animals "commonly kept as pets." Curiously, however, the law still disregards pet turtles, rabbits, and pigeons, which are commonly eaten by Anglos. As Vietnamese-born editorial writer Andrew Lam claimed, the legislation implied that "[t]he yellow horde is at it again, that the eating habits of South East Asians, specifically the Vietnamese, are out of control" while "[i]t remains chic in a French restaurant to eat squab, as it is an accepted ritual for American fraternity boys to swallow live goldfish. And rabbit is nice in red wine."[7]

Horses heading for a fall

Several localities and states have recently banned horse tripping, an event traditionally performed in *charreadas* or Mexican-style rodeos.[8] *Charreadas* have been staged throughout Mexico for several centuries and are also frequently held throughout the southwestern United States. In this event, the legs of a horse that is galloping across the rodeo arena are lassoed by men who are pursuing on horseback. Once the legs are encircled by the lasso, the rope is pulled tight, throwing the horse to the ground. It is not uncommon for horses felled in this fashion to suffer injuries or even death.

The spreading efforts to ban horse tripping are grounded on the argument that the event is inhumane. But more to the point, horse tripping violates the deeply contradictory human-animal borders in force within dominant Anglo culture. It is difficult to underestimate the importance of horses to Anglo-European culture,[9] including Hispanic-origin societies. But in the US today, horses are seen both as pets (the number of working horses is now vanishingly small) and as perhaps the premier animal symbol of freedom, nobility, beauty, grace, and power. While it is acceptable to derive money from equine suffering and death (after all, no one is seriously trying to ban horse racing, picket the horse slaughterhouses that supply Alpo or Purina, or prevent the export of horse flesh to France), how could civilized people derive entertainment pleasure from watching such a glorious animal be thrown violently to the ground? Also, the method—tripping an innocent, noble, unsuspecting individual—is so sneaky and underhanded. It might be OK for cattle to be "hazed" (that is, roped, thrown, and hog-tied), but then again, they're *cattle*.

The people who perform the horse tripping are *charros* or *vaqueros*. Historically, *vaqueros* were simply Mexican cowboys who worked throughout the western borderlands. But as the Anglo land grab of the frontier proceeded, they were displaced by American cowboys who went on to become the most revered figures of the American West. Hollywood subsequently recast the *vaquero* in racialized and heavily masculinized terms, to become the image of a cruel, macho Mejicano, a mustachioed bandit-figure digging his razor-sharp spurs into his horse's sides until they bleed.

The blood of the lamb

In April 1987, a church practicing Santeria announced plans to open a house of worship in Hialeah, Florida. This announcement, along with a spate of angry calls from residents reacting to "whole piles of animals, stinking and with flies" that had been left behind following a sacrifice, prompted the Hialeah City Council to hold an emergency meeting in June. At that time, the council adopted a resolution noting that the Santeria religious group was potentially threatening public morals, peace, and safety, and passed an ordinance extending Florida's animal cruelty laws to cover ritual sacrifice, thus imposing criminal sanctions on the activity. The attorney general of Florida also expressed an opinion that religious animal sacrifice was not a "necessary" killing and so it was prohibited by state law. A few months later, the council adopted an ordinance that went further, prohibiting the possession, sacrifice, or slaughter of an animal with the intent to use such animal for food purposes. The prohibition applied, however, only if the animal was killed in any type of *ritual* regardless of whether the animal was in fact consumed for food. This left as legal the killing of animals in properly zoned and licensed establishments.[10]

The Hialeah ordinance was followed by bans in other cities. Los Angeles, for example, became the first city in the nation to outlaw ritual sacrifice (termed "torture-killings" in one headline) under any circumstances; and San Francisco followed with a ban of its own, amid news reports about an estimated one thousand cases of ritual slaughter, "disemboweled chickens" and "decapitated native songbirds" left as evidence. In San Francisco, where Santeria "high priest" Pete Rivera claimed that only high priests extensively trained in ritual sacrifice techniques were allowed to kill four-legged animals and that "[t]he gringo doesn't understand our religion," the ban created a furor and prompted the city council to allow sacrifices if the resulting meat was to be used primarily for purposes of consumption.

Ernesto Pichardo, founder and priest of Hialeah's Santeria church, located on a former used-car lot, took the city to court. In the face of protests by animal rights and humane society groups (as well as local Catholic and Baptist clergy), Pichardo argued that the city had violated the church's rights under the free exercise of religion clause of the Constitution's First Amendment. Santeria sacrifices were integral to key Santeria religious ceremonies (birth, death, marriage events) and were used to intervene with *orishas* or minor gods that are believed to have powers to help people with certain kinds of problems. An action for declaratory, injunctive, and monetary relief was filed in the US District Court, which ruled for the city on the grounds that the jurisdiction had a right to prevent health risks and emotional injury to children, protect animals from cruel and unnecessary killing, and restrict the slaughter of animals to areas zoned for slaughterhouses. The US Supreme Court, however, thought otherwise and on 11 June 1993 ruled that the city had not demonstrated a "compelling interest" in implementing the ban and had unfairly targeted a religious practice, sacrifice, used only by Santeros. The Court thus declared the ordinance void under the First Amendment.

This ruling was hardly surprising. To do otherwise could easily have opened up a Pandora's box for the Court and indeed the nation, since a finding of cruelty, for example, would have threatened such long-standing religious practices as Kosher slaughter, and could even have raised serious questions about the "humaneness" of conventional killing techniques practiced on the slaughterhouse floor. The visceral reaction at the local level thus necessarily faded away in the weighing of such national interests. The question remains, however, why the local response was so swift and so vehement.

The animals killed in Santeria practice include a wide range of domestic animals, including lambs, goats, and chickens, but also turtles, snakes, and (according to some reports) dogs. Most of these animals are eaten, but it is also traditional to leave remains at a major crossroads, leading one observer to note that it is "not uncommon to find decapitated chickens at the intersection of 98th Street and Broadway [in New York City]." Such practices thus violate the human-animal border in the dominant culture (where the killing of animals occurs on a vast scale but is almost completely hidden from view, and offal is "processed" rather than left on the roadway to stink in the sun). But more critical is the perception that the people doing the killing and their reasons for killing are deeply suspect, associated with some of the most threatening populations and illegitimate purposes.

These suspicions come through in most media and even scholarly accounts, in which the Santeria religion is described as a fusion of traditional African religious elements (mostly Yoruba) with parts of Roman Catholicism (the *orishas* are named after the Catholic saints or *santos*). Such descriptions imply that Santeria was "imported" from Africa, and more recently the Caribbean, and thus is not indigenous to the US. But both the history and the geography of these accounts are misleading. Emerging during America's slave-trade past, Santeria is in fact as "American" as Mormonism and less of an "import" than Catholicism or Presbyterianism, neither of which have adapted as much as Santeria to the cultural context within which they are practiced. The religion is also typically referred to as a "sect" or "cult" rather than a religion, despite an estimated 75–100 million practicing Santeros worldwide; Santeria is the majority religion in Brazil, Haiti, and Cuba. In the US, Santeria has spread rapidly among Latinos from Central and South America and among African Americans. Although the religion is not a form of *voodun*, it has certain common historical origins, thus conjuring up images of Satanism, demonism, and (literally) black magic. The frequent addition of the adjective "ritualistic" to descriptions also implies that Santeria is somehow primitive and simultaneously that "modern" religions, such as Catholicism, are *not* ritualistic. In addition, the African roots of Santeria link the religion in many minds to adjectives such as barbaric, backward, primitive, and irrational. All of these aspects of Santeria practice, as described or exaggerated in media accounts, deny it the legitimacy of a "world religion," thus throwing its rituals into question and also reducing the status of Santeria priests in comparison with other clergy.

To celebrate the Supreme Court victory, Hialeah-based Santeria priest Rigoberto Zamora held a public sacrifice. The local newspaper described the event, at which Zamora "poked a steak knife" into the throats of goats and rams, then "sawed through vocal cords and arteries until blood spurted," and finally took small birds and "twisted off their heads." How uncivilized, especially compared to battery-caged chickens, crated veal, factory-farmed hogs, and BST-laced milk from downer cows.

Postcolonial Racialization and the Human-Animal Divide

Our cases illustrate how, in the contemporary US, racialization of others is fostered by postcolonial interpretations of the human-animal boundary or divide, under time-space conditions of postmodernity. Many forms of racialization have, in fact, long relied upon a discourse about human-animal boundaries, namely the dichotomous division of sentient beings into categories of "human" and "animal." The most basic and durable criteria used to fix the boundary have involved differences in *kind*. But although humans and animals do manifestly differ (a point that is universally recognized), the interspecific divide is not solely a behavioral or biologically determined distinction. Rather, like so many other common categorizations (such as race or ethnicity), it is also a place-specific social construction subject to renegotiation over time. Moreover, the reasons for assigning one human group to one side of the boundary or another may also change between times and places.

From its earliest beginnings, Christian theology identified the soul as the defining feature of humanity. Even with the advent of Enlightenment ideas about animals, such as Descartes' identification of animals with machines, the boundary rested on the presence/absence of souls. With the rise of a more secular Western science, the key differences in kind became biological and behavioral characteristics; criteria such as language or intentionality were employed to maintain the borders.[11] But Darwin's theory of evolution cast a fundamentally new light on the issue. The boundary distinguishing humans and animals was reinterpreted in the West to involve not only differences in kind but also differences in *progress* along an evolutionary path. This path began with "lower" life forms, proceeded through intermediate stages inhabited by "higher" animals, and reached its pinnacle with (white) "man."

This scientific, evolutionary recasting fit squarely within an interconnected set of understandings about the human geography of the colonial world, in which the "discovery" of "races" raised complex questions of human taxonomy. Categorizing exotic-looking peoples from distant lands as lower on the evolutionary scale and thus closer to animals echoed and relied upon a myriad of similar divisions used to separate some humans from others: primitive versus modern, civilized versus savage, heathen versus Christian, cannibal versus noncannibal. In turn, the human-animal division construed as a continuum of *both* bodily form/function and temporal stage in evolutionary progress was used to reinforce these intrahuman categorizations and interpret them in temporal, evolutionary terms rather than in solely social or geographic ways. The stubborn and threatening heterogeneity of the colonies was contained and disciplined not only by branding them socially or geographically different from Europe but also, as Ann McClintock suggests, "temporally different and thus as irrevocably superannuated by history":

> [I]mperial progress across the space of empire is figured as a journey backward in time to an anachronistic moment of prehistory. By extension, the return journey to Europe is seen as rehearsing the evolutionary logic of historical progress forward and upward to the apogee of the Enlightenment in the European metropolis. Geographical difference across space is figured as a historical difference across time.[12]

In postcolonial, Western capitalist space, the idea of a human-animal divide as reflective of both differences in kind and in evolutionary progress has retained its power to produce and maintain racial and other forms of cultural difference. The dominant uses of human-animal distinctions during the colonial epoch relied upon representations of similarity to animals to dehumanize and thus racialize particular cultural groups. Contemporary arguments, in contrast, are primarily characterized by a focus on animal practices employed by subdominant cultural groups as cruel, savage, criminal, and *inhuman*: the literal blood-letting of animals, the slicing up of their bodies. But although the precise terms of reference have shifted over time, the postcolonial moment is one that continues to use putative human-animal boundaries to inscribe totems of difference (savage, barbaric, heathen, or archaic versus civilized, Christian, or modern).

We will say more about each of these uses of the human-animal boundary in the racialization process in the sections of the chapter that follow. Here, we want to understand why, in postcolonial times, animal practices have become a key aspect of the human-animal boundary used to racialize and produce difference. We locate our explanation in radically changing time-space relations that epitomize postmodernity. Just as in the colonial period, when the dimensionality of the world as perceived through Western eyes suddenly expanded in the wake of European exploration and "discovery," time-space relationships have altered dramatically during the course of the twentieth century, particularly over the last two decades. Indeed, a compression of time-space, or shrinking of the world's time-space fabric, is a hallmark of postmodernity. This compression creates what Fredric Jameson terms "*postmodern hyperspace,*"[13] which brings visible difference "home" instead of restricting it to a distant, exotic colonial space. Those seeking to produce racial difference are no longer separated by vast continents and long journeys from groups they wish to dehumanize. Instead, the targets live next door (figuratively and, not uncommonly, literally), inviting an inspection of their unsettling otherness. This implies that fanciful representations of people-as-beasts are less potent than images of people-acting-beastly toward animals. Since older evolutionary interpretations of racial difference persist within postmodern culture, not only immigrants but native-born people of color are sometimes constructed by their animal practices.

Animals and the Body Politic

Racialization is far from a monolithic or static process but instead is situational and shaped by racial ideologies and stereotypes.[14] Exactly how and why does this postcolonial, postmodern form of racialization around animals occur? We argue that in the present instance, animal bodies have become one site of political struggle over the construction of cultural difference and maintenance of American white supremacy. By scrutinizing and interpreting subaltern practices on animal bodies (or simply "animal practices") through their own lenses, dominant groups in the US simultaneously construct immigrant others as uncivilized, irrational, or beastly, and their own actions as civilized, rational, and humane.

In general, animal bodies can be used to racialize, dehumanize, and maintain power relations in three key ways. First, animals serve as absent referents or models for human behavior.[15] Being treated "like an animal" is typically interpreted as a degrading and dehumanizing experience, and such treatment is therefore a powerful tool for subjugation of others. The specific "treatments" in mind here are not the many loving forms of human-animal interaction, but rather involve abuse or violation, physical and/or emotional. The key aspect of such violent treatment that makes it dehumanizing, however, is not just the abuse or violation: it is the fact that victims are *objectified* and used like animals, who are commonly objectified and used without second thought.[16] Abusive treatment of slaves by masters, for example, was modeled on how people use animals without consideration of their subjectivity.[17]

Second, people are dehumanized by virtue of imputed similarities in behavior or bodily features and/or associations with the animal world in general or certain animals in particular. (Human identities also derive meaning and an enormous range of positive values from imputed similarities, of course, such as bravery, speed, and cunning.) Imputations are often made on the basis of associational representations of both humans and the animals to which they are being linked: colonial images of Africans as "ape-people" come readily to mind. Similarities can also be drawn on the basis of theories of human-animal continuity. For example, in Western thought, women's bodies have been deemed "like" animals due to their biological role, seemingly uncontrolled passions, and perceived irrationality. Using this logic turned on its head, queer bodies are deemed transgressive because they engage in "unnatural" behavior; animals are (supposedly) all heterosexual, thus queerness constitutes a perversion of nature. Turning to race, as we have argued, people of color (especially Africans) were historically situated by Westerners as lower on the "chain of being" and thus in closer evolutionary and behavioral proximity to nonhuman animals (especially the great apes). Colored bodies were thus both more primitive and uncivilized, and closer to animals and their unbridled biological urges and passions. Such associations persist and are often made explicit; in contemporary pornography, for example, it is most often people of color depicted in sex scenes involving intercourse with animals.

The third and least explored manner in which animals play a role in the social construction of racial difference, and the one which we argue characterizes the postcolonial, postmodern moment, involves specific human practices on animal bodies. Such practices have been used to construct other groups as well: in Medieval Europe, for instance, women who harbored feline "familiars" were often regarded as witches. And, as illustrated so vividly by Frederick Simoons,[18] taboos about which animal bodies to eat (and which body parts) are common amongst contemporary peoples, with the result that outsider groups not observing such taboos may be viewed with disgust and distain. The many other sorts of practices on animal bodies—such as those described in our animal stories—that can constitute powerful weapons for the devaluation and dehumanization of people of color have been less remarked. We turn now to an analysis of why certain animal bodies and body practices are taken up in this fashion.

Animal Practices and Dehumanization

What makes one animal practice acceptable and another a potent symbol of savagery that can be used to dehumanize those who engage in it? We have argued that every human group defines the boundary between humans and other animals in part on the basis of their treatment of animal bodies or animal practices. Specific forms of human-animal interactions, legitimized and rationalized over time, are part and parcel of the repertoire of "civilized" behavior that defines the human-animal divide. Those who do not stay within this field fall over the human-animal boundary or at least into the netherworld of "savagery"; if the practices are too far over the line, they can be interpreted as cannibalism, the ultimate act of inhumanity. Policing the human-animal boundary through the regulation of animal practices is necessary to maintain identity as humans and, not coincidentally, to sustain the legitimacy of animal practices of dominant groups.

It is widely recognized that in most societies certain types of animal practices are taboo. Taboo practices involve sexual relations with animals (bestiality is rarely sanctioned, although sometimes tolerated). Beyond bestiality, the killing and eating of the "wrong" species or categories of animals (especially totemic species or those seen as too similar to humans) can also be forbidden. For example, the consumption of apes is widely interpreted as tantamount to cannibalism, since simians occupy an ambiguous position along the human-animal boundary. They are not fully inside the human camp: one would not marry King Kong or have sex with Bonzo (even at bedtime)! But apes are seen almost literally as "inferior" humans because of their physiological similarity to humans. Eating them is thus strictly taboo. Similarly, in societies where pets are perceived to be members of the family and household, they can also come to occupy ambiguous or intermediate positions. Eating them, like the Cambodian men in our story did, becomes out of the question for civilized people.

Despite the importance of animal species or category in determining which animal practices fall beyond the bounds of humanity in any given society, practices are rarely considered (un)acceptable on the basis of species alone. Species is only one part of the immediate context through which animal practices are interpreted. Often, it is only this immediate context that separates "civilized" from "savage," thus revealing both the presence of unspoken criteria for judging animal practices and the extreme cultural relativity of these judgements.

Specifically (as our animal stories illustrated), there are at least four other key elements of context which define the human-animal borderline. One is reason or rationale for harm. Was a specific harmful practice necessary for survival or to minimize human or animal pain/death? Few humans raise objections to killing and eating taboo animals if the alternative is starvation; the most commonly stated reason for killing laboratory animals (even "pet" species such as dogs and cats) is to prevent suffering or death; and "euthanasia" of companion animals is justified as a way to reduce animal suffering. When the rationale for harm is seen as unnecessary or irrational, or the results are defined as damaging, however, practices may be condemned. Just what is unnecessary or irrational or damaging varies from group to group.

Another important aspect of context is the social location of the perpetrator: was the person(s) involved in the harmful practice "appropriate"? For example, if an animal was killed for purposes of human consumption, did a butcher or slaughterhouse worker perform the act? Or if a companion animal was killed, was a veterinarian presiding? As our cases illustrate, problems arise when the human actor does not have the role and/or training deemed necessary by the dominant group to legitimize the act. Religious functionaries, for example, are no longer normally linked with animal sacrifices: Christian clergy are trained to deal in immortal souls, not corporeal affairs; and rabbis only serve to insure that kosher methods of killing are used. Thus, as religious specialists, neither Hmong shamans or Santeria priests are seen to have the credentials to sacrifice food animals, much less companion animals. Similarly, where the actual killing of animals has become industrialized, professionalized, and removed from the course of everyday life, lay people (such as the Cambodian men charged with pet eating at home) have no legitimacy as animal killers.

A further contextual element revolves around the means or methods of harm: how was the harm inflicted? What techniques or tools were utilized, and did they fall within the range of local convention? Or were methods seen as archaic, barbaric, or brutally employed? A puppy can legitimately lose her head in a laboratory decapitator, but bludgeoning her to death is deemed too brutal. Similarly, bolt-guns are acceptable for dispatching a lamb led to (professional) slaughter, but the kitchen knife is no longer seen as civilized or humane. Certainly "twisting off the heads" of small birds is completely beyond the pale, and hunting to injure rather than swiftly kill is apt to be defined not only as cruel and inhumane but both unmanly and unsporting.

Lastly, the site of harm is perhaps the most crucial aspect of context in determining the legitimacy of an animal practice. Was an animal killed in a slaughterhouse or in the backyard barbeque pit next to the pool? Were rats killed in the lab or were they disemboweled in the living room? The issue of site has two dimensions. One is whether the harmful action is carried out in purpose-built quarters or reserved places (slaughterhouses, labs, shelters, forests during hunting season) or "out of site" in unspecialized spaces more typically used for other purposes or banned for the animal practice in question (residential areas, posted lands). A second site-related issue is whether the action occurs "out of sight" in abattoirs or factory farms banished from the city or in labs behind locked doors, or in highly visible places of everyday life such as homes, street corners, or church. Although in traditional societies the killing and death of individual animals was (and in many places remains) a quotidian experience, keeping mass, mechanized, and industrialized violence toward animals "out of sight" is necessary to legitimize suffering on the vast scale required by the mass market's demand for meat and medicine.

Place and the Borders of Humanity

Human-animal borders and human practices on animals vary according to place. In representational politics that seek to dehumanize people by associating them with certain animals, place is often used to reinforce such associations.[19] Places are imbued with negative characteristics because they harbor (or are thought to harbor) certain feared or disliked animals, and then these places are linked to people who take on the dirty, polluted, or dangerous aspects of the place (and its animals). For example, "jungles" are dangerous places in the Western popular imagination, conjuring up images of dense foliage beneath which poisonous snakes slither and vicious beasts wait to pounce on unsuspecting humans. More concretely, marginalized groups such as gypsies are often relegated to residual places in urban areas (such as dumps), often inhabited by "dirty" and "disease-ridden" animals, for example, rats. Thus a "dirty—unsafe—rats—gypsies" association arises, linking a so-called pest-species to a particular subaltern group. This associational process has long been used to connect poor people, "dirty" animals, and dirt more generally.

In the case of animal practices, however, place plays both more straight-forward and more nuanced roles. At a basic level, specific repertoires of animal practices evolve and become normalized *in place*. Such repertoires are in part environmentally determined, since the diversity of animal species available to kill, eat, or otherwise use is shaped by environmental factors, as are particular modes of subsistence linked to specific animals (for example, pastoralism). In addition, however, cultural ideas about animals (like other aspects of culture) evolve in place over time due to social or technological change generated within a society, or by externally driven events such as migrations or invasions. Thus values and practices concerning cosmological, totemic, or companionate relations between people and animals, and the material uses of animals as food or clothing, medicines or aphrodisiacs, shift as a result of social dynamics, technological change, or culture contact. The result is a shifting but place-specific ensemblage of animals, valued and used according to particular, legitimized codes. Transgressions of such place-specific codes or boundaries of practice *by definition* situate an individual or group as "outsider," "savage," or "subhuman."

What happens when the coding of animal bodies and the codes of animal practice shared by people dominant in one place are broken or challenged by people from another place, who do not share these codes but share the same space? When people are uprooted and brought to new places, they encounter different human-animal boundary constructions and if they persist in their indigenous practices are much more likely to transgress the border than locals. During much of (pre)history, the pace of such culture contact was relatively slow, allowing both host and newcomer groups to adjust; in earlier international migration waves to the US, origins of immigrants were sufficiently similar to host populations that conflict on the basis of animal practices does not appear to have been rife. With the economic globalization, escalating geopolitical instabilities and conflicts, and vast international population flows that characterize the postmodern condition, the "empire" has come home. Newcomers from a wide variety of radically different environments and cultural landscapes are suddenly living cheek by jowl. Typically, immigrants must move into the territories of a more powerful host community. Adjustment possibilities are foreshortened; for the largest immigrant groups, the need to adjust may be obviated by the emergence of relatively self-contained immigrant districts, such as "ethno-burbs."[20] Thus in the contemporary US, immigrants whose indigenous animal practices clash with the codes of dominant society are at the greatest risk of racialization and dehumanization.

Nevertheless, non-immigrant people of darker (versus lighter) color can also be at risk on the basis of their animal practices. Here, place plays a more nuanced role, by exoticizing the imaginary places of origin of such groups. Risk in this case arises not only because dominant norms of animal practice are contravened, but also because of the deeply engrained evolutionary connotations of the primitive, exotic, racialized "homelands" lurking in the Western imagination just below the surface of contemporary race relations. Thus cock fighting among Native Americans or Chicanos, the adoption of Santeria on the part of many Chicanos and African Americans, or the keeping of aggressive, vicious dogs (or, worse, dog-fighting) among youth in inner-city communities of color can place such subaltern groups on the far side of the human-animal boundary. When problematic practices occur in racialized and marginalized places, such as "ghetto" areas that are already indirectly and sometimes even explicitly linked to Africa (by virtue of names like "The Jungle"), prospects of racialization on the basis of animal practices may rise still higher.

Lastly, there may be time-space displacement of one group's animal practices onto another group located in a different place. With globalization of environmental degradation and the rise of international efforts to prevent species extinction, local groups may risk racialization by virtue of animal practices occurring in their ancestral or natal-origin countries or regions rather than their *own* behavior toward animals. By a quick twist in the logic of postmodern hyperspace, they can in effect be held suspect while being thousands of miles away from the action. For example, the rhinoceros faces extinction due to poaching and the subsequent sale of their pulverized horns as an aphrodisiac to Asian consumers. Such practices contravene dominant Western environmental values as well as acceptable reasons for animal harm and may be used to devalue and dehumanize Asian Americans or Asian immigrants regardless of whether or not they support the market for such substances.

Toward *Le Pratique Sauvage*

Our purpose in attempting to explicate the links between race, place, and animal practices has been to show how deeply engrained ideas about people and animals have been used to produce cultural difference and devalue subaltern groups. In the US, such differences play into a multifaceted and dynamic process of racialization in which immigrants who appear to threaten dominant cultural identities are powerfully marked as outside the project of becoming American, and thus excluded from its associated benefits. This exploration reveals the extreme relativity of legitimate animal body codes and practices with respect to time, place, and culture. Ironically, however, our consideration also exposes the universality of human violence toward animals. We are left with a dual challenge: how to break the links between animals and racialization, and stop the violence done to people racialized on the basis of their animal practices; and how to make the links between animals and people, and stop the violence directed at animals on the basis of their nonhuman status. Lynda Birke captures the first aspect of this challenge nicely, when she suggests that:

> [I]t may be true that animals suffer less if they are stunned before killing, so that animals killed according to certain religious practices are likely to suffer. But that cannot justify (for example) anti-Semitic attacks. No human culture is free of animal suffering; and slaughterhouses that stun are hardly repositories of kindness and compassion. We need to find ways of expressing concern about what happens to the animals that do not express some kind of cultural imperialism.[21]

We maintain that making the links between animals and people requires a rejection of "dehumanization," as a basis for cultural critique. For the connotations of the very term "dehumanization" are deeply insidious. They imply human superiority and thus sanction mastery over animals and nature, and also suggest that violent or otherwise harmful treatment is acceptable as long as the targets are nonhuman beings. Thus dehumanization not only stimulates violence toward people, it implicitly legitimizes violence toward animals.

This does not mean that the human-animal boundary should simply be banished for good. For, as Val Plumwood argues, the denial of difference can be as harmful as its production.[22] Instead, difference—whether amongst humans or between humans and animals—must be respected and valorized. Stopping the violence means neither dismissing difference nor using it to legitimize harm or domination. Rather, in our view, stopping the violence requires adopting recipes for "wild practice" and extending them to embrace animals as well as people.

What changes in human thought and practice does *le pratique sauvage* imply? We see three basic shifts as necessary. One is that humans, especially dominant groups, accept rather than deny some of the vulnerability that animals have always known and reject the illusion that a devaluation of others (human or animal) either empowers or offers protection from harm. Another is that all humans need to abandon the drive for overarching control and instead choose a position of humility or marginality with respect to the Earth that balances needs for safety and security with consideration for the needs of other life-forms. Such marginality must be internally imposed (as opposed to the marginality that humans impose on each other to oppress or gain power) and its costs must be fairly borne. Finally, this sort of *pratique sauvage* implies that people must actively engage in a radically inclusive politics which considers the interests and positionality of the enormous array of animal life and lives, as well as the lives of diverse peoples. Neither human nor animal lives can ever be fully known, of course. We are obliged, however, to discern them as best we are able, through both the practices of interaction and exchange, and the exercise of all our powers of empathy and imagination.

Acknowledgements

We are deeply grateful to Laura Pulido for her comments and suggestions on an earlier version of this chapter. All errors, omissions, and confusions remain entirely our own.

Notes

1. Brigitte Bardot, as quoted in Reuters Ltd, 26 April 1996.

2. See Jan Penrose, "Reification in the Name of Change: The Impact of Nationalism on Social Constructions of Nation, People and Place in Scotland and the United Kingdom," in Peter Jackson and Jan Penrose, eds, *Constructions of Race, Place and Nation*, Minneapolis, Minn.: University of Minnesota Press, 1994, pp. 27–49.

3. We draw this phrase (originally introduced by Althusser), from Gayatri Chakravorty Spivak, *The Post-Colonial Critique: Interviews, Strategies, Dialogues*, New York: Routledge, 1990, p. 54. For a useful discussion of her ideas, see Edward Soja and Barbara Hooper, "The Spaces that Difference Makes," in Michael Keith and Seve Pile, eds, *Place and the Politics of Identity*, London and New York: Routledge, 1993, pp. 183–205.

4. See Max Arax, "Hmong's Sacrifice of Puppy Reopens Cultural Wounds," *Los Angeles Times*, 16 December 1994, pp. A1,5; and Anonymous, "Sacrifice of Dog Highlights Clash of Cultures in Central Valley," *San Francisco Chronicle*, 19 December 1994, p. A22.

5. Frank B. Williams, "Four Men Sentenced to Jail in Cruel Poaching of Deer," *Los Angeles Times*, 13 May 1994, pp. B1,2.

6. See David Haldane, "Culture Clash or Animal Cruelty?," *Los Angeles Times*, 13 March 1989, sect. II, p. 1; David Haldane, "Judge Clears Cambodians Who Killed Dog for Food," *Los Angeles Times*, 15 March 1989, sect. II, p. 1; Clay Evans, "Bill Outlawing Eating of Pets Clears Senate," *Los Angeles Times*, 29 August 1989, sect. I, p. 20; Paul Jacobs, "Governor Signs Pet Protection Bill But Opposes Penalties," *Los Angeles Times*, 19 September 1989, p. B3; Greg Lucas, "Governor Signs Bill Outlawing Dining on Pets," *San Francisco Chronicle*, 19 September 1989, p. A8; Katherine Bishop, "U.S.A.'s Culinary Rule: Hot Dogs Yes, Dogs No," *New York Times*, 5 October 1989, sect. A, p. 22.

7. Andrew Lam, "Cuisine of a Pragmatic People", *San Francisco Chronicle*, 9 August 1989, p. A17.

8. See Teresa Puente, "Tripping of Horses Faces Reins: Mexican Tradition May Be Riding for a Fall," *Chicago Tribune*, 28 July 1995, sect. 2c, p. 1; Carla Rivera, "Supervisors Ban Horse Tripping," *Los Angeles Times*, 19 April 1995, sect.B, p. 2; Anonymous, "States Move to Outlaw Mexican Rodeo Event of Horse Tripping," *New York Times*, 2 April 1995, sect. 1, p. 23; and Dave McKinney and David Hoffman, "House Bill Reins in Mexican Rodeos," *Chicago Sun-Times*, 15 May 1996, p. 20.

9. On the role of horses in the spread of pastoralism, agriculture, diffusion of innovation, and conquest, see Paul Shepard, *The Others: How Animals Made Us Human*, Washington D.C.: Island Press, 1995.

10. The Santeria case is United States of America, *U.S. Supreme Court Reports*, 1993. Church of the Lukumi Babalu Aye, Inc., and Ernesto Pichardo v. City of Hialeah, 124 L Ed 472. Media reports include Paul M. Barrett, "Court to Test Religion Rights in Sacrifice Case," *Wall Street Journal*, 18 October 1992, sect. B, pp. 1, 7; Joan Biskupic, "Animal Sacrifices Ban Tests Religion Rights," *Washington Post*, 1 November 1992, sect. A, pp. 1, 8, 9; Anonymous, "Santeria Priest Performs Sacrifices," *Sunday Telegram*, 27 June 1993, sect. A, p. 5. Earlier reports described Santeria in general (for example, Rick Mitchell, "Out of Africa: An Ancient Nigerian Religion Comes through Caribbean to the United States," *San Francisco Chronicle*, 1 May 1988, magazine 15, p. Z25); and the bans in Los Angeles and San Francisco are described in Anonymous, "L.A. Animal Torture-Killings Blamed on Sect," *San Francisco Chronicle*, 22 July 1988, sect. A, p. 14; Anonymous, "Ritualistic Animal Sacrifice Is Outlawed by L.A. Council," *San Francisco Chronicle*, 3 October 1990, sect. A, p. 17; Suzanne Espinosa, "Resistance to S. F. Ban on Animal Sacrifice," *San Francisco*

Chronicle, 24 July 1992, sect. A, p. 24; Elaine Herscher, "Panel Compromises on Animal Sacrifice," *San Francisco Chronicle*, 19 August 1992, sect. A, p. 20.

11. In a fascinating departure from such evolutionary taxonomies, the surrealists Breton and Aragon denounced the natural sciences for classification mania and argued against reductive typologies privileging exterior detail—anatomic or ethologic in the narrow and individualized sense of the term. Breton and his associates adopted, instead, the idea of analogic thought—a means of viewing the relations of an entire biotype with its milieu (for example, all Mexican animals were named specifically). In addition, mimicry or mimetism was one of Breton's requirements in biological classification: the nuptial dance among birds was like passion among humans. The result was a sort of surrealist totem. See Claude Maillard-Chary, "Le Sentiment de la nature chez les surréalistes," *L'Homme et la société*, nos 91–2, pp. 157–72.

12. Ann McClintock, *Imperial Leather: Race, Gender and Sexuality in the Colonial Contest*, New York: Routledge, 1994, p. 40.

13. Fredric Jameson, *Postmodernism, or the Cultural Logic of Late Capitalism*, Durham, N.C.: Duke University Press, 1991.

14. Laura Pulido, "A Critical Review of the Methodology of Environmental Racism Research," *Antipode* 28, 1996, pp. 142–59.

15. See Carol J. Adams, *The Sexual Politics of Meat: A Feminist-Vegetarian Critical Theory*, New York: Continuum, 1990.

16. A long line of ethical reasoning, perhaps epitomized by Kantian theory, argues that humans have only indirect moral duties to animals because they lack subjectivity and thus can be treated like inanimate objects.

17. Marjorie Spiegel, *The Dreaded Comparison: Human and Animal Slavery*, Denmark: Heretic Books, 1988.

18. See Frederick J. Simoons, *Eat Not This Flesh: Food Avoidances from Prehistory to the Present*, Madison, Wisc.: University of Wisconsin Press, 1994.

19. David Sibley, *Geographies of Exclusion*, London: Routledge, 1995.

20. See Wei Li, *Chinese Ethnoburbs of Southern California*, Ph.D. diss., Department of Geography, University of Southern California, 1997.

21. Lynda Birke, "Exploring the Boundaries: Feminism, Animals and Science," in Carol Adams and Josephine Donovan, eds, *Animals and Women*, Durham, N.C.: Duke University Press, 1995, p. 49.

22. Val Plumwood, *Feminism and the Mastery of Nature*, London: Routledge, 1993.

5

Are You Man Enough, Big and Bad Enough? Wolf Eradication in the US

Jody Emel

Introduction

When I was growing up in Nebraska, my best friend's father shot coyotes for $2 a pair of ears. He didn't do it for the money; he was a middle-class Republican with a boat and a cabin at the lake. He wasn't a redneck either, in fact, he, more than anyone I knew, represented the most acceptable, middle-class, Wasp masculinity of that place and time. He discriminated against practically everyone—"niggers," "queers," "bohunks," "big dumb swedes," women, failures of "manhood"—and he hunted. Every year, he and his friends packed their jeeps and other suitable rough-and-ready vehicles with guns, ammunition, khaki, and food to drive from our town in Nebraska to Wyoming for antelope or to Colorado for elk. The goal, of course, was to pull down the buck with the biggest rack.

I despised these men, who shot wild animals, laughed at slanderous jokes, and dispensed spiritual and cultural deprivation along with class values. For me, in my Nebraska farm town, animal cruelty and racism were the most obvious and repellent aspects of the persona represented by people like my neighbor. Although I experienced them, the attendant sexism and classism were less shocking, probably because they were so widely acceptable. Furthermore, we lived at the edge of the fields. There was nothing urban or urbane about this part of the country—no labor unions, no strikes, only one factory. My alienation, shaped by the rurality of the place, gave birth to a nascent radical ecofeminism that I could only name many years later.

Learning as an adult that my environmentalism or animal advocacy was considered elitist, bourgeois, misanthropic, and possibly worse gave me pause.[1] Yet discovering a number of feminist writers who saw the congruence of racism, sexism, classism, the domination of nature, and the abuse of animals provided the impetus for looking further into these relationships.[2] This essay, then, is informed by a version of left-green ecofeminism that holds that, in addition to economic modes of production, deep-seated cultural attitudes shaped gender in ways that gave rise to patriarchal, nature-fearing, militaristic, and exploitative political-economic systems.[3] The argument claims that oppressions of humans and animals are linked in complex ways.[4] Indeed, in her preface to Marjorie Spiegel's book, *The Dreaded Comparison: Human and Animal Slavery*, Alice Walker writes of the similarities between the enslavement of black people in the past (and by implication other enslaved peoples) and the enslavement of animals:

> [It] is a comparison that, even for those of us who recognize its validity, is a difficult one to face. Especially so, if we are the descendants of slaves. Or of slaveowners. Or of both. Especially so if we are also responsible in some way for the present treatment of animals. Especially so if we, for instance, participate in or profit from animal research ... or if we own animals or if we eat animals or if we are content to know that animals are shut up "safely" in zoos. In short, if we are complicit in their enslavement and destruction, which is to say if we are at this juncture in history, master.[5]

How are the oppressions linked? Through the multiple processes of identity and hierarchy creation bolstered by the dualisms of human-animal, civilization-wilderness, society-nature, male-female, mind-body, white-black, and so on. To use Donna Haraway's terms, "Western dualisms reflect the One's domination of the Others: women, lower class people, nonwhites, and all those whose task is to mirror the unitary self."[6] What it means to be human can never be determined without the animal other. Who we are, and who and whether they are, are mutually reinforcing. As the last link in the "great chain of being" (which begins with the privileged white male), their fates are inseparable from our own. Thus, the contest over "nature" is a fight over who we (human beings) are and will be.[7]

Representation and identification are instrumental to oppression and resistance. How we represent and identify ourselves and others, whether they be animals or people, means everything for what and how we feel or don't feel. If we are taught to believe or have "rationalized" that an animal is "vermin" and deserves to be killed, a feeling of sympathy can be suppressed or altogether replaced with hatred, rage, anger, or detachment. How we come to identify ourselves—as hunters, masters, victims, prey—leads us to specific practices and forms of expression. And it is the human capacity to distance, background, deny, stereotype, and devalue the other that has led to the great atrocities of history.[8] The cultivation of "alibis for aggression" in the name of progress has tangled ideological roots in various mediums of material and cultural conditions, and it is this cultivation that I wish to examine with respect to animal oppression.

"Of Wolves and Man"

In this essay, I look at the eradication of the wolf in the United States to illustrate the interrelatedness of sexism, racism, animal abuse, and economic practices. The wolf occupies a special place in the history of human and nature relations for a number of reasons, including its skill as a predator, its social behavior within the pack, its resemblance to "man's best friend," and its physical tenacity. Throughout history, the wolf has been a complex symbol embodying numerous projected normative attributes, and for that, it has suffered more than most predator species.[9] It was virtually eradicated from the continental United States (with two or three geographical exceptions) and most of Western Europe. In England, Germany, and France the wolf was particularly hated, and immigrants from those places to North America brought with them a very negative cultural representation of the wolf.[10] Early writings reveal that, in the ninth century, wolf hunting was a part of the young nobleman's training. The breeding of dogs specially developed and trained to hunt wolves was practiced in Britain. Italians, in contrast, were more ambivalent toward the wolf (recall the positive Romulus and Remus symbolism), and Italy today has a native wolf population in the Apennines—even where sheep are raised. Wolves also exist in the Balkans and the eastern Alps, not only because these regions were less densely populated than other parts of Europe, but because they lacked or never formed institutionalized systems of incentives for extermination.[11] Why these differences? Numerous stories have been told about wolves encircling plague-ridden European cities, waiting for fresh corpses to devour, or of witches or devils having wolflike characteristics, or of Christian lambs or sheep being sent into the midst of wolves (the connection between wolves and Roman aristocrats is important here). No child of European or American descent has missed the tale of the Three Little Pigs or Little Red Riding Hood. Dante puts those condemned for the "sins of the wolf" (seducers, hypocrites, magicians, thieves, and liars) in the eighth circle of Hell.[12] Wolves were indeed a threat to medieval peasants, whose means of existence could be wiped out in one night by a pack of wolves. However, the same could be said of peasants in other places in Europe where wolves were not so deeply hated. Whatever the reasons, the wolf created by Western European culture—with its connotations of the wild, the savage, the diabolic, the lustful—was resolutely destroyed until a new wolf was imagined in the late twentieth century.

Today, the negative symbolism of the wolf has changed for many North Americans and Europeans. As a result of environmental movements during the past two decades, wolves have been reintroduced to areas of Idaho and Montana, and other populations receive some measure of protection. The importance of this "recovery" must not be understated. The wolf today is for many a potent symbol of the wild, the free, the uncommodifiable. Its reintroduction to a place where the myths of progress, private property, and rugged individualism reign supreme is, to my mind, a highly transgressive act. But the battle is not over and the legislation that protects endangered species and wilderness areas (not only in the US but in Italy and other countries) is subject to revocation, underfunding, and changes that would thwart such acts of recovery. Thus, the fight for the existence of the material wolf and a very important symbol of freedom and the wild continues.

Wolf Ecology and Behavior

Young and Goldman published the first major scientific treatise on North American wolves in 1944. At the time that they were writing, most of the remaining wolves in the United States were in the national forests. These authors, in the employ of the US Biological Survey, noted the presence of wolves throughout most of northern Mexico and sparse populations in the western United States—including Minnesota, Wisconsin, Michigan, Arkansas, Missouri, Washington, Oregon, Oklahoma, Louisiana, Texas—and in the Canadian prairie provinces and far eastern Canada. Ontario probably had the largest representation of wolves. Larger numbers were estimated in Alaskan territory (estimates of up to 7,000) and the Canadian Northwest Territories. Wolves were also found on Vancouver Island and in the northern interior of British Columbia, particularly the Peace River country. Some 36,000 wolves were thought to follow the migratory caribou herds of northern Canada, ranging over some 600,000 square miles. Wolves also ranged along the southern borders of Arizona and New Mexico, where they continually entered from the states of Sonora, Chihuahua, and Coahuila in Mexico.

Not until the late 1940s did anyone take a serious, "scientific" look at wolves. In the 1960s and 1970s a number of studies on boreal wolf ecology and behavior became available.[13] Also informative but with a naturalist slant are works such as those by Mowat.[14] A designed study of southwest wolf ecology or behavior has never been conducted and is now no longer possible.[15] Recent work has been done on the remaining wolves in Alaska, Canada, Isle Royale, and the wolf recovery efforts in the southeastern US and northern Mexico.[16]

We know very little of wolves for certain because they are elusive and have been exterminated over most of their former territories. David Mech, perhaps the most renowned North American wolf biologist, has come upon only a dozen or so that he didn't first spot with airplanes or track down with radio collars in thirty years of field research. Wolf researchers (who more recently are also conservation biologists) agree that wolves are amazing animals.[17] Lopez writes:

> In the winter of 1976 an aerial hunter surprised ten gray wolves traveling on a ridge in the Alaska Range. There was nowhere for the animals to escape to and the gunner shot nine quickly. The tenth had broken for the top of a spur running off the ridge. The hunter knew the spur ended at an abrupt vertical drop of about three hundred feet and he followed, curious to see what the wolf would do. Without hesitation the wolf sailed off the spur, fell the three hundred feet into a snowbank, and came up running in an explosion of powder.[18]

There are hundreds, if not thousands, of stories like this one. Many of them are told by the hunters who killed the wolves (more on that later). It's no wonder people admire them so. They spend an average of eight to ten hours out of every twenty-four on the move. They travel great distances and have tremendous stamina. One observer in British Columbia tracked two wolves for twenty-two miles as they broke trail through five feet of snow. The animals paused but never lay down to rest.[19] Naturalist Adolph Murie, watching a pack in Alaska, witnessed a regular daily round of forty miles in search of food while the female was denning.[20]

Everyone who has studied and written on wolves notes their friendliness to each other and to the pups within a pack. The solidarity of the pack is maintained, at least in part, by play. Playing includes games of keep away with sticks or bones (even antlers and hides), engaging in mock combat, ambushing one another, scruff biting, and chasing. At the conclusion of play, a greeting ceremony occurs around the dominant wolves—licking jaws, nosing heads, and paw poking. Allen says that all field workers see it frequently, and there is clearly a paying of "fond" respects involved.[21] Howling is another form of social behavior and, of course, the wolf howl (like other social behaviors) has provided considerable grist for the mill of the human imagination. Wolves howl to assemble the pack, to pass on an alarm—especially at the den site—to locate each other, and to communicate across great distances (six miles or more in still air).

Wolf diets are about 98 percent meat (they eat about five to ten pounds a day) with the remainder constituted by grass, insects, and other matter. Hunting of large animals is taught. Allen describes these hunting skills as being beyond our human ken. He says, "[t]hese are made possible by inborn capacities effectively tuned and developed in the young animal through an apprenticeship that only the capable survive."[22] Hunting of mice and rabbits seems to be more natural. Young females, because they are not so heavy in the chest, are the best hunters under some circumstances. Supposedly, wolves can smell prey from two miles away. They can run for miles behind fleeing game and then speed up to charge. They use ambush strategies in some cases: two wolves herd prey into an ambush. Sometimes they chase animals, corner them ("test" them), and then walk away. Fresh tracks may be examined and then disregarded for no apparent reason. No one knows for sure why.

Killing animals as large as elk, caribou, moose, or musk ox is dangerous and many studies have found wolves with fractured skulls, broken limbs, and so on. A wolf that is crippled or hurt loses its status or rank within the hierarchy. In some cases the crippled animal may be cared for, but more generally the other animals move up one notch in the hierarchy and the crippled animal is killed or driven from the pack. In fact, killing large (and wild) ungulates is so dangerous and dependent upon a group that lone wolves have often killed livestock because of the comparative ease. The irony here is that the more wolves were destroyed, the more likely livestock would be prey.

Other than humans, adult wolves have no predators.

The Eradication of Wolves in the United States

At one time, wolves lived in nearly every part of North America, particularly in the forests and on the plains. Although the wolf had been bountied from the 1660s—the Massachusetts Bay Company established the first bounty at a penny per wolf—by the early part of the eighteenth century they were still prevalent. Possibly their numbers increased for a time because of the new domesticated animals brought into their midst. Native Americans were encouraged to hunt them for the bounties, and, like the fur trade, the effect of the bounty was to create a market. The wolf was bountied in most of the eastern states during the 1600s and 1700s. In New Jersey, the bounties were established in 1697, allowing 10 shillings to Negroes and Indians and twice that to "Christian" killers. In addition to the bounties, there were collective hunts and activities geared to clear out areas that might shelter the animals. Cronon remarked, "[b]ecause, unlike Indians, wolves were incapable of distinguishing an owned animal from a wild one, the drawing of new property boundaries on the New England landscape inevitably meant their death."[23]

> Through the night I was kept awake by what I conceived to be a jubilee of dogs, assembled to bay the moon. But I was told in the morning that what disturbed me was only the howling of wolves which nobody there regarded. When I entered the Hall of Justice, I found the Squire giving judgment for the reward of two wolf whelps a country man had taken from the bitch. The judgment seat was shaken with intelligence that the she wolf was coming— not to give bail . . . but to devote herself, or rescue her offspring. The animal was punished for this daring contempt, committed in the face of the court, and was shot within an hundred yards of the tribunal.[24]

By 1800 the wolf's destruction in New England and in eastern Canada had been largely effected. Connecticut had already withdrawn its bounty, and the last wolf in New England was killed in Maine in 1860. In the Midwest and West, the eradication took less time. Lopez points out that while the European wolf hunter of 1650 might kill twenty to thirty wolves in a lifetime, a single American wolfer could, in the late 1800s, kill four to five thousand in ten years.[25] No other animal control effort ever achieved in geographic scope or economic scale the war waged against wolves in the nineteenth and early twentieth centuries in the US and Canada.

Wolf pelts became commodities in the mid nineteenth century, marketed mostly to Russia. Records of an upper Missouri trading company indicated that in 1850 they shipped twenty wolf pelts down river; by 1853, the total had risen to three thousand.[26] Buffalo hunters were killing the *lobos* who followed the herds. The buffalo-hunting era was brief. It began after tanners learned how to make good leather from buffalo skins and lasted only a dozen years or so (1871–83), seldom more than four years in any given place on the Plains.[27] That was long enough to slaughter the enormous herds that had roamed the grazing lands of the West. In total, between 1850 and 1880, some 75 million buffalo were killed—usually only for the hams, the tongue, and the skin. Carcasses left to rot in the sun attracted wolves and other carnivores, whom the buffalo hunters shot for sport and skinned when convenient.

It is difficult to imagine the magnitude of slaughter and death that occurred during the latter part of the nineteenth century. Descriptions of the prevalence of buffalo are remarkable when juxtaposed with a knowledge of the animals' rapid elimination. Captain Benjamin Bonneville, standing on a high bluff near the North Fork of the Platte River in 1832, saw a country "blackened by innumerable herds."[28] The same year and in nearly the same place J. K. Townsend wrote: "our vision, at the least computation, would certainly extend ten miles; and in the whole of this vast space, including about eight miles in width from bluffs to the river bank, there apparently was no vista in the incalculable multitude."[29]

But by the early 1880s the buffalo hunters had to go north to Wyoming and Montana for a living: it was all over on the southern Plains. The large herds were gone and the small, remnant herds that remained were not enough to make hide taking economically worthwhile. Some of the hunters went north, where some larger buffalo herds still existed, but even those were largely wiped out by 1883. The kill rate was astonishing: a pair of experienced hunters could kill forty or fifty buffaloes in a day. Buffalo Bill Cody killed nearly five thousand in eight months. In the winter of 1881–82, John Edwards downed seventy-five buffaloes in a single stand; a companion of his downed eighty-five.[30] One year later, when the white hunters went out again, they found no buffaloes.

Native Americans and wolves, who had been dependent on the buffalo herds, resorted to other means of survival. The warrior societies who ran and traded thousands of ponies and were at one time considered the "Lords of the Southern Plains" became semi-starved reservation dwellers or died fighting. The disappearance of the buffalo made these Plains "natives" prisoners of the reservation because of the promise of rations. It ensured that tribes like the Comanches and Kiowas would continue raiding for horses, mules, and cattle. This made them "thieves" and the rage they felt toward the whites for destroying their way of life made them "murderers" when they resisted. A number of attacks on buffalo hunters occurred because of the alarm among the Comanches and others regarding the rate of buffalo killing. The demise of the buffalo meant no hide for tepees and forced use of canvas, which was a government annuity or purchase item. It meant little to exchange for things like cooking utensils, canned goods, and ammunition. Worst, there was not enough to eat. The Plains people were warriors and nomads, not farmers. Toward the end of the nineteenth century, a buffalo could hardly be located for religious ceremonies. In 1881 the Kiowas did finally find two buffalo, a bull and a cow, and the bull's head was the focus of their summer ceremonial. One of their medicine men promised to bring up from the earth all the buffalo they needed if they

Figure 5.1 Female wolf with her litter, all killed by strychnine

From Stanley P. Young and Edward Goldman, *The Wolves of North America*,
New York: Dover Publications, 1944

used only bows and arrows for killing; as early as 1859 the Lakota tried
through spiritual means to bring about the return of the vanishing buffalo.
But the buffalo used in the Kiowa sun dance of 1887 had to be purchased
from Charles Goodnight, Texas hero and owner of one of the largest
ranches. Two years later the Kiowa paid him $100 (a fortune) for a
ceremonial buffalo.[31]

While the Native Americans were being pushed onto reservations, most
of the wolfers were working for commercial cattlemen, and from 1875 to
1895 the slaughter of wolves on the Plains reached its peak. During this
period of "strychnine insanity," range dogs died, children died, "everything
that ate meat died."[32] Wolf hunting was easier than buffalo hunting because
it could be done with strychnine. Wolfers would lace buffalo carcasses with
the poison, and the next day ten to twenty wolves would be found dead in
the area (Figure 5.1). Other animals that touched the meat would be dead
as well. Because the wolves salivated upon the grass before dying, the Indian
ponies and other animals that ate the grass would also die. Stanley Young,
who published one of the first major pieces on the wolf in North America
from his predator control research with the Division of Wildlife Research
(Fish and Wildlife Service, Department of the Interior), wrote:

> Destruction by this strychnine poisoning campaign that covered an empire
> hardly has been exceeded in North America, unless by the slaughter of the
> passenger pigeon, the buffalo and the antelope. There was a sort of unwritten
> law of the range that no cowman would knowingly pass by a carcass of any

kind without inserting in it a goodly dose of strychnine sulphate, in the hope of eventually killing one more wolf. The hazard to other forms of wildlife involved by this lavish use of strychnine was not taken into consideration by stock interests at the time. Kit foxes, so prevalent at the time on the plains, were poisoned by the thousands, for they were generally the first to take the poisoned meat. The predominant thought was "to get the wolf by any and all means."[33]

Although no one knows how many animals were killed on the Great Plains during the latter half of the nineteenth century during the greatest animal slaughter ever, Lopez ventures that, counting buffaloes, antelopes, passenger pigeons, Indian ponies, and wolves, there were perhaps over five hundred million. Of these, maybe one to two million were wolves. During 1884, 5450 wolves were turned in for bounty in Montana alone; in 1885, 2224. By the 1930s, the wolf was eliminated from the Great Plains and the ranching country of Montana and Wyoming. From 1883 to 1918, in Montana alone, 80,730 wolves were killed for $342,764 in bounties.[34]

In the southwestern United States, the development of the cattle and sheep industries ensured the demise of the wolf as well, although the wolves were never present in such large numbers as they had been in the North and East, where the buffalo and elk herds were large. By the late 1880s, the Southwest was "one large livestock ranch" and overgrazing was already a problem because of the huge numbers on the range.[35] Oversupply and bad weather conditions led to the collapse of the livestock industry during this period. And, again, as in Montana, the survival margin of the rancher was said to be so slim that every cow counted.

By 1905, the wolves had been thinned out in the open country of Texas, New Mexico, and Arizona. Only remnants remained in the Gila National Forest, the Pecos Valley, and a few other areas.[36] But ranchers and their congressmen were still dissatisfied and insisted the US government, specifically the Forest Service, take some responsibility for remedying the predator problem. The ranchers were paying government grazing fees and argued that the grazing right included wolf elimination. As early as 1897, ranchers in New Mexico had asked the US Biological Survey (then part of the Department of the Interior) to issue "responsible men" in every county for the express purpose of distributing poison free of charge to the owners of stock.[37]

Pursuant to this plea, Vernon Bailey, biologist and hunter for the US Biological Survey, wrote bulletins in 1907 for the Department of Agriculture, entitled "Wolves in Relation to Stock, Game and the National Forest Reserves" and "Directions for the Destruction of Wolves and Coyotes." Bailey was intent upon their complete extermination from the western range.[38]

The Forest Service joined the Biological Survey in the predator control effort. Forest rangers and guards were the first government predator control agents (see Figure 5.2). In total, over 1800 wolves and 23,000 coyotes were

Figure 5.2   A government wolf hunter's camp, showing wolf and coyote
catch from one range in eastern Wyoming
From Stanley P. Young and Edward Goldman, *The Wolves of North America*,
New York: Dover Publications, 1944

killed in the western United States in 1907 under the auspices of co-
operative stockmen, the rangers and special hunters on the forest reserves,
along the lines recommended in the wolf publications of the survey.

Around 1914, game protective associations were organized in New Mexico
and Arizona. Sportsmen now joined the livestock interests in advocating a
federal control program as the only real solution to the predator problem.
J. Stokley Ligon, a "conservationist" and protégé of Bailey, was joined by the
young ranger, Aldo Leopold, author of the now famous *Sand County
Almanac*, in the promotion of predator control, scientific game laws, and
game refuges throughout the southwest.[39] Leopold and Ligon believed that
predator control was an essential route to game abundance. Hunters and
mountain lions were perhaps more responsible than wolves for reducing
the numbers of available bucks, but, as Flader wrote in her history of
Leopold's career, "wolves, feared throughout history as killers of domesticated animals and even of people, had become both symbol and scapegoat
of the predatory species and were thus more zealously eradicated."[40] The
men's efforts were successful, and on 30 June 1914 Congress made the US
Biological Survey responsible for experiments and demonstrations in
destroying wolves, prairie dogs, and other animals injurious to agriculture
and animal husbandry. Ligon got the Predatory Animal and Rodent Control
(PARC) branch of the survey organized and functioning. An expert hunter
and trapper himself, he hired some three hundred hunters, including

several known "wolfers." Ligon and Leopold got the sportsmen to throw all their support behind the survey and insist that the job be completely finished—to the last wolf and lion, lest these "vermin" regain their range.[41] Thus, wolf killing was formally institutionalized and bureaucratized. Western rangelands were organized into control districts, each with a supervisor and field personnel. Ligon was the inspector for the New Mexico–Arizona district.

During the first two years of operation, the PARC took over 150 adults and pups.[42] Just as a poor consumer market for beef was reason to slaughter wolves, so was the good market during World War I. Cattle losses meant dollars and cents taken from the war effort, according to the PARC. Thus, control efforts intensified, and in 1918 eighty-one adult wolves and thirty pups were taken in the district. Between 6 February and 30 June 1918, another twelve wolves were poisoned by a special force employed by New Mexico A and M College. Nothing short of total extermination was the game plan. Ligon stated in 1918:

Very few wolves lived to get away from the dens the past spring in New Mexico. This signifies that there will never be any more young maturing in the State unless it be along the Mexican border, and this line will be carefully guarded in the future for animals that may drift in from Mexico.[43]

By 1925 the wolf had ceased to be a major predator in the southwest United States. In his annual report for the fiscal year 1925, Inspector Ligon wrote that "[t]he passing of the wolf in New Mexico, as well as in other western states, is every year becoming more apparent . . . The 'Lobo's' final exit from New Mexico has long been heralded. His stay, which has been far too long, seems fast drawing to a close."[44]

Once the bureaucracy was involved, the methodical, year-in, year-out extermination of the wolves was guaranteed. The hard-to-get wolves had reputations and so did the men who took them; considerable efforts were made to obtain the "last wolf." One of the wolves that W.C. Echols, the famous US–Mexico border hunter, took in 1926 had as a companion a dog with a collar and name plate identifying its owner in Cochise County, Arizona. Echols was indefatigable in hunting down border-crossing wolves and no cost was spared. He swore that "as long as they come, I shall catch 'em."[45]

The wolf's fate was hastened by the development of new chemical predacides such as Compound 1080. A dwindling number were taken through the 1950s and 1960s. Perhaps one last southwestern wolf, in Arizona's Aravaipa Canyon (a private wildlife refuge), was seen or heard in the mid 1970s. Some people believed this animal might offer a chance for the Mexican wolf to stage a comeback and be saved from extinction. But as Brown recounts,

Wolf life history, and southwest tradition, dictated otherwise. The "wolf" was "quietly" taken by a private trapper for a reputed bounty of $500 put up by local stockmen. U.S. Fish and Wildlife Service photographs of the skull allegedly from the Aravaipa wolf indicate that the animal was a true wolf. If so, it is likely the last wolf taken in the U.S. half of the Southwest.[46]

And with that wolf taking, the animal was gone from the southwest and nearly all of the western range.[47]

Man, Wolf, Good, and Evil

Like the Native American, the wolf was killed to secure land and investment. No less importantly, it was killed to sustain big game animals so that human hunters could kill them. It was killed for pelts, for data, for science, and for trophies. It was also tortured, set on fire, annihilated:

> They poisoned them with strychnine, arsenic, and cyanide, on such a scale that millions of other animals—raccoons, black-footed ferrets, red foxes, ravens, red-tailed hawks, eagles, ground squirrels, wolverines—were killed incidentally in the process. In the thick of the wolf fever they even poisoned themselves, and burned down their own property torching the woods to get rid of wolf havens. In the United States in the period between 1865 and 1885 cattlemen killed wolves with almost pathological dedication. In the twentieth century people pulled up alongside wolves in airplanes and snowmobiles and blew them apart with shotguns for sport. In Minnesota in the 1970s people choked Eastern timber wolves to death in snares to show their contempt for the animals' designation as an endangered species.[48]

What is this all about? It is not solely about protecting livestock, because as we shall see the slaughter went on long after the economic threat ended. It continues to this day when almost no one living in the continental United States has seen a wolf. As cultural ecofeminists contend, cultural phenomena and economic factors interact with each other in a complex dialectic. So while much of the story about wolf eradication has to do with class and economy, there is an intertwining causality stemming from a dominant construction of masculinity that is predicated upon mastery and control through the hunt. A fear of the "wild," the "irrational," or the "different" is also part of the construction. Wild animals, and particularly predators like the wolf, have represented longings, needs, and urges that were suppressed in the particular construction of masculinity that dominated during the late nineteenth and early twentieth centuries. They have been targets for hatred, the same hatred that launched armies and lynch mobs against human "others." A Massachusetts law of 1638, for example, stated that "[w]hoever shall shoot off a gun on any unnecessary occasion, or at any game except an Indian or a wolf, shall forfeit 5 shillings for every shot."[49]

Killing for necessity (hardening the heart)

Unquestionably, wolves were a problem for livestock owners during the latter part of the nineteenth century. With the buffalo gone and the land to the east settled by farmers, many wolves became dependent upon stock when elk, deer, and other wild ungulates were unavailable. The institution of private property meant that gains and losses were not shared. Thus, each individual owner felt losses keenly. R.M. Allen, general manager of Standard Cattle Company in Ames, Nebraska wrote in a letter of 3 April 1886:

> The loss is incalculable. I was told by one man who had 11 colts running in a pasture with 11 mares that he lost all of the 11 colts and one of the mares. The Continental Cattle Co., on the Little Missouri in Montana, who have a yearly brand of colts of some 700 head, lose, as I hear, annually about one third of their colts, and doubtless a great percentage of their calf brands as well.[50]

It was a war between the ranchers and the wolves. Senator John B. Kendrick of Wyoming, one of the pioneer stock producers of the Rocky Mountain area, testified:

> Our fight on the ranges over which I had supervision and management at the time began in the fall of 1893.... [A]ll told on this one cattle ranch, covering territory of probably 30 or 35 miles square, we had a record when I left the ranch, and lost track of it, of about 500 gray wolves that we had killed. And the coyotes were threw in for good measure; they numbered hundreds, but we had no disposition to either count them or keep track of them.[51]

But there was more to it than money. The war was fueled by masculine stereotypes that provided negative representations of the wolf and sympathy for the prey. There were standards for "manhood" or desirable forms of masculinity, as evinced through literature, cultural organizations like the Boy Scouts, and cinema. Considering the imperialism and colonialism of the Victorian era, and the attendant killing and death associated with both the colonized people and the destruction of animals in the path of settlement, these standards necessarily had much to do with behavior relative to hunting and death. For example, as illustrated in R.S.S. Baden-Powell's best-selling *Scouting for Boys* (first published in 1908), the exemplary masculine idols were the "peace scouts":

> The "trappers" of North America, hunters of Central Africa, the British pioneers, explorers, and missionaries over Asia and all the wild parts of the world, the bushmen and drovers of Australia, the constabulary of North-West Canada and of South Africa—all are peace scouts, real men in every sense of the word, and thoroughly up on scout craft, i.e., they understand living out in the jungles, and they can find their way anywhere, are able to read

meaning from the smallest signs and foot-tracks; they know how to look after their health when far away from any doctors, are strong and plucky, and ready to face any danger, and always keen to help each other. They are accustomed to take their lives in their hands, and to fling them down without hesitation if they can help their country by doing so.[52]

The killing had to be done right. In MacKenzie's work on the imperial pioneer and hunter, he describes the etiquette of hunting or the "sporting code" that was laid out in the vast hunting literature of the period. He says that Baden-Powell frequently alluded to the code and in particular admonished readers never to kill an animal without some real reason for doing so, and in that case the kill should be quick and effective, to give as little pain as possible.[53] The wolf failed to live up to European and North American expectations of the proper hunter; it was considered wasteful and unsportsmanlike. Senator Kendrick, speaking before the United States Senate Committee on Agriculture and Forestry, described the wolf as "vicious in its cruelty" in that "[h]is prey is literally eaten alive, its bowels torn out while it is still on its feet in many cases."[54] A renowned American natural historian, William Hornaday, wrote that "[o]f all the wild creatures . . . none are more despicable than wolves. There is no depth of meanness, treachery or cruelty to which they do not cheerfully descend. They are the only animals on earth which make a regular practice of killing and devouring their wounded companions, and eating their own dead."[55] A true hunter was supposed to be humane, to make the kill quick and clean. And certainly, you weren't to eat your own mates. Remarking on horse destruction, Alexander Ross, an early fur-trader, wrote that wolves "do not always kill to eat; like wasteful hunters, they often kill for the pleasure of killing, and leave the carcasses untouched. The helplessness of the horse when attacked by wolves is not more singular than its timidity and want of action when in danger by fire."[56] The true hunter was also supposed to hunt alone. Ganging up on the prey was inappropriate sporting form. (Native Americans failed on this count as well.) Ligon scientifically described the "savage energy" of a group of wolves he thought brought down an "unfortunate bruin" in southeastern Alaska.[57] A military man in Wyoming wrote in his 1843 journal of a "fine Durham Bull (probably lost by some of the Emigrants) . . . who fought manfully" but was outnumbered and overpowered by wolves in the end.[58]

The construction of the wolf as a merciless killer of innocent livestock is quite interesting considering the slaughter of cattle and buffalo that went on at the hands of the wolf's killers (not to mention the slaughtering methods of the wolf hunters themselves). Of course, the wolf *is* merciless, I suppose, but the abhorrence for the wolf's killing technique that was shown during the 1930s when the national predator control group (PARC) was authorized and funded existed alongside a congressional unwillingness to pass a law against lynching African Americans. Mutilations of black men— the removal of fingers, toes, and penises—not to mention murders, went unpunished. But wolves who killed livestock and other ungulates, generally for survival, were methodically rubbed out of existence.

On top of it all was the wolf's so-called cowardice. Once wolves had experienced gunfire, they ran at the sight of guns and humans. Cowardice was perhaps the most disdained violation of the sporting code and the American frontier. You were supposed to "be a man" and stand up to the challenge. At bottom, a "real man" was supposed to take the pain, not cry out or show care, and stand to fight. In her analysis of the Western novel (and later cinema), which created a model for men who came of age in the twentieth century, Jane Thompkins argues that most Westerns describe the same man, "a man whose hardness is one with the hardness of nature."[59] She cites the cover of *Heller with a Gun*, which reads: "[h]e was as merciless as the frontier that bred him." If a man showed sentiment or indulged in "excessive or unnecessary feeling" then he would be "soft, womanish, emotional, the very qualities the Western hero must get rid of to be a man."[60] Heroism was conceived upon the notion of self-mortification, which demanded that the suffering of animals (or others) could not be recognized without jeopardizing the admirability of such self-sacrificial heroism. For if it was considered callous and unmerciful to inflict pain upon sentient beings, then self-sacrifice could not be considered so gallant.

Dying, to be a man
During the latter half of the nineteenth and the early twentieth centuries, hunting and killing fierce animals was one of the highest forms of sport, an indicator of virility and prowess. For men of "frontier masculinity," manhood and death went hand in hand. Thompkins argues that the masculine profile was secular, materialist, and anti-feminist; that it focused on conflict in public space and was obsessed by death. America's obsession with the frontier had much to do with the appeal of untamed animals and, as Young and Goldman claimed, no animal stood less in awe of man than the wolf. The dramatic possibilities of conquering a fierce beast equaled or surpassed those of a battle with the elements alone. Surviving a storm was one thing, but looking your savage prey in the eye was surely more exhilarating.

It was permissible within the hunting or virility code to show admiration or sympathy for the foe—particularly if he (it was always a "he") was a worthy opponent. For example, Ernest Thompson Seton was a hunter of wolves who admired them to the extent that he wrote a book largely devoted to them (see *Great Historic Animals, Mainly about Wolves*). He killed the mate of a wolf he had been trying to trap for months: "[W]e each threw a lasso over the neck of the doomed wolf, and strained our horses in opposite directions until the blood burst from her mouth, her eyes glazed, her limbs stiffened and then fell limp."[61] When he found the female wolf's male partner dead in one of his traps the next day, Seton put their dead bodies together in his shed because he was so deeply moved that the male would abandon all of his previous caution in his determination to find his mate.

In the writings of wolfers and government biologists who were also hunters, there seems to be considerable respect for the hard-to-get animals they so methodically wiped out. Stanley Young, for instance, devotes several pages of his book on North American wolves to reminiscing about "renegade" wolves who were tough to take. He writes that "[w]ith every hand turned against them, their wisdom was respected by the stockmen upon whose cattle they depredated, as well as by the wolf trappers who finally eliminated them at the cost of much time, money and patience."[62]

It was, in part, their great sense of duty that allowed the hunters to kill those they considered such worthy opponents: a worthy foe, a worthy kill, a continued sense of mastery. Lopez tells the story of an Alaskan trapper named Carson who tracked a wolf that had dragged one of his traps more than twenty miles. The hunter found the wolf hung upside down by the dragline on a steep hillside.

> He disentangled the wolf for the purpose of taking pictures then shot him in the head. "Lobo died as he had lived, in defiance of all things that would dare to conquer him. His bloody career was ended, but even in death his fiery eyes and truculent jaws opened in a look of unremitting hate. Lobo, king of his domain—and rightly a king he was called—was dead."[63]

This ability to admire what one has murdered requires a curious detachment. In some way it may be the same as a Comanche homage to the buffalo. It is certainly similar to the emulation of the "noble savage" as reflected in the writings and words of General Crook, reputedly the greatest Indian fighter in American history.[64] Crook fought the Apaches, the Lakota, the Cheyenne, the Arapaho, and many other tribes. He was once asked if it was not hard to go on another Indian campaign. His famous answer was, "[y]es, it is hard. But, sir, the hardest thing is to go and fight those whom you know are in the right."[65] Nevertheless, he performed his duty. One wolf hunter said of himself and his partner as they strangled another den of fluffy, playful pups, "[w]e both felt somewhat ashamed and guilty . . . but it was duty."[66]

A learned capacity to cut off feelings in order to facilitate death or degradation is problematic. In his essay "Liberal Society and the Indian Question," Michael Rogin, after Hannah Arendt, argues that this capacity has disturbing connections to totalitarianism and raises questions that cannot be resolved by viewing Indian removal as "pragmatic and inevitable." He shows, through an analysis of Indian removal policy, how the worse the policy got in terms of human rights violations, the more extreme the process of denial. Conceptions of human rights collapsed, with "civilized atrocities" committed as defenses against "savage atrocities." There was casual talk and sometimes practice of tribal extermination, perceived impossibility of cultural coexistence, and all or nothing conflicts over living space.[67] Rogin reviews the history of Cherokee removal wherein four thousand of the estimated fifteen thousand members of the eastern branch of the tribe were killed in the process. The secretary of war (Lewis Cass) described the state of affairs as a "generous and enlightened policy ... ably and judiciously carried into effect. . . . Humanity no less than good policy dictated this course toward these children of the forest."[68] Native American destruction (like slavery and animal extermination) became an "abstracted and generalized process removed from human control and human reality."[69] In his second annual message to Congress, Andrew Jackson, the "father" of Indian removal policy, justified himself in "the image of the creator," as a "real tool in the hands of" a divine father, "wielded, like a mere automaton, sometimes, without knowing it, to the worst of purposes."[70] In Rogin's words, "[t]o be a man meant to participate, emotionally separated from the actual experience, in a genocide."[71]

Emulation, mastery, and masculinity

Lopez says that "man" wants to be the wolf. Clearly, lots of people throughout history and within different cultures have admired and wanted to emulate the wolf. Killing a wolf for its skin or other body part was something Native Americans of various cultures did. For Native Americans who were not ranchers or sheep herders, the wolves were admired for their hunting and other social skills. They were the Plains peoples' spirit talkers.[72] Wolf tales about life and how to live, hunt, save, behave, and so forth are abundant among the legends and tales of Native Americans. Plenty-Coups, a chief of the Crows in Montana's Yellowstone River country, once described to historian Frank Linderman the way he and other boys of his tribe were taught to hunt like wolves:

> off would go our shirts and leggings. There was no talking, no laughing, but only carefully suppressed excitement while our teacher painted our bodies with the mud that was sure to be there. He made ears of it and set them on our heads, so that they were like the ears of wolves ... our teacher would cover our backs with the wolf skins we had stolen out of our fathers' lodges. Ho! Now we were a real party of Crow Wolves and anxious to be off. . . . Slipping quietly through camp, stealing and then sharing bits of meat taken from the drying racks of aunts and grandmother, swimming in ice-cold creek water, learning to be tough and clever as the wolf, the boys prepared for the first bowe and their first antelope hunt, which would come soon.[73]

In many of the stories the wolf teaches the way to live by example. The wise and enduring Wolf is present, alongside Lucky-Man, in the Arikara explanation for the creation of earth in its present state. One way the Plains people taught each other the right way to live was through the Medicine Wheel, which comprises a series of points, each one representing a point of view—a way of seeing and experiencing that is signified by a particular animal.

> Wolf, with his endurance and caring for his family, is found at one of the points, associated also with the clouds or wind. The young, seeking guidance

and knowledge of themselves, take a journey in quest of a vision and a guiding spirit to assist them; those who choose Wolf as their particular Spirit will be lucky; but all can be stronger and wiser by seeking from Wolf's point of view.[74]

The wolf ritual was the dominant ritual of the Nootkans and their neighbors along the Northwest Coast (near the Queen Charlotte Islands). Both men and women were initiated into a secret society in which they received certain powers from the "Wolves". The wolf was the tutelary animal because it was considered the bravest and fiercest. Patterns of conduct aiming at tribal (as opposed to individual) welfare were publicly reinforced using wolves and their pack behavior as an ideal.[75]

But something went awry with the desire to *be* the wolf in white American masculinity. Lopez tells a story of a white man who had killed some thirty wolves himself from a plane and flown hunters who had killed nearly four hundred more:

> [f]or him the thing was not the killing; it was that moment when the blast of the shotgun hit the wolf and flattened him—because the wolf's legs never stopped driving. In that same instant the animal was fighting to go on, to stay on its feet, to shake off the impact of the buckshot. The man spoke with awed respect of the animal's will to live, its bone and muscle shattered, blood streaking the snow, but refusing to fall. "When the legs stop, you know he's dead. He doesn't quit until there's nothing left." He spoke as though he himself would never be a quitter in life because he had seen this thing, four hundred times.[76]

Legitimated by the "rationales" of predator control, conquering wilderness, and aiding helpless indigenous people (the "children of the forest"), a lust for violent mastery was let loose. Jack O'Connor, a former gun editor of *Outdoor Life*, wrote about the great satisfaction he got from killing a wolf in British Columbia:

> It was a lovely sight to see the crosshairs in the 4X settle right behind the wolf's shoulder. Neither ram nor wolf had seen me. The wolf's mouth was open, his tongue was hanging out, and he was panting heavily. . . . When my rifle went off, the 130 grain .270 bullet cracked that wolf right through the ribs and the animal was flattened as if by a giant hammer.[77]

Lopez says O'Connor shot at every wolf he ever saw and was no different from many such men of the 1920s, 1930s, and 1940s (and later) who supported aerial hunting of wolves. At bottom was a "distorted sense of manhood"; killing wolves was a way to prove to other men that they were no fools, that they were tough.[78] In a similar vein, Thompkins argues that the abuse of animals in Westerns is part of a sadomasochistic impulse central to the masculinist profile which "aims at the successful domination of the emotions, of the fleshly mortal part of the self, and of the material world outside the body."[79] This mastery, of self and other, in the construction of identity through killing is recognizable in other contemporary white American heroes as well. In his work on masculinity in the twentieth century, Schwenger wrote about male initiation rites through the kill.[80] He cites a version in Norman Mailer's *Why Are We in Vietnam?* that involved a wolf kill in Alaska. In the novel, one of the character's (Rusty Jethroe) "executive vanity" demands that he bring back a grizzly from an Alaskan hunting trip ("Rusty was sick. He had to get it up. They had to go for grizzer now"). Meanwhile his son's friend Tex has shot a wolf and the hunting guide enacts an initiation rite.

> Well, he got down and gave us each a cup of blood to drink and that was a taste of fish, odd enough, and salt, near to oyster sauce and then the taste of wild meat like an eye looking at you in the center of a midnight fire, and D.J. [the son] was on with the blood.... D.J. next thing was on his hands and knees, looking into the upper Yukon wolf mouth, those big teeth curved like a tusk, and put his nose up close to that mouth, and thought he was looking up the belly of a whale, D.J. was breathing wolf breath, all the fatigue of the wolf running broken ass to the woods and the life running the other way from him, a crazy breath, wild ass odor, something rotten from the bottom of the barrel like the stink of that which is unloved, whelp shit smell, wild as wild garlic, bad, but going all the way right back into the guts of things, you could smell the anger in that wolf's heart (fucked again! I'll kill them!) burnt electric wire kind of anger like he'd lived to rip one piece of flesh from another piece, and was going to miss it now, going to miss going deep into that feeling of release when the flesh pulls loose from the flesh, and there D.J. was sweating, cause he was ready to get down and wrestle with the wolf, and get his teeth to its throat, his teeth had a glinty little ache where they could think to feel the cord of the jugular, it was all that blood he'd drunk, it was a black shit fuel, D.J. was up tight with the essential animal insanity of things.[81]

The hero lusts to kill what he loves. "Love and murder are intermingled and confused."[82] Certainly, the sexual undertones of this passage and of hunting in general are explicit. MacKenzie remarks that hunting can readily be interpreted as sexual sublimation. The hunting literature is full of descriptions of the physical agonies of the Hunt, of the tensions induced by the great risk and the "ecstasy of the release when the hunter prevails and stands over his kill."[83]

That hunting and killing are definitive of at least some constructions of masculinity, both now and in the past, is common currency. As Cartmill writes, "the connection of hunting with masculinity runs deep, and both hunters and their critics often comment on it."[84] In a recent survey of blacks, whites, and Lumbee Indians in rural Scotland County, North Carolina, 91 percent of the men interviewed agreed with this statement:

"(h)unting provides an opportunity for a boy to identify with the world of men, which is the most important influence of hunting on a boy."[85] The further affirmation of virility associated with hunting has been remarked upon by both hunters and anti-hunting activists. Some hunters accuse animal rights activists of being "limp-wristed sissies and aging hippies" who are governed by "large women and matriarchal mores."[86] Anti-hunting activists, conversely, accuse hunters of fearing they are not "man enough" unless they are proving themselves through the hunt. Robert Bly, new-age poet, self-proclaimed expert on fathering, and alleged misogynist, admonishes mothers against raising their sons to be too kind-hearted to animals. He suggests killing small animals should be part of every boyhood.[87]

Identifying with the prey
In direct contrast to the European hunter, who is a representative of a privileged class, the hunter in some African American folk tales is made ridiculous. From the point of view of those who have been treated as prey, the hunt is evil. Zora Neale Hurston relates the story of the "nigger" who was taken hunting by his master and told to shoot a deer. The deer went tearing past him, but "[h]e didn't make a move to shoot de deer." When the white man asked him if he killed the deer, "de nigger" claimed he hadn't seen it. "All I seen was a white man with a pack of chairs on his head and Ah tipped my hat to him and waited for de deer."[88] When a hunter levels his gun to shoot at a pond of three thousand ducks, the lake freezes and the ducks fly away. Better still, the prey controls the predator: a catfish pulls the fisherman into the lake. A boll weevil asks to drive a car. Devoid of feet or claws, even the snake complains to God that everything steps on him: "Ah ain't got no kind of protection at all." Accordingly, God apologizes and gives him poison to put in his mouth.[89]

So too in Richard Wright's work, the animals serve as scapegoats for humans, and many of his African American characters empathize and identify with the animals. In *Pagan Spain*, Wright understands the bullfight as a ritual for the overcoming of fear. But, unlike Hemingway, Wright does not identify with the matador, "neither in the expression of desire to kill a bull nor in the triumph experienced in a public display of courage. The crowds press forward with '*Bravo hombre . . .!*' but Wright does not join in; instead he watches the bull."[90] Wright's identification is with the black bull—the black, uncontrolled bull who is the victim of the cold, manipulative matador. "What someone else might see as the discipline of emotions is presented here as the absence of them."[91] The account of the bullfighting ends in a crude village bullfight in which the bull is killed and the people rush to his testicles and begin "kicking them, stamping them, spitting at them, grinding them under their heels," with an "excited look of sadism."[92] But, of course, the people love the bull. Unlike the white hunters and matadors who kill what they love, the protagonists of Wright's stories find no glory in hurting innocent animals. "They have shared too much with them to make a game of that kind of torture."[93]

Another more contemporary case of the wolf as a symbol of opposition or outsider status is found in the name and music of *Los Lobos*, a *Chicano* band from East Los Angeles who play rhythm and blues as well as traditional Mexican and Norteño music. Their first release was entitled "How Will the Wolf Survive?" The wolf in the title song is both a real wolf and an illegal immigrant from Mexico who is trying to survive in an alien land, hunted and alone.[94] The music of this group speaks of inner-city struggles with subtle emotional strains and "wafting idealism."[95] Similarly, in her book *Women Who Run with Wolves*, Clarissa Pinkola-Estes uses the wolf as a metaphor for marginalized women who have been oppressed and damaged by societal norms stemming from a number of sources. The wolf is a valuable metaphor in both of these cases because it is an outsider, but not a victim. It is running and hunted, but fierce and resourceful. The recovery of the wolf as a positive symbol by marginalized others stands in sweet opposition to the negative cultural coding.

Conclusions

On a Saturday afternoon in Texas a few years ago, three men on horseback rode down a female red wolf and threw a lasso over her neck. When she gripped the rope with her teeth to keep the noose from closing, they dragged her around the prairie until they'd broken her teeth out. Then while two of them stretched the animal between their horses with ropes, the third man beat her to death with a pair of fence pliers. The wolf was taken around to a few bars in a pickup and finally thrown in a roadside ditch.[96]

In February 1972, an Oglala from Pine Ridge [Reservation in South Dakota] named Raymond Yellow Thunder, aged fifty-one, was severely beaten for the fun of it by two white brothers named Hare, then stripped from the waist down and paraded before a patriotic gathering at an American Legion dance in Gordon, Nebraska; the injured man was thrown into the street, after which the brothers stuffed him into a car trunk and rode him around town for forty-five minutes before dumping him out at a Laundromat.[97]

As a white, working-class girl I could not wholly identify with the hero of the Western, the rancher, or my hunter-neighbor. I identified with the wolf, the cattle, the Native Americans, and other "outsiders." Thompkins suggests that we all identified to some extent with the hero in Westerns—but the hero is clearly white, Anglo-Saxon, and male, and there is a hierarchy. Women, people of color, and animals are all down the ladder. I felt sorry for the wolf "bitches" who had their dens robbed year after year so that bounty hunters could secure incomes and identities. Like Richard Wright, I did not identify with the matador but with the bull. I have no empathic resonance with the idea of "killing what you love," nor have I had to suppress my emotions to live up to a masculine ideal. I have not had to "harden my heart," though I have been mocked for not doing so. For me, the wolf represents a yearning for the wild *against* the "rational" (as it has come to be limitedly defined), the unemotional, the oppressive. Perhaps these are the reasons why I have come to be a radical environmentalist and a feminist with an interest in wilderness and wildlife preservation. People like me are charged with being "green bigots" who place the interests of wildlife and lands over the legitimate needs of the impoverished masses of humanity. Preservationists are stereotyped as people who have "made it" and are more interested in protecting birds and wildflowers than ameliorating the plight (health care, housing, education, and income) of the underprivileged.[98] This is undoubtedly true in some cases. But for many of us, wolves and wildernesses are symbols of resistance. These animals are metaphors for oppositional ways of thinking and feeling. Passion for their survival results from *not* anesthetizing oneself to the oppression of animals and its links to other forms and sites of oppression.

Sentiment and feeling are necessary for struggle. Adrienne Rich writes about how "poetry can open locked chambers of possibility, restore numbed zones to feeling, recharge desire—and how sensual vitality is essential to the struggle for life." For many of us, nature is not dead, dull, or in any way meaningless. It is a miracle composed of countless beings, processes, and things. To ignore animals other than ourselves is to avoid questioning the moral foundations of our economy and the constructions of "self" to which we adhere. To leave unexamined the structures, be they gender, race, class, or culture, that teach us to slice off or repress empathy and to distance ourselves from the "other," invites oppression, brutality, holocaust. We are all complicitous in and victimized by such ignorance.

Lopez writes, "[w]e are forced to a larger question: when a man cocked a rifle and aimed at a wolf's head, what was he trying to kill? And other questions. Why didn't we quit, why did we go on killing long after the need was gone? And when the craven and deranged tortured wolves, why did so many of us look the other way?"[99] As we have seen, there are many reasons. There was killing out of duty or "necessity." There was killing to *be* the animal, the savage, the wild. There was also the idea that killing, with skill and expertise, could make one more of a man in the hunting-code tradition—the Rooseveltian virility tradition. There was killing to stomp out the hated or envied: freedom, difference, a place in the world, "savagery." Killing from depravity occurred as well. The license to hate and aggress guaranteed by both racism and speciesism is written in one paranoid and sadistic hand.

Thompkins argued, against other historians, that the Western was not so much about the American dialectic between civilization and nature, as about the fear of losing mastery and therefore identity.[100] Killing is a way to preserve mastery, but is it more terrifying when it is accomplished out of rage and depravity, or when it is done out of duty and desert—tied to a methodical, rationalistic, technological pattern of elimination at which those who do well are skilled, praiseworthy, capable, or perhaps chivalric outdoorsmen? It is the conjuncture of these factors, these overdeterminers, that makes it possible and confusing. Constructions of masculinity, cruelty, regimes of bureaucracy, commodity production, class relations, myth, and superstition, all determined the wolf's demise. Altogether, they supported and mutually defended one another.

No wolves remained in Nebraska when I was growing up. I have not seen a live one except in a zoo. My favorite exhibits in the small natural history museum in my town were the two stuffed wolves set in a twilight blue winter scene and the "Indian" room with its pit of bones belonging to the also mostly eradicated Lakota and Pawnee. Both the wolves and the "Indians" depended upon the buffalo, also completely gone except for a few in the Black Hills and Yellowstone National Park. The dogged determination to pound the wolf out of existence was impressive. It was the same determination that characterized the bureaucratic and military maneuvering which eliminated free, or non-reservation, Native Americans. Reading the military journals of officers tracking the last small groups of free Comanches along the canyons of the Llano Estacado in West Texas is remarkably like reading the accounts of government hunters tracking the last remaining south-western wolves. Both eliminations were undertaken as "necessities"; both beg the question of alternative possibilities.

Acknowledgements

This is a modified version of an article published originally in *Environment and Planning D, Society and Space*, vol. 13, no. 6, December 1995, pp. 707–34. Grateful thanks are extended to its publishers, Pion Limited, for permission to reprint.

Notes

1. See, for example, Richard Lichtman, "Critical Discussion III: Humans Must Be So Lucky: Moral Prejudice, Specieism, and Animal Liberation," *Capitalism, Nature, Socialism*, vol. 3, no. 2, 1992, pp. 114–17; Steven Rose, "Critical Discussion IV: Humans Must Be So Lucky: Moral Prejudice, Specieism, and Animal Liberation", *Capitalism, Nature, Socialism*, vol. 3, no. 2, 1992, pp. 117–20; Anna Bramwell, *Ecology in the 20th Century: A History*, New Haven, Conn. 1989; Michael Heiman, *The Quiet Evolution: Power, Planning, and Profits in New York State*, New York 1988.

2. Carol J. Adams, *Neither Man Nor Beast: Feminism and the Defense of Animals*, New York 1994; Val Plumwood, *Feminism and the Mastery of Nature*, London and New York 1993; Donna Haraway, *Simians, Cyborgs, and Women: The Reinvention of Nature*, New York 1989; Alice Walker, "Am I Blue? 'Ain't These Tears in These Eyes Tellin' You?'" in Irene Zahava, ed., *Through Other Eyes: Animal Stories by Women*, Freedom, Calif. 1988, pp. 1–6; Karen J. Warren, "Feminism and Ecology: Making Connections", *Environmental Ethics*, vol. 9, 1987, pp. 3–20; Carolyn Merchant, *The Death of Nature*, New York 1980; Susan Griffin, *Women and Nature: The Roaring Inside Her*, New York 1978.

3. Joni Seager, *Earth Follies: Coming to Feminist Terms with the Global Environmental Crisis*, New York, 1993.

4. Warren, "Feminism and Ecology," pp. 3–5.

5. Alice Walker, "Preface," in Marjorie Spiegel, *The Dreaded Comparison: Human and Animal Slavery*, London and Philadelphia 1988, p. 9.

6. Haraway, *Simians, Cyborgs, and Women*, p. 177.

7. For an exceptionally well-wrought and fascinating example of this complex and

changing relationship, see Matt Cartmill's *A View to a Death in the Morning: Hunting and Nature through History*, Cambridge, Mass. 1993.

8. See Peter Gay, *The Cultivation of Hatred*, New York and London 1993. For fuller elaboration of the concepts "background," "distance," and "stereotype," see Plumwood, *Feminism and the Mastery of Nature*, pp. 41–68.

9. Stanley P. Young and Edward Goldman, *The Wolves of North America*, New York: Dover Publications 1944; Michael W. Fox, *The Soul of the Wolf: A Meditation on Wolves and Man*, New York 1992.

10. Luigi Boitani and Peggy Bruton, "Interview," *Defenders* May/June 1989, p. 209.

11. Boitani and Bruton, "Interview."

12. Barry H. Lopez, *Of Wolves and Men*, New York 1978, p. 205.

13. See, for example, B. L. Burkholder, "Movements and Behavior of a Wolf Pack in Alaska," *Journal of Wildlife Management*, vol. 23, 1959, pp. 1–11; L. David Mech, "The Wolves of Isle Royale," *US Department of Interior National Park Service Fauna Series*, vol. 7, 1966, pp. 1–210; Mech, *The Wolf*, Garden City, N.Y. 1970; Mech, "Productivity, Mortality, and Population Trends of Wolves in Northeastern Minnesota," *Journal of Mammology*, vol. 58, 1977, pp. 559–74.

14. Farley Mowat, *Never Cry Wolf*, Boston 1963.

15. See David E. Brown's *The Wolf in the Southwest: The Making of an Endangered Species*, Tucson, Ariz. 1983, for an excellent recounting of existing information on southwestern wolves.

16. Mike Link and Kate Crowley, *Following the Pack: The World of Wolf Research*, Stillwater, Minn. 1994.

17. Link and Crowley, *Following the Pack*; Fox, *Soul of the Wolf*.

18. Lopez, *Of Wolves and Men*, p. 3.

19. Lopez, *Of Wolves and Men*, p. 25.

20. Lopez, *Of Wolves and Men*, p. 25.

21. Durward Allen, *The Wolves of Minong: Their Vital Role in a Wild Community*, Boston, Mass.: Houghton Mifflin, 1979.

22. Allen, *Wolves of Minong*.

23. William Cronon, *Changes in the Land: Indians, Colonists, and the Ecology of New England*, New York 1983.

24. George Washington quoted in Young and Goldman, *Wolves of North America*, p. 375.

25. Lopez, *Of Wolves and Men*, p. 169.

26. Lopez, *Of Wolves and Men*, p. 177.

27. Wayne Gard, *The Great Buffalo Hunt: Its History and Drama, and Its Role in the Opening of the West*, Lincoln, Neb. 1959, pp. 133–53.

28. Gard, *Great Buffalo Hunt*, p. 5.

29. Gard, *Great Buffalo Hunt*, p. 5.

30. Gard, *Great Buffalo Hunt*, p. 271.

31. William T. Hagan, *United States–Comanches Relations: The Reservation Years*, New Haven, Conn. and London 1976.

32. Lopez, *Of Wolves and Men*, p. 180.

33. Young and Goldman, *Wolves of North America*, pp. 336–7.

34. Lopez, *Of Wolves and Men*, pp. 180–83.

35. Brown, *Wolf in the Southwest*, p. 41.

36. Brown, *Wolf in the Southwest*, pp. 46–7.

37. Young and Goldman, *Wolves of North America*, pp. 361–3.

38. Vernon Bailey, "Destruction of Wolves and Coyotes—Results Obtained During 1907." *US Department of Agriculture Bureau of Biological Survey Circular*, no. 63, 1908, p. 1. Bailey had made the wolf something of an obsession in his work.

39. Bailey, "Destruction of Wolves," p. 52. In later years, as head of the Wisconsin Game Commission, Leopold voted to reinstate a bounty and control program against the handful of wolves remaining in that state, despite his public acknowledgement that they posed no threat to the state's burgeoning deer herd, which he was then attempting to reduce.

40. Susan Flader, *Thinking Like a Mountain: Aldo Leopold and the Evolution of an Ecological Attitude toward Deer, Wolves and Forests*, Madison, Wis. 1974.

41. Brown, *Wolf in the Southwest*, pp. 52–7.

42. Brown, *Wolf in the Southwest*, p. 56.

43. Brown, *Wolf in the Southwest*, p. 59.

44. J. S. Ligon, "Predatory Animal Control. New Mexico District," *Annual Reports*, US Department of Agriculture Bureau of Biological Survey 1919. Quoted in Brown, *Wolf in the Southwest*, p. 72.

45. Brown, *Wolf in the Southwest*, p. 78.

46. Brown, *Wolf in the Southwest*, p. 115.

47. Today perhaps 1200 wolves survive in remote northern Minnesota, a forested area unsuitable for agriculture. Some fifty remain in Michigan and Wisconsin, and maybe a few more in northwestern Montana. The gray wolf was "listed" in 1967 under a predecessor to the now-extant Endangered Species Act. Listing did not require protection. The Mexican gray wolf is now a formal endangered species, even though it is extinct in the U.S. Minnesota wolves were listed as endangered under the act, making it illegal to kill or harass them. However, there was so much opposition that the status was reduced in 1978 to "threatened," thus allowing them to be killed by federal agents if there is a problem with livestock. Farmers are also compensated for lost livestock.

Alaska continues to have a large wolf population—some five to six thousand animals. The wolves are managed by the Alaska Department of Fish and Game, which hunts them with aerial-hunting programs. Wolves are tranquilized and collared with radio collars, then tracked by the planes and shot down. Of course, there is no livestock at risk, but the sport hunters want the caribou and moose to themselves.

48. Lopez, *Of Wolves and Men*, p. 137.

49. Lopez, *Of Wolves and Men*, p. 170.

50. Young and Goldman, *Wolves of North America*, p. 270.

51. John B. Kendrick, "Control of Predatory Animals." *Hearings before the Committee on Agriculture and Forestry*, US Senate, 71st Cong., 2nd and 3rd sess., on S. 3483, 8 May 1930 and 28 and 29 January 1931, p. 6.

52. R. S. S. Baden-Powell, *Scouting for Boys*, London 1908, p. 5.

53. John M. MacKenzie, "The Imperial Pioneer and Hunter and the British Masculine Stereotype in Late Victorian and Edwardian Times," in J. A. Mangan and James Walvin, eds, *Manliness and Morality: Middle-Class Masculinity in Britain and America 1800–1940*, New York 1987, p. 186.

54. Kendrick, "Control of Predatory Animals," p. 7.

55. William T. Hornaday, *The American Natural History*, New York 1904, p. 36.

56. Quoted in Young and Goldman, *Wolves of North America*, p. 271.

57. J. Stokley Ligon, "When Wolves Forsake Their Ways," *Nature Magazine*, vol. 7, 1926, pp. 156–9.

58. Talbot was with General Fremont on the Platte River in 1834. He is quoted in Young and Goldman, *Wolves of North America*, p. 265. This identification with the poor bull and the unfortunate bear is interesting given that bulls and their companions were usually thought of as "livestock" or live property and bears were themselves shot as predators and for sport. Little sentiment was elicited on their behalf; the cattle, in particular, existed to die and become meat.

59. Jane Thompkins, *West of Everything: The Inner Life of Westerns*, New York and Oxford 1993, p. 73.

60. Thompkins, *West of Everything*, p. 121.

61. Ernest Thompson Seton, *Great Historic Animals: Mainly about Wolves*, New York 1937, p. 73.

62. Young and Goldman, *Wolves of North America*, p. 285.

63. Lopez, *Of Wolves and Men*, p. 163.

64. Peter Matthiessen, *In the Spirit of Crazy Horse*, New York 1983, 1991, p. 7.

65. Matthiessen, *Spirit of Crazy Horse*, p. 11.

66. Lopez, *Of Wolves and Men*, p. 191.

67. Michael Paul Rogin, *Ronald Reagan, the Movie: And Other Episodes in Political Demonology*, Berkeley, Calif. 1987, p. 140.

68. Rogin, *Ronald Reagan*, p. 167.

69. Rogin, *Ronald Reagan*, p. 168.

70. Rogin, *Ronald Reagan*, p. 168.

71. Rogin, *Ronald Reagan*, p. 168.

72. Lopez, *Of Wolves and Men*, p. 177.

73. Jeanette Ross, "The Wolf in Native American Tales," in Wolves in American Culture Committee, eds, *Wolf!*, Ashland, Wis. 1986, p. 40.

74. Ross, *Wolf in American Tales*, p. 43.

75. Alice Henson Ernst, *The Wolf Ritual of the Northwest Coast*, Eugene, Oreg. 1952.

76. Lopez, *Of Wolves and Men*, p. 166.

77. Jack O'Connor, "Wolf!" *Outdoor Life*, vol. 127, no. 4, 1961, p. 75.

78. Lopez, *Of Wolves and Men*, p. 162.

79. Thompkins, *West of Everything*, p. 107.

80. Peter Schwenger, *Phallic Critiques: Masculinity and Twentieth-Century Literature*, London, Boston, Melbourne, and Henley 1984.

81. Norman Mailer, *Why Are We in Vietnam?* London 1969, pp. 69–70.

82. Thompkins, *West of Everything*, p. 95.

83. MacKenzie, *Imperial Pioneer*, p. 180.

84. Cartmill, *View to a Death*, p. 233.

85. Stuart A. Marks, *Southern Hunting in Black and White: Nature, History, and Ritual in a Carolina Community*, Princeton, N.J. 1991, p. 276.

86. Cartmill, *View to a Death*, p. 237.

87. Kenneth Clatterbaugh, *Contemporary Perspectives on Masculinity: Men, Women, and Politics in Modern Society*, Boulder, Colo. 1990; Robert Bly, *The Pillow and the Key: Commentary on the Fairy Tale of Iron John, Part One*, St. Paul, Minn. 1987.

88. Zora Neale Hurston, *Mules and Men*, New York 1990, p. 74.

89. See Richard M. Dorson, *American Negro Folktales: Collected with Introduction and Notes by Richard M. Dorson*, Greenwich, Conn. 1967.

90. Mary Allen, *Animals in American Literature*, Urbana, Ill. 1983, p. 148.

91. Allen, *Animals in American Literature*, p. 148.

92. Richard Wright, *Pagan Spain, Richard Wright Reader*, Ellen Wright and Michel Fabre, New York 1978, pp. 142–3.

93. Allen, *Animals in American Literature*, p. 149.

94. I have read and been told several stories about how some of the old wolf "runways" over the Mexican-American border are now used by Mexicans to slip across unseen.

95. Josef Woodard, "The Wolf Prospers," *Down Beat*, vol. 57, no. 12, 1990, pp. 26–7.

96. Lopez, *Of Wolves and Men*, p. 152.

97. Matthiessen, *Spirit of Crazy Horse*, p. 59.

98. Max Oelschlaeger, *The Idea of Wilderness: From Prehistory to the Age of Ecology*, New Haven, Conn. and London 1991.

99. Lopez, *Of Wolves and Men*, p. 138.

100. Thompkins, *West of Everything*, p. 45.

PART II

NEGOTIATING THE
HUMAN-ANIMAL BORDERLANDS

6

Zoöpolis

Jennifer Wolch

[W]ithout the recognition that the city is of and within the environment, the wilderness of the wolf and the moose, the nature that most of us think of as natural cannot survive, and our own survival on the planet will come into question.[1]

1. Introduction

Urbanization in the West was based historically on a notion of progress rooted in the conquest and exploitation of nature by culture. The moral compass of city builders pointed toward the virtues of reason, progress, and profit, leaving wild lands and wild things—as well as people deemed to be wild or "savage"—beyond the scope of their reckoning. Today, the logic of capitalist urbanization still proceeds without regard to nonhuman animal life, except as cash-on-the-hoof headed for slaughter on the "disassembly" line or commodities used to further the cycle of accumulation.[2] Development may be slowed by laws protecting endangered species, but you will rarely see the bulldozers stopping to gently place rabbits or reptiles out of harm's way.

Paralleling this disregard for nonhuman life, you will find no mention of animals in contemporary urban theory, whose lexicon reveals a deep-seated anthropocentrism. In mainstream theory, urbanization transforms "empty" land through a process called "development" to produce "improved land," whose developers are exhorted (at least in neoclassical theory) to dedicate it to the "highest and best use." Such language is perverse: wildlands are not "empty" but teeming with nonhuman life; "development" involves a thorough denaturalization of the environment; "improved land" is invariably impoverished in terms of soil quality, drainage, and vegetation; and judgements of "highest and best use" reflect profit-centered values and the interests of humans alone, ignoring not only wild or feral animals but captives such as pets, lab animals, and livestock, who live and die in urban space shared with people. Marxian and feminist varieties of urban theory are equally anthropocentric.[3]

Our theories and practices of urbanization have contributed to disastrous ecological effects. Wildlife habitat is being destroyed at record rates as the urban front advances worldwide, driven in the First World by suburbanization and edge-city development, and in the Second and Third Worlds by pursuit of a "catching-up" development model that produces vast rural to urban migration flows and sprawling squatter landscapes.[4] Entire ecosystems and species are threatened, while individual animals in search of food and/ or water must risk entry into urban areas, where they encounter people, vehicles, and other dangers. The explosion of urban pet populations has not only polluted urban waterways but led to mass killings of dogs and cats. Isolation of urban people from the domestic animals they eat has distanced them from the horrors and ecological harms of factory farming, and the escalating destruction of rangelands and forests driven by the market's efforts to create/satisfy a lust for meat. For most free creatures, as well as staggering numbers of captives such as pets and livestock, cities imply suffering, death, or extinction.

The aim of this paper is to foreground an urban theory that takes nonhumans seriously. Such a theory needs to address questions about (1) how urbanization of the natural environment impacts animals, and what global, national, and locality-specific political-economic and cultural forces drive modes of urbanization that are most threatening to animals; (2) how and why city residents react to the presence of animals in their midst, why attitudes may shift with new forms of urbanization, and what this means for animals; (3) how both city-building practices and human attitudes and behaviors together define the capacity of urban ecologies to support nonhuman life; and (4) how the planning/policy-making activities of the state, environmental design practices, and political struggles have emerged to slow the rate of violence toward animals witnessed under contemporary capitalist urbanization. In the first part, I clarify what I mean by "humans" and "animals," and provide a series of arguments suggesting that a trans-species urban theory is necessary to the development of an eco-socialist, feminist, anti-racist urban praxis. Then, in the second part, I argue that current considerations of animals and people in the capitalist city (based on US experience) are strictly limited, and suggest that a trans-species urban theory must be grounded in contemporary theoretical debates regarding urbanization, nature and culture, ecology, and urban environmental action.

2. Why Animals Matter (Even in Cities)

The rationale for considering animals in the context of urban environmentalism is not transparent. Urban environmental issues traditionally center around the pollution of the city conceived as human habitat, not animal habitat. Thus the various wings of the urban progressive environmental movement have avoided thinking about nonhumans and have left the ethical as well as pragmatic ecological, political, and economic questions regarding animals to be dealt with by those involved in the defense of endangered species or animal welfare. Such a division of labor privileges the rare and the tame, and ignores the lives and living spaces of the large number and variety of animals who dwell in cities. In this section, I argue that even common, everyday animals should matter.

The human-animal divide: a definition

At the outset, it is imperative to clarify what we mean when we talk about "animals" or "nonhumans" on the one hand, and "people" or "humans" on the other. Where does one draw the line between the two, and upon what criteria? In many parts of the world beliefs in transmogrification or transmigration of souls provide a basis for beliefs in human-animal continuity (or even coincidence). But in the Western world animals have for many centuries been defined as fundamentally different and ontologically separate from humans, and although explicit criteria for establishing human-animal difference have changed over time, all such criteria routinely use humans as the standard for judgement. The concern is, can animals do what humans do? rather than, can humans do what animals do? Thus judged, animals are inferior beings. The Darwinian revolution declared a fundamental continuity between the species, but standing below humans on the evolutionary scale, animals could still be readily separated from people, objectified and used instrumentally for food, clothes, transportation, company, or spare body parts.

Agreement about the human-animal divide has recently collapsed. Critiques of post-Enlightenment science,[5] greater understanding of animal thinking and capabilities, and studies of human biology and behavior emphasizing human-animal similarities have all rendered claims about human uniqueness deeply suspect. Debates about the human-animal divide have also raged as a result of sociobiological discourses about the biological bases for human social organization and behavior, and feminist and anti-racist arguments about the social bases for human differences claimed to be biological. Long-held beliefs in the human as social subject and the animal as biological object have thus been destabilized.

My position on the human-animal divide is that animals as well as people socially construct their worlds and influence each other's worlds. The resulting "animal constructs are likely to be markedly different from ours but may be no less real."[6] Animals have their own realities, their own worldviews; in short, they are *subjects*, not objects. This position is rarely reflected in ecosocialist, feminist and anti-racist practice, however. Developed in direct opposition to a capitalist system riddled by divisions of class, race/ethnicity, and gender, and deeply destructive of nature, such practice ignores some sorts of animals altogether (for example, pets, livestock) or has embedded animals within holistic and/or anthropocentric conceptions of the environment and therefore avoided the question of animal subjectivity.[7] Thus, in most forms of progressive environmentalism, animals have been objectified and/or backgrounded.

Thinking like a bat: the question of animal standpoints
The recovery of animal subjectivity implies an ethical and political obligation to redefine the urban problematic and to consider strategies for urban
praxis from the standpoints of animals. Granting animals subjectivity at a
theoretical, conceptual level is a first step. Even this first step is apt to be
hotly contested by human social groups who have been marginalized and
devalued by claims that they are "closer to animals" and hence less
intelligent, worthy, or evolved than Anglo-European white males. It may
also run counter to those who interpret the granting of subjectivity as
synonymous with a granting of rights and object either to rights-type
arguments in general or to animal rights specifically.[8] But a far more
difficult step must be taken if the revalorization of animal subjectivity is to
be meaningful in terms of day-to-day practice. We not only have to "think
like a mountain" but also to "think like a bat," somehow overcoming Nagel's
classic objection that because bat sonar is not similar to any human sense, it
is humanly impossible to answer a question such as "what is it like to be a
bat?" or, more generally," what is it like to be an animal?"[9]

But is it impossible to think like a bat? There is a parallel here with the
problems raised by standpoint (or multipositionality) theories. Standpoint
theories assert that a variety of individual human differences (such as race,
class, or gender) so strongly shape experience and thus interpretations of
the world that a single position essentializes and silences difference, and
fails to challenge power relations. In the extreme, such polyvocality leads to
a nihilistic relativism and a paralysis of political action. But the response
cannot be to return to practices of radical exclusion and denial of
difference. Instead, we must recognize that individual humans are embedded in social relations and networks with people similar or different upon
whom their welfare depends.[10] This realization allows for a recognition of
kinship but also of difference, since identities are defined through seeing
that we are similar to, and different from, related others. And through
everyday interaction and concerted practice, and using what Haraway terms
a "cyborg vision" that allows "partial, locatable, critical knowledge sustaining
the possibility of webs of connection called solidarity,"[11] we can embrace
kinship as well as difference and encourage the emergence of an ethic of
respect and mutuality, caring and friendship.[12]

The webs of kinships and difference that shape individual identity involve
both humans and animals. This is reasonably easy to accept in the abstract
(that is, humans depend upon a rich ecology of animal organisms). But
there is also a large volume of archeological, paleoanthropological, and
psychological evidence suggesting that concrete interactions and interdependence with animal others are indispensable to the development of
human cognition, identity, and consciousness, and to a maturity that accepts
ambiguity, difference, and lack of control.[13] In short, animals are not only
"good to think" (to borrow a phrase from Lévi-Strauss) but indispensable to
learning how to think in the first place, and how to relate to other people.

Who are the relevant animal others? I argue that many sorts of animals matter, including domestic animals. Clearly, domestication has profoundly altered the intelligence, senses, and life ways of creatures such as dogs, cows, sheep, and horses so as to drastically diminish their otherness; so denaturalized, they have come to be seen as part of human culture. But wild animals have been appropriated and denaturalized by people too. This is evidenced by the myriad ways wildlife is commercialized (in both embodied and disembodied forms) and incorporated into material culture. And like domestic animals, wild animals can be profoundly impacted by human actions, often leading to significant behavioral adaptations. Ultimately, the division between wild and domestic must be seen as a permeable social construct; it may be better to conceive of a *matrix* of animals who vary with respect to the extent of physical or behavioral modification due to human intervention, and types of interaction with people.

Our ontological dependency on animals seems to have characterized us as a species since the Pleistocene. Human needs for dietary protein, desires for spiritual inspiration and companionship, and the ever-present possibility of ending up as somebody's dinner required thinking like an animal. This aspect of animal contribution to human development can be used as an (anthropomorphic) argument in defense of wildlife conservation or pet keeping. But my concern is how human dependency on animals was played out in terms of the patterns of human-animal interactions it precipitated. Specifically, did ontological dependency on animals create an interspecific ethic of caring and webs of friendship? Without resurrecting a 1990s version of the Noble Savage—an essentialized indigenous person living in spiritual and material harmony with nature—it is clear that for most of (pre)history, people ate wild animals, tamed them, and kept them captive, but also respected them as kin, friends, teachers, spirits, or gods. Their value lay both in their similarities with and differences from humans. Not coincidentally, most wild animal habitats were also sustained.

Re-enchanting the city: an agenda to bring the animals back in
How can animals play their integral role in human ontology today, thereby helping to foster ethical responses and political practices engendered by the recognition of human-animal kinship and difference? Most critically, how can such responses and practices possibly develop in places where everyday interaction with so many kinds of animals has been eliminated? Most people now live in such places, namely cities. Cities are perceived as so human-dominated that they become naturalized as just another part of the ecosystem, that is, the human habitat. In the West, many of us interact with or experience animals only by keeping captives of a restricted variety or eating "food" animals sliced into steak, chop, and roast. We get a sense of wild animals only by watching "Wild Kingdom" reruns or going to Sea World to see the latest in a long string of short-lived "Shamus."[14] In our apparent mastery of urban nature, we are seemingly protected from all nature's dangers but chance losing any sense of wonder and awe for the nonhuman world. The loss of both the humility and the dignity of risk

results in a widespread belief in the banality of day-to-day survival. This belief is deeply damaging to class, gender, and North-South relations as well as to nature.[15]

To allow for the emergence of an ethic, practice, and politics of caring for animals and nature, we need to renaturalize cities and invite the animals back in, and in the process re-enchant the city.[16] I call this renaturalized, re-enchanted city *zoöpolis*. The reintegration of people with animals and nature in zoöpolis can provide urban dwellers with the local, situated, everyday knowledge of animal life required to grasp animal standpoints or ways of being in the world, to interact with them accordingly in particular contexts, and to motivate political action necessary to protect their autonomy as subjects and their life spaces. Such knowledge would stimulate a thorough rethinking of a wide range of urban daily life practices: not only animal regulation and control practices, but landscaping, development rates and design, roadway and transportation decisions, use of energy, industrial toxics, and bioengineering—in short, all practices that impact animals and nature in its diverse forms (climate, plant life, landforms, and so on). And, at the most personal level, we might rethink eating habits, since factory farms are so environmentally destructive *in situ*, and the Western meat habit radically increases the rate at which wild habitats are converted to agricultural land worldwide (to say nothing of how one feels about eating cows, pigs, chickens, or fishes once they are embraced as kin).

While based in everyday practice like the bioregional paradigm, the renaturalization or zoöpolis model differs in including animals and nature in the metropolis rather than relying on an anti-urban spatial fix like small-scale communalism. It also accepts the reality of global interdependence rather than opting for autarky. Moreover, unlike deep ecological visions epistemically tied to a psychologized individualism and lacking in political-economic critique, urban renaturalization is motivated not only by a conviction that animals are central to human ontology in ways that enable the development of webs of kinship and caring with animal subjects, but that our alienation from animals results from specific political-economic structures, social relations, and institutions operative at several spatial scales. Such structures, relations, and institutions will not magically change once individuals recognize animal subjectivity, but will only be altered through political engagement and struggle against oppression based on class, race, gender, and species.

Beyond the city, the zoöpolis model serves as a powerful curb on the contradictory and colonizing environmental politics of the West as practiced both in the West itself and as inflicted on other parts of the world. For example, wildlife reserves are vital to prevent species extinction. But because they are "out there," remote from urban life, reserves can do nothing to alter entrenched modes of economic organization and associated consumption practices that hinge on continual growth and make reserves necessary in the first place. The only modes of life that the reserves change are those of subsistence peoples, who suddenly find themselves alienated from their traditional economic base and further immiserated. But an interspecific ethic of caring replaces dominionism to create urban regions where animals are not incarcerated, killed, or sent off to live in wildlife prisons, but instead are valued neighbors and partners in survival. This ethic links urban residents with peoples elsewhere in the world who have evolved ways of both surviving and sustaining the forests, streams, and diversity of animal lives, and enjoins their participation in the struggle. The Western myth of a pristine Arcadian wilderness, imposed with imperial impunity on those places held hostage to the International Monetary Fund and the World Bank in league with powerful international environmental organizations, is trumped by a post-colonial politics and practice that begins at home with animals in the city.

3. Ways of Thinking Animals in the City

An agenda for renaturalizing the city and bringing animals back in should be developed with an awareness of the impacts of urbanization on animals in the capitalist city, how urban residents think about and behave toward animal life, the ecological adaptations made by animals to urban conditions, and current practices and politics arising around urban animals. Studies that address these topics are primarily grounded in empiricist social science and wildlife biology. The challenge of trans-species urban theory is to develop a framework informed by social theory. The goal is to understand capitalist urbanization in a globalizing economy and what it means for animal life; how and why patterns of human-animal interactions change over time and space; urban animal ecology as science, social discourse, and political economy; and trans-species urban practice shaped by managerial plans and grassroots activism. Figure 6.1 lays out a metatheoretical heuristic device that links together the disparate discourses of the trans-species urban problematic. This device does not seek to privilege a particular theoretical perspective, but rather highlights multiple sources of inspiration that may be fruitful in theory development.

Animal town: urbanization, environmental change, and animal life chances
The city is built to accommodate humans and their pursuits, yet a subaltern "animal town" inevitably emerges with urban growth. This animal town shapes the practices of urbanization in key ways (for example, by attracting or repelling people/development in certain places, or influencing animal exclusion strategies). But animals are even more profoundly affected by the urbanization process under capitalism, which involves extensive denaturalization of rural or wild lands and widespread environmental pollution. The most basic types of urban environmental change are well-known and involve soils, hydrology, climate, ambient air and water quality, and vegetation.[17] Some wild animal species (for example, rats, pigeons, cockroaches) adapt to and/or thrive in cities. But others are unable to find appropriate food or shelter, adapt to urban climate, air quality, or hydrological changes, or

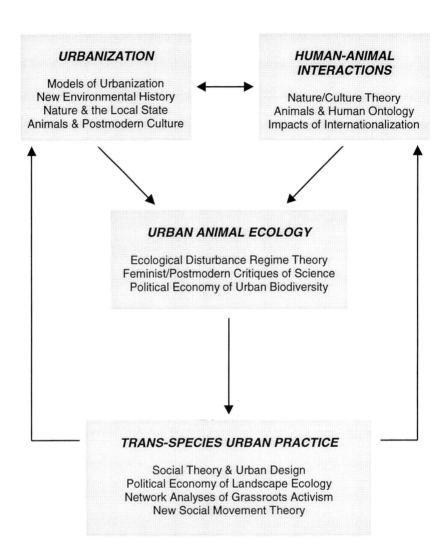

Figure 6.1 Conceptual framework for linking the disparate discourses of
the trans-species urban problematic

tolerate contact with people. Captives, of course, are mostly restricted to homes, yards, or purpose-built quarters such as feed lots or labs, but even the health of pets, feral animals, and creatures destined for dissecting trays or dinner tables can be negatively affected by various forms of urban environmental pollution.

Metropolitan development also creates spatially extensive, patchy landscapes and extreme habitat fragmentation that especially affects wildlife. Some animals can adapt to such fragmentation and to the human proximity it implies, but more commonly animals die *in situ* or migrate to less fragmented areas. If movement corridors between habitat patches are cut off, species extinction can result as fragmentation intensifies, due to declining habitat patch size,[18] deleterious edge effects,[19] distance or isolation effects, and related shifts in community ecology.[20] Where fragmentation leads to the loss of large predators, remaining species may proliferate, degrade the environment, and threaten the viability of other forms of wildlife. Weedy, opportunistic and/or exotic species may also invade, to similar effect.

Such accounts of urban environmental change and habitat fragmentation are not typically incorporated into theories of urbanization under capitalism. For example, most explanations of urbanization do not explicitly address the social or political-economic drivers of urban environmental change, especially habitat fragmentation.[21] By the same token, most studies of urban environments restrict themselves to the scientific measurement of environmental-quality shifts or describe habitat fragmentation in isolation from the social dynamics that drive it.[22] This suggests that urbanization models need to be reconsidered to account for the environmental as well as political-economic bases of urbanization, the range of institutional forces acting on the urban environment, and the cultural processes that background nature in the city.

Efforts to theoretically link urban and environmental change are at the heart of the new environmental history, which reorients ideas about urbanization by illustrating how environmental exploitation and disturbance underpin the history of cities, and how thinking about nature as an actor (rather than a passive object to be acted upon) can help us understand the course of urbanization. Contemporary urbanization, linked to global labor, capital, and commodity flows, is simultaneously rooted in exploitation of natural "resources" (including wildlife, domestic and other sorts of animals) and actively transforms regional landscapes and the possibilities for animal life—although not always in the manner desired or expected, due to nature's agency. Revisiting neo-Marxian theories of the local state as well as neo-Weberian concepts of urban managerialism to analyze relations between nature and the local state could illuminate the structural and institutional contexts of for example, habitat loss/degradation. One obvious starting place is growth machine theory, since it focuses on the influence of rentiers on the local state apparatus and local politics;[23] another is the critique of urban planning as part of the modernist project of control and domination of others (human as well as nonhuman) through rationalist city building and policing of urban interactions and human/animal proximities in the name of human health and welfare.[24] Finally, urban cultural studies may help us understand how the aesthetics of urban built environments deepen the distanciation between animals and people. For instance, Wilson demonstrates how urban simulacra such as zoos and wildlife parks have increasingly mediated human experience of animal life.[25] Real live animals can actually come to be seen as less than authentic since the terms of authenticity have been so thoroughly redefined. The distanciation of wild animals has simultaneously stimulated the elaboration of a romanticized wildness used as a means to peddle consumer goods, sell real estate, and sustain the capital accumulation process, reinforcing urban expansion and environmental degradation.[26]

Reckoning with the beast: human interactions with urban animals
The everyday behavior of urban residents also influences the possibilities for urban animal life. The question of human relations with animals in the city has been tackled by empirical researchers armed with behavioral models, who posit that, through their behavior, people make cities more or less attractive to animals (for example, human pest management and animal control practices, urban design, provision of food and water for feral animals and/or wildlife). These behaviors, in turn, rest on underlying values and attitudes toward animals. In such values-attitudes-behavior frameworks, resident responses are rooted in cultural beliefs about animals, but also in the behavior of animals themselves—their destructiveness, charisma and charm, and, less frequently, their ecological benefits.

Attitudes toward animals have been characterized on the basis of survey research and the development of attitudinal typologies.[27] Findings suggest that urbanization increases both distanciation from nature and concern for animal welfare. Kellert, for example, found that urban residents were less apt to hold utilitarian attitudes, were more likely to have moralistic and humanistic attitudes, suggesting that they were concerned for the ethical treatment of animals, and were focused on individual animals such as pets and popular wildlife species.[28] Urban residents of large cities were more supportive of protecting endangered species; less in favor of shooting or trapping predators to control damage to livestock; more apt to be opposed to hunting; and supportive of allocating additional public resources for programs to increase wildlife in cities. Domestic and attractive animals were most preferred, while animals known to cause human property damage or inflict injury were among the least preferred.

Conventional wisdom characterizes the responses of urban residents and institutions to local animals in two ways: (1) as "pests," who are implicitly granted agency in affecting the urban environment, given the social or economic costs they impose; or (2) as objectified "pets," who provide companionship, an aesthetic amenity to property owners, or recreational opportunities such as bird-watching and feeding wildlife.[29] Almost no systematic research, however, has been conducted on urban residents' behavior toward the wild or unfamiliar animals they encounter or how behavior is shaped by space or by class, patriarchy, or social constructions of race/ethnicity. Moreover, the behavior of urban institutions involved in urban wildlife management or animal regulation/control has yet to be explored.[30]

How can we gain a deeper understanding of human interactions with the city's animals? The insights from wider debates in nature/culture theory are most instructive and help put behavioral research in proper context.[31] Increasingly, nature/culture theorizing converges on the conviction that the Western nature/culture dualism, a variant of the more fundamental division between object and subject, is artificial and deeply destructive of Earth's diverse life-forms. It validates a theory and practice of human/nature relations that backgrounds human dependency on nature. Hyperseparating nature from culture encourages its colonization and domination. The nature/culture dualism also incorporates nature into culture, denying its subjectivity and giving it solely instrumental value. By homogenizing and disembodying nature, it becomes possible to ignore the consequences of human activity such as urbanization, industrial production, and agroindustrialization on specific creatures and their terrains. This helps trigger what O'Connor terms the "second contradiction of capitalism," that is, the destruction of the means of production via the process of capital accumulation itself.[32]

The place-specific version of the nature/culture dualism is the city/country divide; as that place historically emblematic of human culture, the city seeks to exclude all remnants of the country from its midst, especially wild animals. As we have already seen, the radical exclusion of most animals from everyday urban life may disrupt development of human consciousness and identity, and prevent the emergence of interspecific webs of friendship and concern. This argument filters through several variants of radical ecophilosophy. In some versions, the centrality of "wild" animals is emphasized, while the potential of tamer animals, more common in cities but often genetically colonized, commodified, and/or neotenized, is questioned. In other versions, the wild/tame distinction in fostering human-animal bonds is minimized, but the progressive loss of interspecific contact and thus understanding is mourned.[33] Corporeal identity may also become increasingly destabilized as understandings of human embodiment traditionally derived through direct experience of live animal bodies/subjects evaporates or is radically transformed. Thus what we now require are theoretical treatments explicating how the deeply ingrained dualism between city (culture) and country (nature), as it is played out ontologically, shapes human-animal interactions in the city.

The ahistorical and placeless values-attitudes-behavior models also miss the role of social and political-economic context on urban values and attitudes toward animals. Yet such values and attitudes are apt to evolve in response to place-specific situations and local contextual shifts resulting from nonlocal dynamics, for example, the rapid internationalization of urban economies. Deepening global competition threatens to stimulate a hardening of attitudes toward animal exploitation and habitat destruction in an international "race to the bottom" regarding environmental/animal protections. Moreover, globalization sharply reveals the fact that understandings of nature in the West are insufficient to grasp the range of relationships between people and animals in diverse global cities fed by international migrant flows from places where nature/culture relations are radically different. Variations on the theme of colonization are being played back onto the colonizers; in the context of internationalization, complex questions arise concerning how both colonially imposed, indigenous, and hybrid meanings and practices are being diffused back into the West. Also, given globalization-generated international migration flows to urban regions, we need to query the role of diverse cultural norms regarding animals in the racialization of immigrant groups and spread of nativism in the West. Urban practices that appear to be linked to immigrant racialization involve animal sacrifice (for example, Santeria) and eating animals traditionally considered in Western culture as household companions.

An urban bestiary: animal ecologies in the city
The recognition that many animals coexist with people in cities and the management implications of shared urban space have spurred the nascent field of urban animal ecology. Grounded in biological field studies and heavily management-oriented, studies of urban animal life focus on wildlife species; there are very few ecological studies of urban companion or feral animals.[34] Most studies tend to be highly species- and place-specific. Only a small number of urban species have been scrutinized, typically in response to human-perceived problems, risk of species endangerment, or their "charismatic" character.

Ecological theory has moved away from holism and equilibrium notions toward a recognition that processes of environmental disturbance, uncertainty, and risk cause ecosystems and populations to continually shift over certain ranges varying with site and scale.[35] This suggests the utility of reconceptualizing cities as ecological disturbance regimes rather than ecological sacrifice zones whose integrity has been irrevocably violated. In order to fully appreciate the permeability of the city/country divide, the heterogeneity and variable patchiness of urban habitats and the possibilities (rather than impossibilities) for urban animal life must be more fully incorporated into ecological analyses. This in turn could inform decisions concerning prospective land-use changes (such as suburban densification or down-zoning, landscaping schemes, transportation corridor design) and indicate how they might influence individual animals and faunal assemblages in terms of stress levels, morbidity and mortality, mobility and access to multiple sources of food and shelter, reproductive success, and exposure to predation.

Scientific urban animal ecology is grounded in instrumental rationality and oriented toward environmental control, perhaps more than other branches of ecology since it is largely applications driven. The effort by pre-eminent ecologist Michael Soulé to frame a response to the postmodern reinvention of nature, however, demonstrates the penetration into ecology of feminist and postmodern critiques of modernist science.[36] Hayles, for instance, argues that our understanding of nature is mediated by the embodied interactivity of observer and observed, and the positionality (gender, class, race, species) of the observer.[37] Animals, for example, construct different worlds through their embodied interactions with it (that is, how their sensory and intellectual capabilities result in their world-views). And although some models may be more or less adequate interpretations of nature, the question of how positionality determines the models pro-posed, tested, and interpreted must always remain open. At a minimum, such thinking calls for self-reflexivity in ecological research on urban animals and ecological tool-kits augmented by rich ethnographic accounts of animals, personal narratives of nonscientific observers, and folklore.

Finally, scientific urban animal ecology is not practiced in a vacuum. Rather, like any other scientific pursuit, it is strongly shaped by motives of research sponsors (especially the state), those who use research products (such as planners), and ideologies of researchers themselves. Building on the field of science studies, claims of scientific ecology must thus be interrogated to expose the political economy of urban animal ecology and biodiversity analysis. How are studies of urban animals framed, and from whose perspective? What motivates them in the first place—developer proposals, hunter lobbies, environmental/animal rights organizations? Sort-ing out such questions requires not only evaluation of the technical merits of urban wildlife studies, but also analysis of how they are framed by epistemological and discursive traditions in scientific ecology and embedded in larger social and political-economic contexts.

Redesigning nature's metropolis: from managerialism to grassroots action
A nascent trans-species urban practice, as yet poorly documented and under-theorized, has appeared in many US cities. This practice involves numerous actors, including a variety of federal, state, and local bureaucra-cies, planners, and managers, and urban grassroots animal/environmental activists. In varying measure, the goals of such practice include altering the nature of interactions between people and animals in the city, creating minimum-impact urban environmental designs, changing everyday practices of the local state (wildlife managers and urban planners), and more forcefully defending the interests of urban animal life.

Wildlife managers and pest-control firms increasingly face local demands for alternatives to extermination-oriented animal-control policies. In the wildlife area, approaches were initially driven by local protests against conventional practices such as culling; now managers are more apt to consider in advance resident reactions to management alternatives and to adopt participatory approaches to decision-making in order to avoid opposition campaigns. Typically, alternative management strategies require education of urban residents to increase knowledge and understanding of, and respect for, wild animal neighbors, and to underscore how domestic animals may harm or be harmed by wildlife. There are limits to educational approaches, however, stimulating some jurisdictions to enact regulatory controls on common residential architectures, building maintenance, garbage storage, fencing, landscaping, and companion-animal keeping that are detrimental to wildlife.

Wild animals were never a focus of urban and regional planning. Nor were other kinds of animals, despite the fact that a large proportion of homes in North America and Europe shelter domestic animals. This is not surprising given the historic location of planning within the development-driven local state apparatus. Since the passage of the US Endangered Species Act (ESA) in 1973, however, planners have been forced to grapple with the impact of human activities on threatened/endangered species. To reduce the impact of urbanization on threatened/endangered animals, planners have adopted such land-use tools as zoning (including urban limit lines and wildlife overlay zones), public/nonprofit land acquisition, transfer of development rights (TDR), environmental impact statements (EIS), and wildlife impact/habitat conservation linkage fees.[38] None of these tools is without severe and well-known technical, political, and economic problems, stimulating the development of approaches such as habitat conservation plans (HCPs)—regional landscape-scale planning efforts to avoid the fragmentation inherent in project-by-project planning and local zoning control.[39]

Despite the ESA, minimum-impact planning for urban wildlife has not been a priority for either architects or urban planners. Wildlife-oriented residential landscape architecture remains uncommon. Most examples are new developments (as opposed to retrofits), sited at the urban fringe, planned for low densities, and thus oriented for upper-income residents only. Many are merely ploys to enhance real-estate profits by providing home-buyers, steeped in an anti-urban ideology of suburban living emphasizing proximity to "the outdoors," with an extra "amenity" in the form of proximity to wild animals' bodies. Planning practice routinely defines other less attractive locations which host animals (dead or alive), such as slaughterhouses and factory farms, as "noxious" land uses and isolates them from urban residents to protect their sensibilities and the public health.

Wildlife considerations are also largely absent from the US progressive architecture/planning agenda, as are concerns for captives such as pets or livestock. The 1980s "costs of sprawl" debate made no mention of wildlife habitat, and the adherents to the so-called new urbanism and sustainable cities movements of the 1990s rarely define sustainability in relation to animals. The new urbanism emphasizes sustainability through high density and mixed-use urban development, but remains strictly anthropocentric in perspective. Although more explicitly ecocentric, the sustainable cities movement aims to reduce human impacts on the natural environment through environmentally sound systems of solid-waste treatment, energy production, transportation, housing, and so on, and the development of urban agriculture capable of supporting local residents.[40] But while such approaches have long-term benefits for all living things, the sustainable cities literature pays little attention to questions of animals per se.[41]

Everyday practices of urban planners, landscape architects, and urban designers shape normative expectations and practical possibilities for human-animal interactions. But their practices do not reflect desires to enrich or facilitate interactions between people and animals through design, nor have they been assessed from this perspective. Even companion animals are ignored; despite the fact that there are more US households with companion animals than children, such animals remain invisible to architects and planners. What explains this anthropocentrism on the part of urban design and architectural professions? Social theories of urban design and professional practice could be used to better understand the anthropocentric production of urban space and place. Cuff, for example, explains the quotidian behavior of architects as part of a collective, interactive social process conditioned by institutional contexts including the local state and developer clients; not surprisingly, design outcomes reflect the growth orientation of contemporary urbanism.[42] More broadly, Evernden argues that planning and design professionals are constrained by the larger culture's insistence on rationality and order and the radical exclusion of animals from the city.[43] The look of the city as created by planners and architects, dominated by standardized design forms such as the suburban tract house surrounded by a manicured, fenced lawn, reflects the deep-seated need to protect the domain of human control by excluding weeds, dirt, and—by extension—nature itself.

Environmental designers drawing on conservation biology and landscape ecology have more actively engaged the question of how to design new metropolitan landscapes for animals and people than have planners or architects.[44] At the regional level, wildlife corridor plans or reserve networks are in vogue.[45] Wildlife networks and corridors are meant to link "mainland" habitats beyond the urban fringe, achieve overall landscape connectivity to protect gene pools, and provide habitat for animals with small home ranges.[46] Can corridors protect and reintegrate animals in the metropolis? Corridor planning is a recent development, and we need case-specific political-economic analyses of corridor plans to answer this question. Preliminary experience suggests that at best large-scale corridors can offer vital protection to gravely threatened keystone species and thus a variety of other animals, while small-scale corridors can be an excellent urban design strategy for allowing common small animals, insects, and birds to share urban living space with people. However, grand corridor proposals can degrade into an amenity for urban recreationists (since they often win taxpayers' support only if justified on recreational rather than habitat-conservation grounds). At worst, corridors may become a collaborationist strategy that merely smooths a pathway for urban real-estate development into wilderness areas.

A growing number of urban grassroots struggles revolves around the protection of specific wild animals or animal populations, and around the preservation of urban wetlands, forests, and other wildlife habitat due to their importance to wildlife. Also, growing awareness of companion-animal wants and desires has stimulated grassroots efforts to create specially designed spaces for pets in the city, such as dog parks.[47] But we have very little systematic information about what catalyzes such grassroots trans-species urban practices or about the connections between such struggles and other forms of local eco/animal activism. It is not clear if grassroots struggles around animals in the city are linked organizationally either to larger-scale environmental activism or green politics, or to traditional national animal welfare organizations, suggesting the need for mapping exercises and organizational network analyses. Ephemeral and limited case-study information suggests that political action around urban animals can expose deep divisions within environmentalism and the animal welfare establishment. These divisions mirror the broader political splits between mainstream environmentalism and the environmental justice movement, between animal rights organizations and environmentalists, and between groups with animal rights and groups with animal welfare orientations. For example, many mainstream groups only pay lip service (if that) to social justice issues, and so many activists of color continue to consider traditional environmental priorities such as wildlands and wildlife—especially in cities—as at best a frivolous obsession of affluent white suburban environmentalists, and at worst reflective of pervasive elitism and racism. Local struggles around wildlife issues can also expose the philosophical split between holistic environmental groups and individualist animal rights activists; for example, such conflicts often arise over proposals to kill feral animals in order to protect native species and ecosystem fragments. And reformist animal welfare organizations such as urban humane societies, concerned primarily with companion animals and often financially dependent on the local state, may be wary of siding with animal rights/liberation groups critical not only of state policies but also the standard practices of the humane societies themselves.[48]

The rise of organizations and informal groups acting to preserve animal habitat in the city, change management policies, and protect individual animals indicates a shift in everyday thinking about the positionality of animals. If such a shift is underway, why and why now? One possibility is that ecocentric environmental ethics and especially animal rights thinking, with its parallels between racism, sexism, and "speciesism," have permeated popular consciousness and stimulated new social movements around urban animals. Other avenues of explanation may open up by theorizing trans-species movements within the broader context of new social movement theory, which points to these movements' consumption-related focus; grass-roots, localist, and anti-state nature; and linkages to the formation of new sociocultural identities necessitated by the postmodern condition and contemporary capitalism.[49] Viewed through the lens of new social movement theory, struggles to resist incursions of capital into urban wildlife habitat or defend the interests of animals in the city could be contextualized within larger social and political-economic dynamics as they alter forms of activism and change individual-level priorities for political action. Such an exercise might even reveal that new social movements around animals transcend both production and consumption-related concerns, reflecting instead a desire among some people to span the human-animal divide by extending networks of caring and friendship to nonhuman others.

4. Toward Zoöpolis

Zoöpolis presents both challenges and opportunities for those committed to eco-socialist, feminist and anti-racist urban futures. At one level, the challenge is to overcome deep divisions in theoretical thinking about nonhumans and their place in the human moral universe. Perhaps more crucial is the challenge of political practice, where purity of theory gives way to a more situated ethics, coalition building, and formation of strategic alliances. Can progressive urban environmentalism build a bridge to those people struggling around questions of urban animals, just as reds have reached out to greens, greens to feminists, feminists to those fighting racism? In time- and place-specific contexts where real linkages are forged, the range of potential alliances is apt to be great, extending from groups with substantial overlap with progressive environmental thinking to those whose communalities are more tenuous and whose focuses are more parochial. Making common cause on specific efforts to fight toxics, promote recycling, or shape air-quality management plans with grassroots groups whose raison d'être is urban wildlife, pets, or farm animal welfare may be difficult. The potential to expand and strengthen the movement is significant, however, and should not be overlooked.

The discourse of zoöpolis creates a space to initiate outreach, conversation, and collaboration in these borderlands of environmental action. Zoöpolis invites a critique of contemporary urbanization from the standpoints of animals but also from the perspective of people, who together with animals suffer from urban pollution and habitat degradation and who are denied the experience of animal kinship and otherness so vital to their well-being. Rejecting alienated theme-park models of human interaction with animals in the city, zoöpolis instead asks for a future in which animals and nature would no longer be incarcerated beyond the reach of our everyday lives, leaving us with only cartoons to heal the wounds of their absence. In a city re-enchanted by the animal kin-dom, the once-solid Enchanted Kingdom might just melt into air.

Acknowledgements

A slightly longer version of this chapter appears in *Capitalism, Nature, Socialism* 7, 1996, pp. 2–48. I am grateful to Guilford Press for allowing it to be reproduced here.

Notes

1. Daniel B. Botkin, *Discordant Harmonies: A New Ecology for the Twenty-First Century*, New York: Oxford, 1990, p. 167.

2. Such commodified animals include those providing city dwellers with opportunities for "nature consumption" and a vast array of captive and companion animals sold for profit

3. For exceptions, see Ted Benton, *Natural Relations: Ecology, Animal Rights and Social Justice*, London: Verso Books, 1993; and Barbara Noske, *Humans and Other Animals*, London: Unwin Hyman, 1989.

4. Maria Mies and Vandana Shiva, *Ecofeminism*, London: Zed Books, 1993.

5. For example, Lynda Birke and Ruth Hubbard, eds, *Reinventing Biology: Respect for Life and the Creation of Knowledge*, Bloomington and Indianapolis, Ind.: Indiana University Press, 1995.

6. Barbara Noske, *Humans and Other Animals*, p. 158; for similar perspectives, see also Donna Haraway, *Simians, Cyborgs, and Women: The Reinvention of Nature*, New York: Routledge, 1991; Val Plumwood, *Feminism and the Mastery of Nature*, London: Routledge, 1993, and, from the perspective of a biologist, Donald Griffin, *Animal Thinking*, Cambridge, Mass.: Harvard University Press, 1984.

7. Progressive environmental practice has conceptualized "the environment" as a scientifically defined system; as "natural resources" to be protected for human use; or as an active but unitary subject to be respected as an independent force with inherent value. The first two approaches are anthropocentric; the ecocentric third approach, common to several strands of green thought, is an improvement, but its ecological holism backgrounds interspecific difference among animals (human and nonhuman) as well as the difference between animate and inanimate nature.

8. A recovery of the animal subject does not imply that animals have rights, although the rights argument does hinge on the conviction that animals are subjects of a life; see Tom Regan, *The Case for Animal Rights*, Berkeley, Calif.: University of California Press, 1986.

9. Thomas Nagel, "What Is It Like to Be a Bat?" *The Philosophical Review* 83, 1974.

10. This argument follows those by Plumwood, *Feminism and the Mastery of Nature*. See also Jessica Benjamin, *The Bonds of Love: Psychoanalysis, Feminism and the Problem of Domination*, London: Virago, 1988, and Jean Grimshaw, *Philosophy and Feminist Thinking*, Minneapolis, Minn.: University of Minnesota Press, 1986.

11. Donna Haraway, "Situated Knowledges: The Science Question in Feminism and the Privilege of Partial Perspective," in *Simians, Cyborgs, and Women*, p. 191.

12. This in no way precludes self-defense against animals such as predators, parasites, or micro-organisms that threaten to harm people.

13. This evidence has perhaps most extensively been marshaled by Paul Shepard in *Thinking Animals: Animals and the Development of Human Intelligence*, New York: Viking Press, 1978; *Nature and Madness*, San Francisco, Calif.: Sierra Club Books, 1982; and, most recently, *The Others*, Washington, D.C.: Earth Island Press, 1996.

14. "Shamu" was the name used for a series of killer whales who performed in a major US marine theme park.

15. Mies and Shiva, *Ecofeminism*.

16. As highlighted in the following section, there are many animals that do, in fact, inhabit urban areas. But most are uninvited, and many are actively expelled or exterminated. Moreover, animals have been largely excluded from our *understanding* of cities and urbanism.

17. Ann Whiston Sprin, *The Granite Garden: Urban Nature and Human Design*, New York: Basic Books, 1984, Michael Hough, *City Form and Natural Process*, New York: Routledge, 1995.

18. O. H. Frankel and Michael E. Soulé, *Conservation and Evolution*, London: Cambridge University Press, 1981; M. E. Gilpin and I. Hanski, eds, *Metapopulation Dynamics: Empirical and Theoretical Investigations*, New York: Academic Press, 1991.

19. Michael E. Soulé, "Land Use Planning and Wildlife Maintenance: Guidelines for Conserving Wildlife in an Urban Landscape," *Journal of the American Planning Association* 57, 1991.

20. M. L. Shaffer, "Minimum Population Sizes for Species Conservation," *BioScience* 31, 1981.

21. See, for example, Michael Dear and Allen J. Scott, *Urbanization and Urban Planning in Capitalist Society*, London: Methuen, 1981.

22. An example is Ian Laurie, ed., *Nature in Cities*, New York: Wiley, 1979.

23. John R. Logan and Harvey L. Molotch, *Urban Fortunes: The Political Economy of Place*, Berkeley, Calif.: University of California Press, 1987.

24. Elizabeth Wilson, *The Sphinx in the City: Urban Life, the Control of Disorder, and Women*, Berkeley: University of California Press, 1991; Christine M. Boyer, *Dreaming the Rational City: The Myth of American City Planning*, Cambridge, Mass.: MIT Press, 1983; Chris Philo, "Animals, Geography and the City: Notes on Inclusions and Exclusions," *Environment & Planning D: Society and Space*, 13, 1995.

25. Alexander Wilson, *The Culture of Nature: North American Landscapes from Disneyland to the Exxon Valdez*, Cambridge, Mass.: Blackwell Books, 1992.

26. Gary Snyder, *The Practice of the Wild*, San Francisco, Calif.: North Point Press, 1990.

27. See the three-part study by Stephen R. Kellert, *Public Attitudes toward Critical Wildlife and Natural Habitat Issues, Phase I*, US Department of Interior, Fish and Wildlife Service, 1979; *Activities of the American Public Relating to Animals, Phase II*, US Department of Interior, Fish and Wildlife Service, 1980; and, co-authored with Joyce Berry, *Knowledge, Affection and Basic Attitudes toward Animals in American Society, Phase III*, US Department of Interior, Fish and Wildlife Service, 1980.

28. Stephen R. Kellert, "Urban Americans' Perceptions of Animals and the Natural Environment," *Urban Ecology* 8, 1984.

29. David A. King, Jody L. White, and William W. Shaw, "Influence of Urban Wildlife Habitats on the Value of Residential Properties," in L. W. Adams and D. L. Leedy, eds, *Wildlife Conservation in Metropolitan Environments*, National Institute for Urban Wildlife, 1991, pp. 165–9, and William W. Shaw, J. Mangun, and R. Lyons, "Residential Enjoyment of Wildlife Resources by Americans," *Leisure Sciences* 7, 1985.

30. For an exception, see William W. Shaw and Vashti Supplee, "Wildlife Conservation in a Rapidly Expanding Metropolitan Area: Informational, Institutional and Economic Constraints and Solutions," in L. W. Adams and D. L. Leedy, eds, *Integrating Man and Nature in the Metropolitan Environment*, National Institute of Urban Wildlife, 1987, pp. 191–8.

31. Donna Haraway, *Primate Visions: Gender, Race, and Nature in the World of Modern Science*, New York: Routledge, 1989; Neil Evernden, *The Social Creation of Nature*, Baltimore, Md.: Johns Hopkins University Press, 1992; Plumwood, *Feminism and the Mastery of Nature*.

32. James O'Connor, "Capitalism, Nature, Socialism: A Theoretical Introduction," *Capitalism, Nature, Socialism* 1, 1988.

33. Paul Shepard, "Our Animal Friends," in S. R. Kellert and E. O. Wilson, eds, *The Biophilia Hypothesis*, Washington, D.C.: Island Press, 1993, pp. 275–300, stresses the wild, while others are more inclusive, such as Noske, *Humans and Other Animals*, and Karen Davis, "Thinking Like a Chicken: Farm Animals and the Feminine Connection," in Carol J. Adams and Josephine Donovan, eds, *Animals and Women: Feminist Theoretical Explorations*, Durham, N.C. and London: Duke University Press, 1995, pp. 192–212.

34. For exceptions, see Alan M. Beck, *The Ecology of Stray Dogs: A Study of Free-ranging Urban Animals*, Baltimore, Md.: York Press, 1974; and C. Haspel and R. E. Calhoun, "Activity Patterns of Free-Ranging Cats in Brooklyn, New York," *Journal of Mammology* 74, 1993.

35. S.T.A. Pickett and P.S. White, eds, *The Ecology of Natural Disturbance and Patch Dynamics*, Orlando, Fla.: Academic Press, 1985; Botkin, *Discordant Harmonies*. In extreme form, the disturbance perspective can be used politically to rationalize anthropogenic destruction of the environment; see Donald Worster, *The Wealth of Nature: Environmental History and the Ecological Imagination*, New York: Oxford University Press, 1993, and Ludwig Trepl, "Holism and Reductionism in Ecology: Technical, Political and Ideological Implications," *Capitalism, Nature, Socialism* 5, 1994. But see also the response to Trepl from

Richard Levens and Richard C. Lewontin, "Holism and Reductionism in Ecology," *Capitalism, Nature, Socialism* 5, 1994.

36. Michael E. Soulé and Gary Lease, eds, *Reinventing Nature? Responses to Postmodern Deconstruction*, Washington, D.C.: Island Press, 1995. For feminist/postmodern critiques of science, see Sandra Harding, *The Science Question in Feminism*, Ithaca, N.Y.: Cornell University Press, 1986; Haraway, *Primate Visions*; and Lynda Birke, *Feminism, Animals and Science: The Naming of the Shrew*, Buckingham: Open University Press, 1994.

37. Katherine N. Hayles, "Searching for Common Ground," in Soulé and Lease, eds, *Reinventing Nature?* pp. 47–64.

38. Daniel L. Leedy, Robert M. Maestro, and Thomas M. Franklin, *Planning for Wildlife in Cities and Suburbs*, Washington, D.C.: US Government Printing Office, 1978; Arthur C. Nelson, James C. Nicholas, and Lindell L. Marsh, "New Fangled Impact Fees: Both the Environment and New Development Benefit from Environmental Linkage Fees," *Planning* 58, 1992.

39. Only a small number of HCPs have been developed or are in progress, and the approach remains hotly contested. See Timothy Beatley, *Habitat Conservation Planning: Endangered Species and Urban Growth*, Austin, Texas: University of Texas Press, 1994.

40. Sim Van der Ryn and Peter Calthorpe, *Sustainable Cities: A New Design Synthesis for Cities, Suburbs, and Towns*, San Francisco, Calif.: Sierra Club Books, 1991; Richard Stren, Rodney White, and Joseph Whitney, *Sustainable Cities: Urbanization and the Environment in International Perspective*, Boulder, Colo.: Westview Press, 1992; Rutherford H. Platt, Rowan A. Rowntree, and Pamela C. Muick, eds, *The Ecological City: Preserving and Restoring Urban Biodiversity*, Minneapolis, Minn.: University of Minnesota Press, 1994.

41. An interesting exception is the green-inspired manifesto for sustainable urban development; see Peter Berg, Beryl Magilavy, and Seth Zuckerman, eds, *A Green City Program for San Francisco Bay Area Cities and Towns*, San Francisco, Calif.: Planet Drum Books, 1986, pp. 48–9, which recommends riparian setback requirements to protect wildlife, review of toxic releases for their impacts on wildlife, habitat restoration, a department of natural life to work on behalf of urban wildness, citizen education, mechanisms to fund habitat maintenance, and (somewhat oxymoronically) the "creation" of "new wild places."

42. Dana Cuff, *Architecture: The Story of Practice*, Cambridge, Mass.: MIT Press, 1991.

43. Evernden, *Social Creation of Nature*, p. 119.

44. R.T.T. Foreman and M. Godron, *Landscape Ecology*, New York: John Wiley and Sons, 1986.

45. Charles E. Little, *Greenways for America*, Baltimore, Md.: Johns Hopkins University Press, 1990; Daniel S. Smith and Paul Cawood Hellmund, *Ecology of Greenways: Design and Function of Linear Conservation Areas*, Minneapolis, Minn.: University of Minnesota Press, 1993.

46. There is also scientific debate about the merits of corridors; see, for instance, Daniel Simberloff and James Cox, "Consequences and Costs of Conservation Corridors," *Conservation Biology* 1, 1987: Simberloff and Cox argue that corridors may help spread diseases and exotics, decrease genetic variation or disrupt local adaptations and coadapted gene complexes, spread fire or other contagious catastrophes, and increase exposure to hunters/poachers and other predators. Reed F. Noss ("Corridors in Real Landscapes: A Reply to Simberloff and Cox," *Conservation Biology* 1, 1987) however, maintains that the best argument for corridors is that the original landscape was interconnected.

47. Jennifer Wolch and Stacy Rowe, "Companions in the Park: Laurel Canyon Dog Park, Los Angeles," *Landscape* 31, 1993.

48. Such practices include putting large numbers of companion animals to death on a routine basis, selling impounded animals to biomedical laboratories, and so on.

49. Alain Touraine, *The Return of the Actor: Social Theory in Postindustrial Society*, Minneapolis, Minn.: University of Minnesota Press, 1988; Alberto Melucci, *Nomads of the Present: Social Movements and Individual Needs in Contemporary Society*, Philadelphia, Pa.: Temple University Press, 1989; Alan Scott, *Ideology and the New Social Movements*, London: Unwin Hyman, 1990.

================= 7 =================

The Cougar's Tale

Andrea Gullo, Unna Lassiter, and Jennifer Wolch

Introduction

What do you do with a mountain lion in the middle of Santa Monica?[1]

The ideology of postwar urban growth in the United States was predicated on the ability of people to control nature and exclude wild animals from cities. Despite the emergence of environmentalism and the animal rights movement, and rising concerns about habitat loss and declining biodiversity, urbanization is still proceeding rapidly in many areas of the country. Ironically, in some of these places the pace of urbanization is so rapid that settlement growth occurs directly through wildlands-to-urban-land conversion, rather than the historical rural-to-urban land-use dynamic. In such border zones, not all of the animals move out, and so more and more people now confront wildlife. Questions about who is responsible for public safety and what sorts of measures (if any) should be mandated to either eradicate or accommodate wildlife have, inevitably, arisen in these places. The presence of wild animals thus often triggers public debate and conflict, lawsuits over wildlife-related injuries, contested hunts and extermination efforts. In short, what *do* you do with a mountain lion in the middle of Santa Monica?

This chapter examines the process by which human ideas about California's cougars (*Felis concolor*, also called mountain lions) have been shaped by patterns of urbanization, by scientific and political debate, and by changing media coverage. We also consider how cougar ideas about people appear to be changing as a result of human encroachment and other habitat pressures. The cougar's tale is particularly compelling. It simultaneously involves conflict between urbanization and environmental protection, hunters and animal rights advocates, and especially lions and suburban residents. Not surprisingly, neither lions nor suburbanites living in the human-animal border zone have much experience in the art of mutual coexistence, but both are faced with the need to learn.

We begin by highlighting the social processes by which animals come to be seen by people as having certain powers and personalities, virtues and

139

vices. We also raise the question of how animals construct people. Clearly, we are very limited in the extent to which such a question can be addressed, but we pose it because of its relevance to the cougar case. Next, the story of Orange County is told. Rapid urbanization and growth induced by the juggernaut of the postwar Southern California economic boom has made it a place where cougars have been brought both to the edge of extinction and into suburban backyards. After detailing these processes, we turn to an examination of media coverage of the cougar controversy since 1985, showing how attitudes reflected in the daily press have shifted along with scientific debates and political struggles. Lastly, we consider how human representations of "cougar character" have evolved as a result of the shifting public discourse around mountain lion management, and how cougar ideas about people seem to have changed in the face of human encroachment into lion country.

Human Ideas about Animals, Animal Ideas about Humans

> Nature and wildlife movies ... [are] one expression of a long human tradition of investing the natural world with meaning. Those meanings are as often as not laden with sexism, colonialism, and species hierarchy—witness the number of cars, tractors, and military machines named after animals.[2]

Following Serpell,[3] Ritvo,[4] Wilson, and many others, we argue that ideas about animals are socially constructed rather than being grounded in direct experience with nature. This is not to say that animals lack materiality or agency, or exist solely in the minds of people, but rather that many *ideas* about animals and what they mean are in some measure social fabrications. In so-called modern societies, the social construction of animals goes largely unmediated by concrete experience, lending the social imaginary even greater constitutive force. It is the form that these social constructions should take in the future, and their implications for policy, that lies at the heart of contemporary debates over relations between people and animals.

How do social constructions of animals take shape? At one level, personal and contextual characteristics of individuals influence the nature of environmental values, ideas about appropriate human-animal relations, and the extent of knowledge of and experience with various types of animals. But animals are also socially constructed/reconstructed by dynamics at the societal and institutional levels. Here, the negotiation of human-animal boundaries is expressed in public discourse about the "character" and behavior of animals, the management problems they present for humans, their ecological and economic roles, and their rights. There are many sites of this public discourse. The print media, TV, and movies are replete with stories about animal *mentalités* and interspecific borders (*Free Willy, Congo,* and *Babe* are some recent examples). Popular writing on environmental philosophy, animal rights, and biodiversity also flourishes, while media campaigns and public policies around specific issues (for example, spotted owl habitat conservation plans, Yellowstone wolf reintroduction programs) bring animal issues to the forefront of public consciousness. Appropriate relations between people and animals are placed explicitly on the bargaining table in national/international debates around endangered species, habitat protection, and wildlife-reserve creation (such as the Rio summit). Numerous actors with dramatically different stakes and powers participate in these debates. Thus the process of negotiation includes scientists, government politicians and policy makers, commercial/industrial interests and their allies, the hunter lobby, nonprofit environmental organizations, community-based groups, and the news/information and entertainment medias themselves.

Together, individual-level factors and the tenor of public discourse create the "animal"—a mediated characterization, imbued with value according to political and social conditions. As the mix of public attitudes shifts in response to (for example) episodic events or new scientific understandings or policy moves, public discourse itself is affected and individuals' ideas about animals and their relationships with the nonhuman world change. Thus, in an iterative cycle, social constructions of animals alter over time and space, and in turn ultimately shape and reshape public policy, which affects animal life chances.

So much for people and their ideas about animals. What about animals and their notions of people? The answer to this question, largely specific to particular kinds of animals, has been sought after by a wide range of people, most of whom are interested in improving their understanding of, or control over, animal behavior. For example, vast numbers of people who raise dogs learn how people fit into normative canine ideas about the appropriate shape of social relations (and how to retain "alpha" dog standing). Others are concerned with either "habituation" (animals becoming accustomed to people) and/or "adaptation" (adaptive behavioral shifts stimulated by human presence). Sometimes such concerns arise out of desires to minimize both habituation and adaptation and thus reduce potentials for conflict (for example, people and bears in Yellowstone), or to permit, for instance, the release of captive animals into the wild, where they need to fear humans in order to survive. In other cases, the motive is to maximize habituation and adaptation; this was obviously a major strategy for people attempting domestication and remains a priority for those who raise and kill livestock.

For many animals we are completely in the dark as to their conceptions of people and how these might change under novel circumstances. There is growing scientific recognition, however, that many birds and mammals are behaviorally flexible due to their intelligence, intentionality, and ability to learn from example and experience. This recognition does not mean that the ways in which particular animals will respond to heightened human-animal interactions are entirely predictable. Yet an understanding of their most likely reactions is crucial for bureaucratic efforts to manage both wildlife and wildlands recreationists, as well as for everyday attempts on the part of people to live in human-animal border zones. Will predators, for example, try to fit unfamiliar animals (such as humans) into a preconceived mental map of prey opportunities and thus see people as prey? Or will they construct a new set of distinctions about danger and diet to guide their behavior? For some animals—including the cougar—the answers to such questions are largely unknown but have become vitally important for both people and animals.

Cougars: A Cautionary Tale

> The real estate boom made [the cougar attack] inevitable.... For lions always will behave as lions, and developers as developers. And there seems to be no outrunning either species.[5]

Orange County, California, is a postsuburban wonderland of theme parks, far-flung edge cities, and all-consuming consumer culture. Less well-known is the fact that cougars also live in Orange County and have become hard-pressed by the county's emergence as the prototypical postwar American "exopolis."

As recently as the 1940s and 1950s, Orange County was a largely rural area with extensive wildlands and a relatively small human population of only 200,000.[6] Rapid decentralization of aerospace firms and (mostly middle-class white) households, however, soon led to a county population explosion of major proportions. Within little more than a decade, traditional towns such as Anaheim had swollen from 14,000 to almost 130,000 residents; Garden Grove, which was not even incorporated until 1956, had over 100,000 residents by 1963.[7] Developers in control of large tracts of land laid out massive new towns, including Irvine, Laguna Niguel, and Mission Viejo. Pro-development forces were either unopposed or (later) unstoppable. By 1990 the county had developed a diverse economic base and heterogeneous population of 2.4 million, sprawled throughout the county's older cities, up-scale beach communities, planned new towns, and exclusive gated enclaves. The absolute loss of wildlands—and cougar habitat—was swift as the county became a dense checkerboard of urbanized areas, farmlands narrowly bypassed by development, and fragments of remnant wilderness.

As people and cities invaded the chaparral wildlands that had once been cougar habitat, there was a dramatic increase in human-cougar interactions. Most of these interactions involved cougar sightings, finds of cougar pawprints, or discoveries of prey remains. Sightings typically occurred in parklands or on/near roads and highways. The number of cougars killed by vehicles in California rose from 4/year in the 1970s to 12/year in 1984, and road kills became the leading cause of death among Orange County's cougar population.[8] Cougars were also sighted in suburban backyards and local shopping malls, and their kills found in residential areas. But only two attacks on humans occurred, neither of which were fatal. In one instance, a six-year-old boy who was alone but within earshot of an adult was attacked;

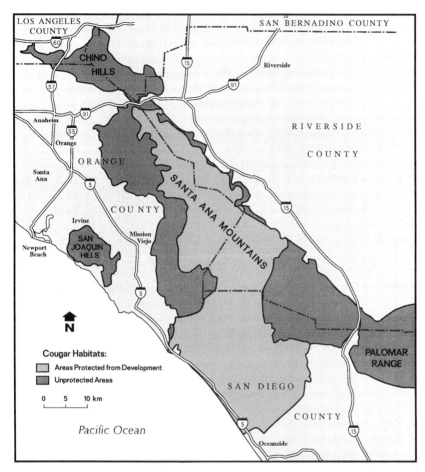

Figure 7.1   Cougar country, Orange County, California
Cartography by Patti Neumann

in a second, a five-year-old girl in the company of an adult was badly clawed and her head partly crushed. Remarkably, no other attacks were recorded for Orange County during the century.

Most of Orange County's cougars make their homes in the Santa Ana Mountain Range—an area of over 2000 square kilometers (Figure 7.1), of which over half is under the jurisdiction of the US Forest Service and US Navy, and thus protected from urbanization. In the Santa Ana Mountains, an estimate tallied 20 adult cougars in the early 1990s, with densities of up to 10 lions per 100 square miles, and average litter sizes of 2.5 kittens.[9] Juveniles typically seek out areas at the fringe of established cougar habitat as their hunting turf (since cougar territories are generally non-overlapping).

Thus their turf is also the suburban fringe with its backyard pools, schools, and shopping centers, and its people, kids, and pets.

Orange County lands west of the Santa Anas were already fully urbanized by 1990 and the eastern zone was growing at a rapid rate. Cougar habitat was thus squarely in the path of urban development. The Orange County cougar population could have been replenished from the Palomar Range to the south, but this range was separated from the Santa Anas by an interstate highway and associated urban developments. Increasing numbers of freeways and residential developments have exposed cougars to greater risks of collisions with cars and have eliminated travel corridors between habitat patches necessary to maintain their breeding population, and thus fragmented cougar territory.

The outlook for Orange County's cougars is unpromising. Efforts to protect cougar habitat "have little power to save land in the pro-development political climate of southern California."[10] Some wildland parcels have been purchased by conservation organizations and removed from the urban land market, and development of other tracts remains subject to planning approval. However, few planning decisions involve areas large enough to affect cougar conservation. With almost thirty local governments making autonomous land-use decisions on a project-by-project basis, major developments in the Chino Hills proceeding, and three highways slated to traverse the county's mountain ranges, the "long term prognosis ... is bleak."[11]

Cat Fights over State Cougar Management

> [I]f the legislature fails to control mountain lions, it will have the blood of these victims on [its hands].[12]

Mirroring the situation in Orange County, reports of encounters between cougars and people increased throughout California during the 1980s and early 1990s. Moreover, some evidence suggested that cougar depredation on pets and livestock had increased. From the 1970s to the early 1990s, the number of depredation permits, which are issued by the California Department of Fish and Game and allow the killing of cougars suspected of pet/livestock killing, steadily climbed from 5 in 1971 to 51 in 1979, 135 in 1985, and by 1993 had reached 192.[13]

Statistically, however, mountain lion attacks remained highly unlikely, "like pianos falling out of the sky."[14] Within the past decade, only four human deaths have been attributed to cougars in the United States (two of these in California in 1994), plus a dozen nonfatal attacks by cougars on humans (seven of which occurred in California). As one ecologist quipped, "cougars are no more dangerous to humans than breathing the Southern California air,"[15] and in fact far more people die each year from bees (forty), dogs (eighteen to twenty), rattlesnakes (twelve), and black widow spiders (three) than from cougars.[16]

But while the actual threat to human safety was minimal, state wildlife officials and park rangers fielding calls from frightened people suggested that the public had become acutely concerned about their outdoor safety. California's Department of Fish and Game, as well as many local and regional wildlands management agencies, faced the complex and contradictory task of balancing political pressures, bioconservation goals, and demands for public safety. Most scientists agreed that a resumption of cougar hunting would not diminish the number of attacks per se, but they did not agree on much else, opening up political space for an escalation of controversy over cougar management. Politicians, in turn, reacting to constituent and special interest concerns, revived a battle over the protectionary status of the cougar. Hunters, gun lobbyists, and their political allies sought a resumption of cougar hunting, while environmentalists, animal welfare advocates, and their political supporters fought to protect the cougars' nongame-animal status.

The science of cougar counting
Management decisions have largely hinged on official cougar-population estimates. These estimates, developed periodically by Department of Fish and Game ecologists, have been hotly contested, however, and thus the scientific debate over cougar counting and what estimates actually mean have fanned the fires of controversy over lion management. The cougar population was estimated at 600 lions in 1920; based on assumptions now viewed as "wild and demonstrably false," this number is considered almost completely unreliable.[17] Estimates by Fish and Game scientists in the early 1970s placed cougar numbers at 2,400, while the most recent official estimates suggested a population of between 4000 and 6000.[18] These latest numbers were held up by ranchers and residents, who blamed this upsurge in the cougar population for what was perceived as a growing boldness on the part of lions, leading them to ravage sheep and cattle herds, and venture into suburban neighborhoods in search of pets to eat. The numbers were also taken up by the Department of Fish and Game's traditional allies and financial supporters—hunters. Hunters cited growing deer mortality and blamed it on excessive cougar predation due to population pressures. Hunters argued, moreover, that depredatory permits were not adequately addressing the perceived problem of increased cougar interactions, and that cougar hunting would be a more effective method of population management needed by virtue of the rebounding cougar numbers. This was a stance that some Department of Fish and Game officials endorsed.

Some cougar ecologists outside the Department of Fish and Game, however, argued that the hypothesis of increased cougar aggressiveness due to population pressures was based on unreliable methods for counting the animals.[19] Moreover, they rejected the proposition that hunting would cause cougar attacks to decline, citing a lack of empirical evidence.[20] Instead, ecologists offered a radically different counter-argument: that an increase in depredatory incidents was more likely a result of habitat loss and fragmentation, which stressed a spatially dynamic cougar population. They also offered a critique of cougar enumeration methods and the interpretation of such population estimates.

In countering the notion of population growth as a cause of increased aggressiveness, some ecologists suggested that stress was a factor in explaining changes in cougar behavior. Moreover, Smallwood, Hopkins, and others argued that estimates regarding the population were inaccurate because they were inferred from unreliable anecdotal evidence (road kills, paid bounties, depredation incidents, sighting reports) and home-range studies. The latter were suspect because they were based on extrapolation from cougar density estimates developed in a small number of intensively studied areas of the state and assumed a unified population of cougars rather than subpopulations with separate dynamics. Different methods of counting considered reliable for estimating long-term population shifts, such as enumerating tracks on a series of cross-state transects, produced very different results.[21] Although not uncontroversial, such estimates called into serious question official Department of Fish and Game figures.

Many cougar ecologists argued that cougar populations were regional in nature because cougar habitat patches were geographically isolated from one another. Smallwood and Wilcox, for example, found differences in regional population trends (with drastic declines in the eastern Sierra Nevada and Southern California, followed by some upsurge in parts of Northern California after 1992).[22] They and other ecologists claimed that management decisions needed to be based on regional-scale population dynamics, accounting for the transient nature of the felines and reflecting the need of this animal to move away from its birthplace to establish a new home range elsewhere. Such biogeographic shifts may well mean that the mountain lion population could reach temporary highs in one area while bottoming out in another, due to increased confinement caused by habitat fragmentation.

Dissident cougar ecologists did not recommend more hunting as a cougar management strategy. Instead, they advocated the speedy public acquisition of critical corridor lands in order to reduce the impacts on cougar populations of urbanization-driven habitat fragmentation.

The politics of cougar management

Blamed for livestock deaths by California ranchers, cougars were bountied in 1907 and carried a price on their heads until the 1960s. Then, in 1963, the cougar's status was changed to a "nonprotected mammal," and although cougar hunting was still allowed no bounties were paid. In 1969, cougar status changed once again to "big game mammals" which gave cougars a level of protection by regulating hunting seasons and creating a bag limit.[23] During the first regulated hunting season (July 1970–June 1971) a total of 4726 tags were sold and 83 cougars were killed by hunters. For the next season (November 1971–February 1972), there was a one-lion limit and 227 tags were sold, with 35 lions killed.[24]

The cougar population in the state, officially estimated to be about 2400, was believed to be falling.[25] An estimated 12,400 lions had been killed since bounties had been placed on the cougar's head.[26] In the wake of rising consciousness about species endangerment (recall that the Endangered Species Act was passed in 1973), growing public concern for California's cougar population was voiced, prompting the passage of Assembly Bill 660. This bill mandated an investigation into the state's cougar population and changed the status of the cougar one more time from "game" animal to protected "nongame" animal. A four-year moratorium banning cougar hunting was enacted and later extended until 1 January 1986. At that point, the bill's reauthorization was vetoed by Republican Governor George Deukmejian. This veto would have returned the cougar to its former game mammal status had it not been for the postponement of the 1986 hunting season by the Fish and Game Commission. Widespread expressions of public concern forced the Department of Fish and Game to conduct a study of the cats' location, density, and population in California to more clearly determine potential effects from hunting. Historically a pro-hunting agency financially dependent upon hunter/angler fees, the department updated a 1981 mountain lion status report and endorsed hunting in Northern California for the 1987 season. But this report was challenged by a coalition of environmental groups in court, and the 1972 hunting ban was upheld. The Department of Fish and Game was thus prohibited from issuing hunting permits not only for the 1987 season but also for the 1988 and 1989 seasons as well, on the grounds that the required environmental impact studies were inconclusive.

The Mountain Lion Foundation, a nonprofit organization dedicated to cougar conservation and opposed to cougar sport-hunting, galvanized opposition to the reinstatement of the cougar hunt. The Mountain Lion Foundation also worked to enact legislation designed to take decision-making power concerning cougar hunting out of the hands of the political appointees on the Fish and Wildlife Commission and bureaucrats at the Department of Fish and Game. The result was passage of Proposition 117 in June 1990 (also known as the California Wildlife Protection Act), which brought back protectionary status for the cougar and allocated $30 million annually to protect and enhance wildlife habitat through the year 2020. One-third of the funds were earmarked to protect critical cougar habitat and cougar prey (for example, deer) habitat, leaving the remaining funds for the purchase of rare and endangered species habitat, wetlands, and riparian and aquatic habitat. The Mountain Lion Foundation was instrumental in the passage of this bill through its direct lobbying of legislators, voter campaigns that used images of charismatic, beautiful cougars to interest people in their cause, and promotion of popular support for cougars through public education and dissemination of cougar research.

By late 1994 and early 1995, however, four cougar-related bills had been introduced in the state legislature. Amid a flurry of shrill public speeches by legislators concerning the evils of a mushrooming cougar population no

longer fearful of humans, support for a change in cougar status appeared
to swell. The proposed bills themselves were ostensibly stimulated by three
issues: (1) the two fatal cougar attacks of 1994; (2) rising reports of
encounters between cougars and humans, pets, and livestock; and (3) the
hypothesized growth in the mountain lion population. The subtext of some
of these bills, however, was clearly about the relegitimation of hunting and
a recasting of the cougar hunter's image from kitten killer to environmental
manager working on the side of scientific ecology.[27]

One of the bills, designed to legalize mountain lion hunting and make it
easier to amend Proposition 117, ultimately went before the voters in a
statewide referendum in March 1996, where it became known as Proposition
197. The voter information package sent to all registered voters in the state
stunningly revealed the polarized nature of political debate around the
cougar issue.[28] Proponents listed in the pamphlet included the father of a
cougar-attack victim, an ecologist, and several former state officials who had
served under Republican governors, such as a past director of the Department of Fish and Game, a former chair of the California Resource
Conservation Commission, and a former undersecretary of the California
Resources Agency. Arguments in favor of the measure emphasized the
cougar's dangerousness to people and pets, its cost to consumers in terms
of livestock depredation, and its population growth, which meant that lions
had "outgrown their food supply" and were thus upsetting California's
ecological balance. "Desperate for food, lions are being forced out of their
natural habitat into existing residential and commercial developments
where children and pets are in severe danger." Not coincidentally, "deer
and elk herds, as well as Big Horn sheep, have been over-killed by lions,"
and thus the measure, which "returns control of the mountain lion to the
experts" was needed to protect people, livestock, and pets and to restore
the "delicate balance of nature."

Among those listed as opponents of Proposition 197 were the Humane
Society of the United States officials, the chair of the California Sierra Club,
and representatives of the California Park Rangers Association. Opponents
did not defend the cougar with arguments about habitat loss, nor did they
point out the minimal risk faced by California residents on account of the
cougar. Instead, they sought to characterize proponents as the "gun lobby"
and named the National Rifle Association, the Gun Owners of California,
and the Safari Club International as backers of the measure. They charged
hunters and their gun lobby allies with wanting to restore cougar trophy
hunting so that they could "kill them for the fun of it." Opponents' major
emphasis was on the cruelty of trophy hunting:

> [a] pack of radio-collared hounds is set on the trail of the big cat until the
> exhausted lion seeks refuge in a tree. The trophy hunter (who often is called
> in from out-of-state after paying huge fees to the houndsmen) then drives in
> to blast the lion off the limb at point blank range. When nursing mothers
> are shot, the kittens starve. Often, the head is severed from the carcass and

becomes the "trophy"—stuffed and put on a wall. THIS IS NOT SPORT; THIS IS SLAUGHTER.

Opponents also blamed Department of Fish and Game bureaucrats for failing to develop an appropriate cougar management plan and for "playing a game of 'chicken' with the public" by "dragging their feet in enforcing existing law that requires them to kill mountain lions perceived to be a threat" and thus stirring up both public demand for the return to sport hunting and political support for allocating extra funds to the department for cougar management, as promised under the proposition.

Covering the Cougar Controversy

Along with the eagle, the bear, the wild river and the far mountain peak, the lion is a symbol of our dwindling wilderness heritage that deserves respect and protection.[29]

It is fast and silent and kills its prey by slicing its fangs between the neck bones, snapping the spinal cord ... With increasing frequency, North America's most efficient four-legged killer, the mountain lion, is prowling areas that human beings think of as their domains: suburban neighborhoods, urban open spaces, sometimes even the shopping mall.[30]

Controversies about counting cougars and alternative policy approaches to cougar management were at the core of dilemmas surrounding human/cougar relationships in California. These controversies, along with episodic attacks, influenced the level and nature of media coverage of the cougar issue. In turn, media representations shaped the social construction of cougars in the public eye.

The *Los Angeles Times* ran seventy-nine items about cougars between the mid 1980s and mid 1990s. As the most widely circulated paper in the state, the *Times* is an especially vital source of understanding about the dynamic renegotiation of nature/society relations, serving to purvey social constructions of animals and thereby shaping public attitudes and policy alike. Like other media texts, *Times* coverage also reflects public attitudes through distribution of the knowledge, attitudes, and positions of public actors such as politicians and scientific experts, and the views of a mix of the general public (neighbors, people attacked by wild animals, animal advocates, and so on).

The tenor of coverage

Articles on cougars tended to reflect an overall position or tone toward cougars, which was positive, negative, or neutral. Of the seventy-nine articles on cougars, the tone of thirty-five (43 percent) were supportive and twenty-eight (35 percent) were negative. About a fifth were neutral, and emphasized cougar legislation and management concerns, and threats to human safety. Negative articles received more visible coverage both in terms of placement—of the eight articles featured on the front page, four were negative while only one was positive—and scope of associated illustrative visual coverage than did positive articles. Nineteen photographs accompanied the negative articles, while only five photographs accompanied supportive articles (Figure 7.2).

Opinion pieces and letters to the editors (30 percent of total coverage) varied dramatically in tone from straight reportage. The vast majority were supportive, in contrast to only 18 percent of the remaining coverage. The most frequent letter/opinion piece themes were legislative issues and threats to cougar safety. The various authors of the letters/editorials included staff writers (ten), residents (seven), animal advocates (five), a legislator and a hunter (a member of the National Rifle Association writing in support of Proposition 117 and arguing that hunting would not prevent recurring attacks).

Respect for cougars and cougar behavior was the hallmark of supportive articles, expressed in over half of them. Some articles denied that cougars were a threat to humans, per se, and blamed loss of habitat due to urbanization for rising rates of cougar encounters. One article quoted ecologist Beier: "[w]hen you look at a map and see how much habitat they [cougars] require, and how much habitat, even in just the last four years, has been lost—basically, when the humans move in, they move out. I'm surprised they are as well behaved as they are."[31] But while not blaming cougars for urbanization-driven episodes, neither did most coverage blame suburbanites for venturing into lion country, nor were developers vilified. Rather, suburbanization was characterized as a natural process coming in the wake of population growth and inner-city turmoil. As a free-lance nature writer, living in the tiny community of Auburn Lakes, where one of the fatal attacks occurred, commented: "[people] would rather take precautions against a possible mountain lion attack, than take precautions against a possible drive-by-shooting."[32]

Half of the supportive coverage expressed direct concern for the safety of cougars. Many articles also expressed fears about the potential for cougar extinction, often using the grizzly bear and the endangered status of the California condor as cautionary examples. Even more common were anti-hunting sentiments, voiced in over 70 percent of supportive coverage. Most involved claims that hunting would not reduce conflicts between people and cougars, and many attacked hunters. As a letter to the editor proclaimed: "[t]rophy hunting is the most blatantly arrogant and vicious type of animal killing; it's done for the pure blood-lust fun of it all."[33] Anti-hunting sentiment was also based on its lack of effectiveness. As a naturalist for the state Department of Parks and Recreation contended: "[w]e're not going to save sheep, cattle, deer or little girls and French poodles by opening up a hunting season on mountain lions."[34]

The cougar was viewed in negative articles as a disruption to urban life, a nuisance to society, or a threat to humans. Unsurprisingly, this latter theme

Figure 7.2   Two rangers photographed moments after they had to kill a cougar

Photographed by Dave Gatley, © 1993, *Los Angeles Times*, and reprinted by permission

of "threat to human safety" was the strongest in negative articles, with the well-being of children, suburban residents, and park users being of primary concern. Worries about safety were expressed in three-quarters of negative coverage, which typically included descriptions of cougar incidents and attacks.

Attacks involving people or domestic animals and cougars were most frequently explained on the basis of increased populations of people and cougars, and development pressure resulting in loss of habitat. Like similar discussions about habitat loss in supportive coverage, these articles seldom questioned the wisdom of suburbanization, but rather took it as a given. And, in language suggesting that cougars somehow knew which territory was "zoned" for them, cougars were sometimes blamed because they had "overrun their habitat."[35]

Support for hunting was the second most common theme echoed in negative articles. This sentiment was reflected in criticisms raised against the 1972 moratorium or Proposition 117 and in comments made in favor of open hunting seasons for cougars. Such articles expressly supported hunting as a means of management and often blamed the rising rates of encounters and attacks on the hunting ban. As the chair of the California Sportsmen's Task Force (and a bow-and-arrow hunter) argued, "[w]hen [Proposition] 117 passed, we knew there were going to be problems in the future," and predicted that "it's just a matter of time before some child is taken."[36] Similarly, a rancher from Eureka claimed that the mountain lions were "annihilating" the deer population and would soon move on to children: "[t]he only thing that is going to turn [the anti-hunting movement] around is when there won't be any deer left and the lion starts coming into the city and going after children."[37]

Changes in cougar coverage
Cougar coverage peaked in 1987 and again in 1995, a year after cougar attacks had occurred (in 1986 and 1994). Similarly, there was an upsurge in editorials in 1987 and 1995. While supportive articles dominated coverage overall, over time the rate of negative articles steadily increased, especially after 1991. Sixty percent of the supportive articles occurred between 1985 and 1989, as did most supportive editorials, during debates on whether to return cougars to game mammal status. Their subsequent decline serves as evidence of decreasing support for lions as the decade progressed, paralleling the rising rate of people-cougar interactions and the upsurge in scientific and political debate over cougar management.

Constructions of Character

It's like moving Jeffrey Dahmer or something next door to you, waiting for the next thing to happen.[38]

To a lion, a child looks like a little fat pig to eat.[39]

Descriptors of cougars highlighted in press coverage and Proposition 197 election materials revealed how ideas about cougar "character" had shifted over the course of the 1980s and 1990s. So, too, had interpretations of how cougar perceptions of people were changing.

Table 7.1   Terms describing cougar character/behavior

Negative terms	Positive terms	Neutral terms
One of nature's finest killing machines	Symbol of dwindling wilderness heritage	Animal
North America's most efficient four-legged killer	Indigenous, rapidly vanishing wild creatures	Creature
Killer-animals	Magnificent wild creature	Predator
Serial killers	Spectacular looking	Mountain lion
Increasingly aggressive population of predators	Elusive and fascinating wild creature	Panther
Prince of predators	Innocent	Cat
Lean, mean, killing machine	Majestic	Big cat
Wayward	Beautiful	Roamer
Big troubles	Proud	German-shepherd-sized lion
Roaming like phantoms		Natural resource
Wildest of the wild		California's resource
Menace		Loner

Changes in representations of cougar character

The descriptions of cougars as well as explanations of cougar behavior based on inferences about their character reflected the changing public judgements about the lion and its moral worthiness for continued protection by the state. Attacks were increasingly attributed to cougar "character" and changes in behavior, which no amount of charisma could offset.

Table 7.1 illustrates the range of descriptive labels used in *LA Times* coverage. Supportive descriptors such as "majestic" linked cougars to aesthetic values. Others such as "a symbol of our dwindling wilderness heritage" conjured up an Edenic, pristine nature embodied by cougars, to be saved for future generations. These terms defined cougars as part of nature, genetically programmed to act the way they do and thus innocent and undeserving of harm from humans (through hunting, for instance). However, most supportive terms also objectified cougars and appropriated them as part of "our" heritage or as a resource available for human use.

In contrast, negative terms for cougars, such as "serial killers," tended to be graphic and alarmist, evoking images of cougars as vicious killers set out to inflict pain and death on people. Such images, linking cougars with premeditated criminal behavior, played on popular worries about rising crime and lawlessness. Thus they also sent implicit messages about how such "criminals" should be dealt with—harshly, and without pity or mercy, as suggested by one commentator, who argued, "[i]n any civilization, killers aren't allowed to run loose."[40] Although unflattering, these terms also granted cougars agency and the ability to think, make, and be held responsible for their decisions.

More subtle negative characterizations also crept into the discussion. For example, a California Department of Fish and Game representative stated: "[sightings] indicate that there is a mountain lion up there that's found an easy way to find a meal. It's easier to snatch a dog than spend all that time fighting over deer."[41] This claim implies that cougars had begun to make a conscious choice to change their historic predatory patterns and secretive lifestyle in order to feast on pets, simply because they suddenly got "lazy" and no longer wished to exert themselves in a challenging hunt. Such characterizations draw on deeply ingrained notions about the value of work, the moral laxity of those suspected of evading labor, and their status as "undeserving" of public support.

As negative coverage of the cougar controversy expanded, negative descriptors of cougars became increasingly common, helping to alter public opinion of cougar character. As reports of cougar-human interactions rose and public fears were fanned by episodic attacks, the images of cougars as charismatic and proud wild animals at home in nature were replaced by terms conjuring danger, death, and criminal intent. The moral status of the cougar fell dramatically, opening the way for efforts to repeal protective status and reinstate trophy hunting.

The campaign for Proposition 197 reveals how the evolution of public discourse around the cougar controversy constrained opponents from making positive claims about lion character; instead, they sought to impugn proponents' characters as cruel and bloody. In the official voter information pamphlet distributed to all registered voters, opponents of the proposition primarily used the neutral term "mountain lion" to refer to cougars. The sole exceptions were the use of the term "big cat" and a reference to mountain lions as being "like any other wildlife resource"—a far cry from "symbol of wilderness heritage."

In contrast, assertions that had become common in the press about cougar character were used by proponents of the measure to sway voter opinion. They emphasized the cougar's practice of "inhumane slaughter" and "over-killing" of prey, inserting a moral judgement about predators as a class and suggesting that cougars had become wanton animals who no longer killed only for food (and thus also implying that they had other reasons, such as killing for fun). Proponents also played upon voter emotions by highlighting how motherhood, children, and pets were now threatened by cougars:

Two small children woke up one morning without a mother because a lion ate her.
A lion preying upon neighborhood pets was found with parts of five different puppies in its stomach.

Some schools have had to cancel recess because lions were found to be lurking around the playground.

Such claims evoke images of literal and sexual predation. Only an animal with a vicious nature could commit such unthinkable crimes or leer at young children. By extension, those seeking to protect cougars were rendered accomplices to crime, their actions tantamount to attacking or molesting mothers, babies, and pets.

Cougar constructions of people
We can only guess about how cougars conceptualize people or about the diversity of their views. Scientific work done in remote areas indicates that in the wilderness cougars hunt at night, are shy and elusive, and keep strictly away from people, who often only realize that they were near a lion later, when tracks are discovered. Cougars have been known to follow or shadow people, but in these instances their behavior was unthreatening; they appeared to be simply curious about people and kept a respectful distance. According to conventional wisdom, if a cougar comes too close for comfort, just making loud noises, waving arms to appear tall, and throwing rocks will make the cat turn tail and run.[42]

However, cougars are considered adaptive and will prey on what is available, alive or not. Much of their behavior is learned, leading ecologists to suggest that they have adapted to sharing habitat with humans in metropolitan border zones.[43] In some locales where urban encroachment has led to the growth of in-town deer herds, cougars may trade off potential danger from humans for the benefits of plentiful prey. For example, studies in Colorado indicate that cougar populations that historically hunted at night, alone, and were only seen during the winter when they followed deer to lowland grazing areas had become active during the day, traveled in groups, and were sighted throughout the year in settled areas where deer populations had increased due to hunting bans, availability of water, and plentiful forage in the form of garden landscaping (and often human hand-outs such as salt licks).

In areas where prey such as deer is not available, cougars may revise their ideas about what is acceptable prey, and come to see people as a potential food source. As cougar biologist Fitzhugh reasoned:[44]

> Confronting a human, the wilderness mountain lion is forced to react to this animal as prey or something to be avoided. . . . It may take several encounters before his [*sic*] instinct tells him [*sic*] what to do . . . But at the edge of the human habitat, where people are common, the lion's curiosity may have been satisfied . . . the lions are now perceiving humans as prey.

The size, movements, and voices of unattended small children in particular may convince cougars that prey is at hand. As one person who had raised and handled cougars suggested, lions did not see children as small humans: "[t]hey see a kid as a big, fat rabbit. They see them as a prey species."[45]

Since several attacks have involved juvenile lions, others hypothesize that these inexperienced hunters have yet to develop accurate ideas of who/ what constitutes prey, and from whom they should stay away. One cougar tracker suggested that a juvenile who killed a Montana child was simply ignorant: "[t]hat one was a kitten and probably got separated from the mother, so it didn't know what it was attacking."[46] Juveniles with the least experience of humans are precisely those who end up in long, thin territories up against the urban edge. And in at least one incident, it was eventually discovered that the attacking cougar had been raised by humans and then was dumped, with no survival skills, in a regional wilderness area to fend for herself.

The changing reactions of at least some cougars to the presence of humans and pets (or among juveniles, their lack of experience with people) are manifest in repeated incidents reported over the past twenty years, during which time there have been more attacks against humans than in the eighty years prior.[47] Cougars appear to have decided that people are viable prey, and/or that humans are interesting and not too threatening. Some cougars are clearly willing to come closer, for example, near school grounds, golf courses, movie theaters, agricultural fields, shopping malls, and suburban neighborhoods where they have been spotted relaxing in backyards and also carrying off domestic animals. Modifying their anti-social habits, such lions "completely ignore lawn mowers, cars and hollering kids."[48]

Ironically, both human constructions of cougars and cougar ideas about people are often misleading. Moreover, they may be working at cross-purposes. While border-zone residents are increasingly convinced that cougars pose major risks (they do not), some cougars on the metropolitan fringe seem to think that people are not dangerous (they are). Such human and feline constructions are thus on a collision course, which could lead to tragic results for both cougars and people.

Conclusions

We want to learn how to limit urban encroachment and how we can enable cougars to survive in an area with this encroachment.[49]

Rapid postwar urbanization in California brought people into cougar habitat. While such urban settlement patterns increased the potential for human-cougar interactions, and may have led some cougars to revise their ideas about humans, the threat posed by cougars remained extremely low. A small number of high-profile incidents during the 1980s and 1990s, however, galvanized fear of cougars among many Californians. The attacks stimulated intense public debate, leading to political pressures to renegotiate human-cougar relationships and a revised social construction of the cougar's character.

The debate over human-cougar relations pitted against each other interest groups with continuing stakes in cougar status, including hunters and their foes, wildlife management officials, ecologists, environmentalists, and animal protection groups. Changing media coverage of the cougar issue reflected and reinforced a backlash against protection for cougars. While coverage initially emphasized the cougar's role in wilderness ecology, the voices of wildlife officials historically allied to hunters, hunters themselves, and politicians linked with their interests became louder. Media reports increasingly reflected the hunter lobby's message that renewed cougar hunting would rescue an overburdened ecosystem. In contrast, the messages of environmentalists and ecologists, who pointed toward the dangers of unfettered urbanization and called for land-use planning solutions, received much less emphasis.

Representations of cougar character which, in the mid 1980s, had highlighted the cougar as a charismatic symbol of wild nature were displaced by a social construction of cougars as cold-blooded killers. The subtext of these messages played upon widespread fears about crime, especially murder and sexual violence, but also about cougar "work ethic." Either criminal or lazy, cougars were by implication unworthy of protection and instead deserved harsh treatment. Such changing representations built support for a return to hunting as a solution to cougar-management problems. At the same time, cougar ideas about people appear to have undergone a change in response to increasing exposure to people and their neighborhoods, making at least some lions less afraid of and thus more willing to enter populated areas. Increasing familiarity, in turn, seems to have led some cougars to view people as prey.

Ultimately, California's voters rejected Proposition 197's endorsement of a return to sport hunting of cougars (with 58 percent opposed). Despite increasing negative media reportage, the strong anti-hunting sentiments of the state's urban-based population, along with a strong editorial in the *Los Angeles Times* against the proposition, may have swayed voters as election day approached.[50] But the proposition's author immediately launched new efforts to change the California Wilderness Protection Act, and the issue is unlikely to go away.

This renegotiation of human-cougar relations must be seen both in its larger political and economic context, and in terms of neighborhood planning practices and ways of everyday life. In a state recently beset by economic recession and largely controlled by pro-development elites, cougar management options (except hunting) involve further restrictions on development and major expenditures for habitat conservation, and so are difficult to promote. Political pressures at both state and federal levels to dismantle environmental regulations such as the Endangered Species Act reinforce and legitimize a critique of environmental planning in general and conservation planning for target species in particular as excessive state intervention into the market, superfluous for environmental well-being, and imposing economic costs on society largely unwarranted by real or perceived benefits. This anti-environmental climate is apt to sharply restrict funds for large-scale habitat acquisition and make urban growth management an uphill battle.

Even in a more favorable political-economic context, however, a growing number of people and cougars will need to learn how to coexist. Developing ways to avoid mutual injury is not only an individual safety issue, but also a critical concern for state and local governments responsible for public safety. But what can be done by people and communities to facilitate coexistence?

On the human side of the equation, there are potential approaches, some of which are already practiced in particular places. Residents can be educated about cougar habits: how to behave in order to avoid attack or how to fend off an attack; when to avoid leaving children or pets unattended; how to store garbage; and how to landscape in order to prevent attracting deer into suburban neighborhoods and public facilities such as parks and other open spaces. Perhaps most importantly, education efforts can raise awareness of the fact that metropolitan border zones are spaces shared with other animals, which entails obligations on the part of people to know about their life ways. Parkland and other open-space management agencies can insure adequate signage and information about cougars, close off areas and restrict entry to adults only if need be. Local and regional land-use plans can take cougar movement paths into account, by defining wildlife corridor zones on the basis of ecological research, acquiring land parcels to reconnect corridor segments, join corridors to larger habitat patches, or create buffer zones around built-up areas to create a green matrix for animal dispersal. Also, designs for infrastructure such as roads and highways can include periodic underpasses to allow cougar (and other wildlife) crossings.

What, if anything, can cougars contribute to solving the dilemmas of coexistence? Once considered largely prisoners of instinct, cougars are now seen to be able to learn from experience and adapt their behavior accordingly; this suggests that purposive efforts to alter their habits may warrant exploration. Indeed, wildlife experts report that lions clearly alter their behavior in response to human activities; for example, they deliberately avoid research teams attempting to radio-collar (and thus "hunt") them, suggesting that cougars could be "hazed" as a means of reorienting their ideas about people and cities.[51] Hazing would involve tracking down cougars who reside in border zones, using dogs to chase them up a tree or pole, and then playing loud audio tapes of talking people, screaming babies, barking dogs, and traffic noises. Cougars could thus learn to associate a frightening experience (being chased) with the sounds of the city, prompting them to fear and avoid such places on their own initiative. While using such a method initially seems stressful for lions, it in fact credits them with the intelligence to learn and the agency for using this new-found knowledge to protect themselves from danger.

If people can learn that life in the human-animal borderlands implies a duty to know their cougar neighbors and how to coexist with them, and if cougars can learn that borderlands survival means keeping a respectful distance, maybe communities of the urban-wildlands fringe can once again rest safe. Then maybe the cougar's tale won't always end with the same sad refrain: "Mountain lion shot and killed today."

Acknowledgements

The authors would like to acknowledge Patti Neumann for her excellent cartographic assistance. Thanks are also due to Jody Emel and Michael Dear for their comments on earlier drafts.

Notes

1. Department of Fish and Game spokesman Jeff Weier, in Renee Tawa, "Mountain Lion Protection Law under Fire," *Los Angeles Times*, 3 April 1995, p. A16.

2. Alexander Wilson, *The Culture of Nature*, Cambridge 1992, p. 120.

3. James Serpell, *In the Company of Animals*, Oxford 1986.

4. Harriet Ritvo, *The Animal Estate: The English and Other Creatures of the Victorian Age*, Cambridge, Mass. 1987.

5. Peter H. King, "The Lion and the Jogger," *Los Angeles Times*, 1 May 1994, p. A3.

6. Rob Kling, Spencer Olin, and Mark Poster, eds, *Postsuburban California*, Berkeley, Calif. 1991, p. 2.

7. C. E. Parker and Marilyn Parker, *Indians to Industry*, Santa Ana, Calif. 1963, p. 84.

8. Kevin Johnson, "Cougars' Worst Foe Is Found to Be Cars," *Los Angeles Times*, 9 August 1993, p. A3.

9. Terry M. Mansfield, "Mountain Lion Management in California," *Transactions of the Fifty-first National Association Wildlife and Natural Resources Conference*, 1986, p. 180.

10. Paul Beier, "Determining Minimum Habitat Areas and Habitat Corridors for Cougars," Conservation Biology, vol. 7, no. 1, 1993, p. 105.

11. Beier, "Determining Minimum Habitat Areas," p. 94.

12. California State Senator Tim Leslie, in Tawa, "Mountain Lion Protection," p. A16.

13. California Department of Fish and Game, Resources Agency, *Report to the Senate Natural Resources and Wildlife Committee and the Assembly Water, Parks and Wildlife Committee Regarding Mountain Lion*, Sacramento 1995.

14. Beier, in Carol McGraw and S. L. Sanger, "Big Cats, Big Trouble," *Los Angeles Times Magazine*, 1 March 1992, p. 28.

15. Beier, in Johnson, "Cougars' Worst Foe," p. A3.

16. R. Weiss, "Researchers Foresee Antivenin Improvements," *Science News*, vol. 138, 1990, pp. 360–62.

17. Bruce A. Hopkins, "California Statewide Estimates and Trend Analysis: Lessons from Diablo Range," Abstract of paper presented at the Fifth Mountain Lion Workshop, 28 February–1 March 1996, San Diego, Calif.: California Department of Fish and Game/ Southern California Chapter of the Wildlife Society, 1996.

18. California Department of Fish and Game, Resources Agency.

19. Bruce A. Hopkins and K. Shawn Smallwood, "Trends in California Mountain Lion Populations," *The Southwestern Naturalist*, vol. 39, no. 1, 1994, pp. 67–72.

20. Paul Beier, "Cougar Attacks on Humans in the United States and Canada," *Wildlife Society Bulletin*, vol. 19, 1991, pp. 403–12.

21. Paul Beier and Stan C. Cunningham, "Power of Track Surveys to Monitor Population Trends," Abstract of paper presented at the Fifth Mountain Lion Workshop, 28 February–1 March 1996, San Diego, Calif.: California Department of Fish and Game/ Southern California Chapter of the Wildlife Society, 1996.

22. Shawn Smallwood and Bruce Wilcox, "Ten Years of California Mountain Lion Track Survey," Abstract of paper presented at the Fifth Mountain Lion Workshop, 28 February–1 March 1996, San Diego, Calif.: California Department of Fish and Game/ Southern California Chapter of the Wildlife Society, 1996.

23. Mountain Lion Foundation, "California Cougar Chronology," Mountain Lion Foundation factsheet, Sacramento, Calif. 1996.

24. Richard A. Weaver, *Status of the Mountain Lion in California with Recommendations for Management*, State of California Resources Agency, Department of Fish and Game, 1982.

25. L. W. Sitton, *Interim Report on Investigations into the Status of the California Mountain Lion, Phase 1 Legislative Report*, California Department of Fish and Game, Sacramento, Calif. 1973, as cited in Mansfield, "Mountain Lion Management," p. 179.

26. Mountain Lion Foundation, "California Cougar Chronology."

27. A brief description of the aims of these bills indicates the terms of political debate around California's cougars. Senate Bill 28 introduced by Tim Leslie (R-Roseville) sought to change the four-fifths vote requirement in the legislature to a majority vote requirement in order to more easily make changes to Proposition 117 and to legalize hunting as a mountain lion management practice under control of the Department of Fish and Game. In an even more extreme move, David Knowles (R-Placerville) presented Assembly Bill 117, recommending a change in cougars' status to game mammal and permitting license tags to be sold for a fee of one dollar. Assembly Bill 87 introduced by Dominic Cortese (R-San Jose) attempted to take a middle ground by establishing a statewide policy and procedure for local authorities that would have further facilitated the removal or taking of mountain lions perceived to pose imminent threat to public health or safety (already legal under Proposition 117). The most progressive of the bills, Senate Bill 582, was introduced by Hilda Solis (D-El Monte) and would have guaranteed more funds for the purchase of land critical to the preservation of Southern California wildlife habitat.

28. California Secretary of State, *California Ballot Pamphlet*, Primary election, 26 March 1996, Sacramento, Calif. 1996, pp. 30–31.

29. Anonymous, "Hunters and the Hunted," *Los Angeles Times*, 10 April 1987, p. II4.

30. Doug Smith, "Too Close to Home," *Los Angeles Times*, 6 September 1994, p. B6.

31. John Needham, "Call of the Wild," *Los Angeles Times*, 13 October 1992, p. A3.

32. Fred Jones, in King, "Lion and the Jogger," p. A3.

33. Dan Pearson, "'Going after the Big Cats,'" *Los Angeles Times*, 15 February 1987, p. V4.

34. Kathleen H. Cooley, "State Postpones Open Season on Mountain Lions," *Los Angeles Times*, 8 April 1986, p. A3.

35. Nicholas Riccardi, "Cougar Is Hunted in Foothills," *Los Angeles Times*, 14 March 1995, p. B1.

36. Smith, "Too Close to Home," B6.

37. Dick Hackett, in Maura Dolan, "Drawing a Bead on Hunting," *Los Angeles Times*, 28 August 1989, p. I1.

38. Cougar attack victim Robert Berrington, in Tawa, "Mountain Lion Protection," p. A16.

39. Dick Hackett, in Dolan, "Drawing a Bead," p. I1.

40. Tony Perry, "Living in Cougar Country," *Los Angeles Times*, 19 December 1994, p. A38.

41. Patrick Moore, in Riccardi, "Cougar Is Hunted in Foothills," p. B1.

42. Lee Fitzhugh, in Steve Emmons and Earl Gustkey, "Answers Are as Elusive as the Lions," *Los Angeles Times*, 29 October 1986, p. I31.

43. Maurice G. Hornocker, "Learning to Live with Mountain Lions," *National Geographic*, vol. 182, 1992, pp. 52–60.

44. Emmons and Gustkey, "Answers Are as Elusive," p. I31.

45. Gary Bogue, in Emmons and Gustkey, "Answers Are as Elusive," p. I1.

46. Bob Weisner, in Margaret L. Knox, "Mountain Lions Thrive in West, Find Civilization," *Los Angeles Times*, 10 January 1991, p. A5.

47. Beier, "Cougar Attacks on Humans," pp. 403–12.

48. Michael Sanders, in McGraw and Sanger, "Big Cats, Big Trouble," p. 28.

49. Beier, in McGraw and Sanger, "Big Cats, Big Trouble," p. 28.

50. Anonymous, "No on Measure to Study Cougars," *Los Angeles Times*, 13 March 1996, p. B8.

51. Dan Lay, Former Conservation Officer (Wildlife Control), Department of the Environment, Province of British Columbia, Canada. Personal communications with Andrea Gullo, March–April 1996.

8

Golden Eagles and the
Environmental Politics of Care

Suzanne M. Michel

> Social theory remains resolutely anthropocentric . . . Animals
> are signifiers, and denied lives of their own. Animals are the
> ultimate Other. Most of us have been afraid (or unwilling) to
> journey across the species divide to construct a more inclusive
> social theory.[1]

Introduction

In January 1994, I visited a Smithsonian Museum traveling exhibit titled Dos
Aquilas—Two Eagles: A Natural History of the Mexico-US Borderlands. This
exhibit displayed the native flora and fauna of the borderland area between
Mexico and the United States. One central theme reiterated in the exhibit
is that the US-Mexico border region is not only a political borderland, but
an ecological borderland. Due to the region's geographical location and
varied topography, numerous different ecosystems collide and intertwine,
producing an incredible amount of biodiversity. A museum display caption
encapsulates this concept of an ecological borderland:

> Barren Wasteland? Look Again!
> The borderlands may well have the greatest biodiversity of any region in
> North America. Its many ecosystems—coasts, riversides, high mountains,
> valleys and arid desert plains—provide habitat for thousands upon thousands
> of plant and animal species.

A borderland delineates an area in which boundaries blur between
nation-states, socioeconomic classes, ecosystems, and so forth. According to
Ingram, Laney, and Gilliam,[2] borderlands are also places where conventional approaches are questioned, stereotypes dissolve, and new understandings emerge. Given this definition, borderlands can be conceived as sites
where we are not afraid to transgress, and we even recognize the interplay
between socially constructed dualisms such as mind-body, rationality-

animality, reason-emotion, or nature-culture.[3] This essay is about traversing the borderlands between nature and culture, more specifically the socially constructed divide between animal and human species. I intend to blur the boundaries between biology, political science, and feminist theory to address the politics of wildlife, or the reproduction rights of the *Aquila chyrsaetos*, the golden eagle in San Diego County, California.

During the past fifty years, golden eagle numbers in San Diego County have declined by at least one third.[4] Raptor ecologists concur that the probable causes of declining eagle numbers are reduced nesting and foraging habitat, and increased human presence near home ranges.[5] In Ramona, a small community in East San Diego County, the loss of habitat to suburban development is one such factor that threatens the local golden eagle population. Thus, to prevent local golden eagle extirpation, certain residents and conservation organizations have decided to contest and obstruct plans to develop land parcels that contain golden eagle nesting sites and foraging ranges. This contestation for space represents ever increasing "hostilities" that are occurring throughout the western United States between those who desire more space for urban growth and those who desire to preserve space for wildlife.

To develop my borderland perspective of wildlife politics, I proceed as follows: first, I briefly explore the realm of the nature-culture dualism, and introduce the concept of borderland identities and possible spheres of political resistance. Second, I examine one political site where human and animal needs for space do conflict: the land-use planning process. Third, I address how discouragement with state-sponsored planning meetings fosters an alternative politics of care at different sites and scales. In this section of my analysis, I examine two particular types of wildlife politics of care: wildlife rehabilitation and the environmental education of families. These political actions occur at the micro-level scale (household, local community) and are conventionally not considered politically relevant in the process of wildlife habitat preservation.

It is my contention that these politics of care are significant because they foster nondualistic thinking, which allows local communities and individuals to become experientially and emotionally connected with the plight of disappearing wildlife. This connection engenders trans-species respect for the lives of nonhuman entities, and political activism that is not grounded in polarized, adversarial politics (such as those present in traditional environmental politics, that is, land-use planning proceedings and legal courts) but in the care of human and animal species. By way of conclusion, I examine how concepts of trans-species relations and politics of care can not only blur the boundaries of nature and culture, but also traverse the boundaries between private and public political activism. Such a traversal occurs within a complex network of political actions which take place at multiple sites (both public and private) and address multiple scales (house-hold to global) linked with destruction of natural entities.

Nature, Culture, and Modernism

Until recently, there has been a silence in geographic literature concerning an understanding of nature as a product of not only physical but social forces.[6] Historically, resistance to accept the natural/nonhuman realm (habitats, ecosystems, species, and so on) as products of social *and* natural histories and forces stems from our extreme reliance upon a particular scientific vision of our natural world. This ontological separation of nature and society is still prevalent in normative discourse and practices validated by experts within the planning, economic, and physical-science disciplines.

The social construction of the nature-culture dualism stems in part from the modernist project that originated during the Age of Enlightenment and has continued for at least three and a half centuries.[7] According to FitzSimmons, the modernist project "reinvented Nature at the same time it reinvented Humanity."[8] This new paradigm of linear progress and empirical thought displaced the Renaissance vision of nature as semiotic and spiritual, and subsequently created nature as a "mechanism—external, universal, primordial and prior to human experience."[9] It is the empiricist thinking and mechanical practices heralded by early scientists and philosophers such as Newton, Galileo, Kepler, and Bacon that has shaped how we presently understand experience and produce nature in modern science.[10]

During the modernist era, the revolutionary paradigm of science positioned nature as an external, material reality to be classified and quantified.[11] Thus, nature became transformed as a discourse of empirical evidence and critical observation that essentially let the "facts speak for themselves."[12] Hence, according to early modernist scientists, culture essentially did not shape our external reality, as Smith ironically notes:

> When he watched the apple fall, Newton did not ask about the social forces and events that led to the planting of the apple tree and the design of the garden, dictating the precise location of the falling apple. Nor did he ask about the domestication of fruit trees that gave the apple its form. He asked rather about the "natural" event, defined in abstraction from its social context.[13]

The evolution of natural-science observation and classification also reinforced the conceptual schism between culture and nature. These practices resulted in the mapping of decisive boundaries between man and nature. Early natural scientists, for example, defined a category man, denoted by "a finite set of attributes specifiable by biological and social scientific discourses."[14] What is important to note here is not whether the boundary between humans and nature is correctly drawn, but that boundaries drawn between human and nonhuman entities are exact, exhaustive, and finite.

This modernist concept of humanity as unique and separate from nature demonstrates a dualistic thought-process known as hyperseparation in which "the master [humans] tries to magnify, to emphasize and maximize the number and importance of differences and to eliminate or treat as inessential shared qualities, and hence to achieve maximum separation [from what is nonhuman or nature]."[15] This insistence upon the separation between nature and culture denies the possibility of continuity between humans and natural entities such as animals. Hence, the denial eliminates sympathy between humans and animals, and this denial also sets up a hierarchical or unequal relationship between the dominating entity (humans) and those dominated (nature).

Within the modernist narrative, hyperseparation is most evident in the scientific, objectified study of nature, which distances the knowing subject, culture (Same), from the subject viewed, nature (Other). Haraway states that the viewer or perpetrator of this distanced gaze is not a detached, passive observer but instead wields a form of "objective power" simply because he/she records a picture of nature that reflects what is really out there.[16] Subsequently, the scientific picture of nature is, more often than not, the only valid and legitimate reality, which in turn delegitimizes alternative visions of nature (other than those grounded in critical science). Within the past two decades, social theorists such as Donna Haraway have contested the concept of scientific objectivity by asserting that passive, neutral images in scientific accounts can never be produced. Instead, Haraway defines scientific observations as "highly specific visual possibilities, each with a wonderfully detailed active, partial way of organizing the world."[17] This assertion challenges the scientific vision of the possibility of distanced, objective observation and even manipulation of natural entities.

The Marxist view: distantiation from nature

Intellectual thought in the modernist era engendered not only the nature-culture dualism but also the concept of humanity as superior or master of nature. According to Passmore,[18] this concept evolved as a combination of the application of science, Judeo-Christian theology, and Greek philosophy. However, according to Marxist theorists, the nature-culture dualism and the subsequent mastery (and devaluation) of nature were also the result of capitalism's concomitant "increasing separation of mental and manual labor and its never ending search for surplus value."[19] Within a Marxist light, capitalism transforms nature into a resource for human exploitation and profit. This vision supports the contemporary view of the use of nature for productive purposes through agriculture, mining, hydroelectric power, and so on, and nature as a commodity object, such as water, timber, and land.[20] These views are prevalent in modern natural-resource management and land-use planning proceedings.[21]

To understand how social labor and economic processes transform nature, we must first examine Marx's unified vision of nature and society. In the book, *Uneven Development: Nature, Capital and the Production of Space*, Neil Smith argues that the mid-nineteenth-century works of Karl Marx clearly "stand out in opposition to the dualistic treatment of nature."[22] This

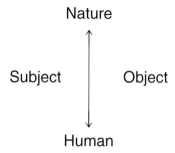

Figure 8.1   Dialectic relationship between nature and humanity

vision of nature recognizes first the vision of humanity embodied within nature; and second that the relationship is not one-sided, but dialectic.[23] In other words, both culture and nature define each other, and the social relations within culture. Expanding upon Smith and Jones,[24] the Marxist vision of nature entails the following propositions: nature as *subject* creates woman/man as *object*; woman/man in turn becomes *subject* by shaping nature as *object*; therefore both nature and woman/man are simultaneously subject and object (see Figure 8.1). Additionally, nature interacts with the creation and reinforcement of human social relations. Humans, who are natural beings, need certain material necessities such as food, housing, and warmth for survival.[25] To satisfy these material necessities, humans produce material goods from nature. This production process creates social relations, new human needs, and further productive interaction with the natural environment. Thus, it is through collaboration with nature "that human beings survive and develop as social beings."[26]

The Marxist narrative of the nature-culture dualism differs from the modernist narrative for two reasons. First, Marx does not accept the concept of an external, primordial nature; instead, he portrays the production of nature and culture as a process of dialectic interaction. Therefore, the Marxist narrative is not a story of nature as an external, pre-existing reality, but of how the separation between humans and nature has evolved through the economic and social relations. In essence, nature is not a pheno-menon that simply is separate from society; instead, society separates itself from nature. Second, the Marxist narrative recognizes that the externaliza-tion of nature within the capitalist mode of production results in the perception of nature as simply a commodity or an instrument for human capital gain. Capitalist societies are portrayed as instrumentalizing or imposing their own needs and ends upon nature, without regard to nature's needs and ends.[27]

Hence, Marxist analyses elucidate how social processes externalize and instrumentalize nature. Yet these analyses concerning the production of nature are at times problematic, simply because most Marxist analyses tend to subsume nature into culture. Marxist literature addressing the topic of the "social construction of nature" remains entrenched within the macro-scale, political-economy framework. Accordingly, if we examine nature solely from a political-economic perspective, nature becomes simply a social product. This social constructionist view of nature strays from Marx's original vision, which stressed the dialectical interactions between nature and culture, as stated by Whatmore and Boucher: "[w]hile nature cannot be (re)produced outside social relations, neither is it reducible to them. Rather the biological and physical dynamics of life forms and processes need to be recognized on their own terms, conceptually independent of human social agency."[28]

This Marxist absorption of nature into culture is a process Plumwood identifies as "incorporation."[29] Incorporation entails that one spectrum of the dualism (nature) is incorporated by the other (culture). Subsequently, incorporation denies the subjectivity of nature by denying difference and treating nature as construction(s) of culture.[30] What is important to note here is that the incorporation of nature not only denies nature as an autonomous entity, but that the relationship between nature and culture again becomes (as with hyperseparation and instrumentalism) unequal. If nature is perceived as purely a social product, then nature cannot exist without culture. Hence, culture becomes the dominant entity within the dualism because culture is the only entity that can define, even control, nature. As with the modernist narrative, nature in Marxist social-constructionist narratives becomes a passive entity subject to the needs and whims of social processes.

From the Nature-Culture Dualism to the Nature-Culture Borderlands

In the above section I have provided a brief exploration of the historical roots of the nature-culture dualism and have discussed in part how this dualism transforms our perceptions and relations with natural entities. I have also pointed out two major problems associated with dualistic thinking: hyperseparation and incorporation. To paraphrase Plumwood,[31] hyper-separation corresponds to the construction of humanity as self-contained, and of nature as alien. Hence, hyperseparation denies interrelationships and continuity between nature and culture. On the other hand, incorpor-ation constructs nature as simply a form of culture, which denies nature difference and subjectivity. Both hyperseparation and incorporation entail instrumentalist and/or unequal relationships between nature and culture.

One solution to overcoming the above problems with dualistic construc-tions of nature and culture is for scientists and social scientists to take the leap and enter the conceptual *borderlands between* nature and culture. The words *borderlands* and *between* are key here because they entail a simultaneous movement for kinship (*borderlands*, that is, blurring of boundaries) and difference (*between* nature and culture).[32] This simultaneous awareness of blurred boundaries and distinctness between human and natural entities such as animals provides us with a tentative framework to look upon identities (both human and animal identities) as *products of interrelations* between distinct humans and natural entities. This reconstruction of identity as a product of interrelations is what Plumwood labels *the relational self* in which the identity (human and nonhuman)

> conceived in terms of mutuality is formed by, bound to an interaction with others through a rich set of relationships which are essential to and not incidental to his or her projects. Nevertheless, he or she can and must remain a distinct individual, separated but not hyperseparated. He or she is not simply at the mercy of these relationships, dissolved, passive and defined by others . . . but is an active participant in them and determinant of them.[33]

However, the above definition, which relates an awareness of mutual, relational selves, is simply the first step in resolving the problems associated with hyperseparation, incorporation, and especially inequality between humans and natural entities. For those of us who deal with widespread species extirpations, it is not enough to simply become aware of the formal problems associated with dualistic thinking and identity-construction processes. In the present world, which often validates the global destruction of ecosystems and associated native wildlife, we must not only recognize a relational awareness between nature and culture, but also embrace goals and actions that encourage a flourishing of the dominated side of the dualism—natural entities/communities.

If we incorporate awareness and action, the metaphor of the borderlands becomes even more significant because we can examine multidimensional interrelations and differences: the borderlands between human and animal entities (bodies); the borderlands between human and animal spaces (communities and habitat); and the borderlands politics or political actions/identities in which the tensions of kinship and difference are played out. One example of this multidimensional borderland is Donna Haraway's metaphor of the cyborg world. The cyborg world, like the borderlands, is a place where people are "not afraid of their joint kinship with . . . and the contradictory standpoints" between humans, animals, and machines.[34] In Haraway's cyborg world, cyborg identities take issue with "relationships for forming wholes from parts [incorporation], including those of polarity [hyperseparation] and hierarchical domination."[35] Yet for Haraway cyborgism not only entails alertness to the problems of hyperseparation, incorporation, and domination, it also entails much-needed political work to resist the worldwide intensification of the domination of nature.[36]

Politics within the nature-culture borderlands

> The cyborg is resolutely committed to partiality, irony, intimacy and perversity. It is oppositional, utopian, and completely without innocence. No longer structured by the polarity of public and private, the cyborg defines a technological polis based partly on revolution of social relations in the *oikos*,

the household. Nature and culture are reworked; one can no longer be the resource for appropriation or the incorporation by the other.[37]

Hence, where can we find examples of borderland relational identities and borderland political resistance to the global dominance of nature? How is this political resistance manifesting itself? Until recently, wildlife preservation analyses and policies have ignored borderland thinking and politics because of our entrenchment within nature-culture dualistic thinking, and because we have not considered different sites, scales, and styles of resistance (that is, resistance that manifests itself beyond government-sponsored proceedings such as land-use planning and the legal courts).[38]

Until recently, wildlife life-preservation studies and policies have focused upon either scientific analyses that emphasize a modernist-scientific knowledge base or structural analyses that only emphasize macro-scale political-economic processes.[39] Such analyses portray the construction of nature as either purely a physical or a social process, thus reinforcing the oppositional and/or unequal relationships between natural entities and culture. In addition, these analyses disregard the interactions between natural entities/processes, political-economic structures, and actions of individuals or agents. In essence, wildlife policies and analyses not only deny the agency of nature, but also fall into the structuralist trap by concentrating solely on political-economic structures/processes. This trap then denies human and nonhuman agency, or the importance of individuals in the creation and transformation of our nature-society relations and landscapes.

Expanding upon Dear and Moos,[40] we must realize that political-economic structures and agency (of human and nonhuman entities) are interconnected things; they recursively (re)produce each other in a ceaseless interplay. Hence, not only nature and culture dialectically reproduce each other, but the same is true for structures and individuals: structures influence people's everyday actions, and individual actions affect structures. Given this notion of dialectic action between nature-culture and structure-agency, it is vitally important that we examine the nature-society relations and politics not only of political-economic structures, but also of local communities, households, and individuals. Such studies can resolve the problem of the structuralist trap by illuminating how both individual courses of action and political-economic structures (such as the state) shape nature-society relations, landscapes, and identities. In addition, such studies break down the barriers between the public-private dualism by elucidating intertwined, dependent relations between public and private political actions.

Therefore, if we expand our analyses to encompass not only scientific/structural processes but the actions of localized communities and individuals, it may be possible to find notions of borderland thinking and politics. As noted above, according to Haraway, it is within the household that we find the historical and gendered emergence of the politics of care, which, as this case study will demonstrate, merges Plumwood's relational self with

Haraway's call for political resistance to the domination of nature. According to feminist Joan Tronto,[41] in Western societies the practice of care, be it caring for children, the infirm, or the elderly, has been relegated almost exclusively to the domain of household work by women or those least well off (slaves, servants). Hence, care has been constructed and conceived in our culture as primarily a feminized concern that takes place in the household or private sphere. Just as nature is devalued, even ignored in the nature-culture dualism, care work is also devalued via its cultural connection with the inferior, feminine sides of the masculine-feminine dualism, which includes the private sphere, dependency, and emotion.

Recently, social theorists, especially feminists, have embraced the practice of care as a political action of resistance to the inequities associated with the nature-culture dualism. For (eco)feminists, the practice of care emulates the relational self by the act of reaching out and caring for another entity other than the self. Care for natural entities entails actions or choices that treat natural entities with respect and compassion rather than as "interchangeable commodities which can be chosen or abandoned at will."[42] In addition, as feminists have argued in political-science literature, social scientists and policy makers can no longer treat care as an apolitical action of the household or private sphere. As this case study will demonstrate, care is political and operates at various sites and scales. Care of natural entities such as wild animals (or even places such as watersheds) engenders not only strong connections with natural entities at the personal, localized scales, but fosters regional, even global, political actions that resist modernist trends to alienate, even destroy, nature.

In the spirit of Haraway's above quote concerning cyborgism, the following case study concerning the politics of golden eagle preservation in Ramona, California, demonstrates that one site of nature-culture borderland political resistance is care for wildlife. First, I delineate the hostilities present in the officially recognized site where contestations for golden eagle preservation occur—the land-use planning process. Second, I examine how these combative, polarized proceedings discourage the participation or voices and actions of those activists whose politics are rooted in an ethic of care. These actions—wildlife rehabilitation and environmental education of families—occur in local communities and households, and are potential sites within Western societies where we can traverse the nature-culture borderlands. It is within these borderland sites that we can recognize the importance of the dialectics between humans and natural entities, and examine how such interactions foster empathy and further political actions of care for human and wildlife communities at various sites and scales.

The Biogeography of Golden Eagle Habitat in Ramona, California

The community of Ramona, as with numerous other suburban communities in the eastern foothills of San Diego County, has experienced rapid population growth over the past twenty years. And it seems that this growth will not stop in the near future. In 1989, a local author cited the Ramona area population between 25,000 and 27,000.[43] Present population estimates are between 35,000 and 39,000.[44] Why is Ramona growing? It seems that prospective residents are attracted to the scenic natural landscapes and rural, country atmosphere present in Ramona. And it is precisely these scenic natural landscapes that are currently at risk due to increased development.

Ramona's biogeographic landscape of chaparral-covered hillsides, native grass flatlands, vernal pools, oak woodlands, riparian corridors, and granite boulder mountains fosters a plethora of avian and mammalian species. It is this panoramic landscape that attracts so many commuters from "down the hill" in San Diego. It is here in Ramona that so many people would like to build that dream house with an ocean view up in those chaparral hillsides. But along with the population increase comes a demand for more housing, more community services, and therefore further community expansion. Hence, local conservationists now wonder, how will this community expansion impact upon local wildlife? Is it possible to have a thriving economy and natural ecosystems including large predators?

Currently, numerous Southern California communities are grappling with the tough dilemma of planning not only for a better quality of life for growing human populations, but also to ensure the survival of native wildlife. In the community of Ramona, the golden eagle is one of the species whose numbers are declining due to reduced nesting and foraging habitat and increased human presence near home ranges—two factors that are results of suburban expansion in eastern San Diego County (Figure 8.2).[45] The conflict between suburban sprawl and wildlife is particularly problematic for the golden eagle and many other large predators such as the mountain lion who require large amounts of space to forage and reproduce.[46] Studies reveal that the foraging range of a California golden eagle is between 19 and 59 square miles.[47] Additionally, these large predators require topographically varied landscapes to breed and forage. According to raptor ecologists Fred Ryser and Thomas Scott,[48] golden eagles nest and reside upon cliffs and mountains, and forage in valleys and flatlands with sparse vegetation cover. Thus, to preserve golden eagle habitat, local raptor ecologists assert that large parcels of land (which include nesting sites and foraging ranges) must be protected.[49] However, in Southern California, where space is a valued commodity, it has been difficult for planners and conservation biologists to justify setting aside large easements for predators such as the golden eagle.[50]

Combative Politics of State-sponsored Land-use Planning

One political site where the contestation between human and animal communities occurs is within city and/or county departments of planning and land use. It is during public planning proceedings, or within the public sphere, that citizens negotiate the conflict between local economic growth

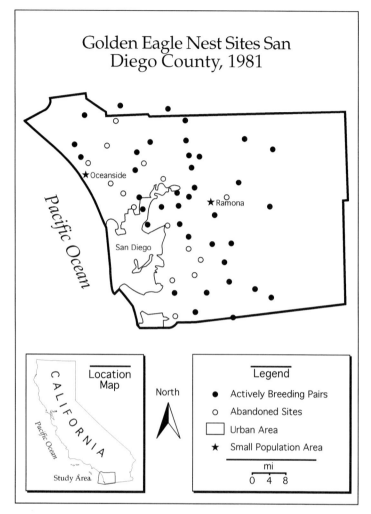

Figure 8.2   Golden eagle nest sites, San Diego County, 1981

Graphics courtesy of Cartography Laboratory at University of Colorado, Boulder;
data provided by Thomas A. Scott

and wildlife rights to space. The public sphere is a concept that originates in modern political theory and is defined by Habermas as a public meeting place where private citizens discuss and debate affairs of the state.[51] Expanding this concept to land-use planning, a planning public sphere is the collection of public meetings in which voluntary members of a territorially based social group peacefully resolve land-use problems that rise to public attention.[52]

Yet within planning public spheres, especially those concerning wildlife habitat spaces, peaceful resolutions between special interest groups are a rarity.[53] As with most government political proceedings, the planning public sphere is one of realpolitik or competitive power politics of accepted real events.[54] Within the realpolitik perspective, the land-use planning process is one that requires participants to use political/economic force and manipulated communications to compete or to battle over land parcels.[55] This mentality encourages polarizations that preclude any agreement between coalitions.[56] The results of these polarizations are usually hostilities between (and even within) factions and little regard for compromise.

Participants within the California state-sponsored planning proceedings (planners, leaders of environmental organizations, representatives of the development community, and politicians) have all relayed to me that planning proceedings in Southern California are, more often than not, a game that participants play to "win at all costs." One reason why the competition is so fierce is because undeveloped land in Southern California is becoming a very rare resource. Thus, during this competition for land, participants from opposing factions assert their unyielding positions through verbal battles. Because the land-use planning process is fraught with numerous polarizations and hostilities, it was no surprise to me when activists indicated that they feel intimidated, even alienated from participating in state-sponsored planning public spheres. Some even refuse to participate at all in antagonistic planning proceedings. Yet even though many activists have given up on the efficacy of the land-use planning process, they have not completely abandoned all hope of political activism. What emerges from this disillusionment with public planning are alternative forms of political activism.

Wildlife Rehabilitation: Alternative Ways of Knowing and Caring for Nature

> I do not know about it; it [the planning process] is over my head. And I am too busy rehabilitating and trying to educate on the native wildlife that we have to get along with in our neighborhood. I think we each have our own little niche. And if somebody's got the fortitude to go to battle on a subject [in planning proceedings], that is their niche.
>
> a wildlife rehabilitator

The realpolitik of government proceedings (such as public planning spheres or the legal courts) is not the only strategy available to participants who wish to make a difference. As indicated by the above quote, certain community activists refuse to operate in government-sponsored planning proceedings and, instead, choose to be active within alternative arenas. It is at this juncture that I shift the site of analysis from planning public spheres, to explore what I label the alternative politics of care. One type of alternative politics that occurs within the local community and the household is the wildlife rehabilitators' actions of care, which foster the day-to-day reproduction of wildlife species. Traditionally, these actions of care (wildlife rehabilitation and educational outreach programs) have been perceived as presumably innocent and apolitical "feel good activities" for women, children, and families.[57] Since these actions of care do not take place within the official planning process (the site where supposedly real politics occur) and do not reflect rational, objective decision-making, they are often dismissed by both academics and politicians as apolitical, affective community activities. However, as this case study demonstrates, alternative politics of care are significant because, first, daily relations with injured wildlife engender a trans-species empathy for the ever-growing animal casualties of our expanding political economies. Second, such empathy, or blurring of boundaries between humans and animals, deconstructs anthropocentric notions of self and fosters the relational identity, which entails kinship and respect for all the wildlife that inhabit the community of Ramona. Third, wildlife rehabilitators' politics of care are nature-culture borderland politics that not only engender kinship with animals, but also result in political resistance at various sites (household, community, ecosystems, and watersheds, for example) and scales (household, local, regional, even global).

Wildlife rehabilitation is a form of community activism in which community volunteers care for animals who are injured, ill, or too young to survive on their own. Usually this care is done in the volunteer's home or, for those birds who need extensive medical assistance, in a rehabilitation center. In fact, Ramona houses one of the largest raptor rehabilitation centers in the United States (Figure 8.3).

It is through this day-to-day care of injured birds of prey that rehabilitators become experientially and emotionally connected to the raptors. During my interviews with rehabilitators, I heard many stories about the individual behaviors of raptors who demonstrate "human-like" traits such as flexibility, intelligence, and pair-hunting strategies.[58] Because of the rehabilitator's daily interaction with animals, the rehabilitator (as with most animal caretakers) is more likely to traverse the nature-culture borderlands between humans and animals by emphasizing human likeness and interconnectedness with animal species rather than differences.[59] Hence, since rehabilitators feel a sense of kinship with animals, it is no surprise that they experience a deep sense of loss as an ever-expanding urbanization process extracts a deadly toll on so many "wildlife families" (as they are described by one rehabilitator).

Daily contact with animals (and animal death) also explains why rehabilitators have a much more emotional and passionate stance on wildlife preservation than, say, decision-makers and experts who participate in planning public spheres. Experts and decision-makers embroiled within the government planning process embody to a greater degree the nature-culture dualism by remaining distanced from the animals they discuss and for whom they make decisions. Participants within the land-use planning process analyze wildlife issues with objectified/scientific reports, cost-benefit

Figure 8.3   A golden eagle in the care of rehabilitators in Ramona; this eagle is
a young fledgling who was found starving in Riverside, California

analyses, and power politics. Life-forms are portrayed in these planning proceedings as objects such as commodities or "minimal viable populations" (the smallest population number possible that still ensures a variable enough gene pool for that species to survive). Experts and decision-makers who participate in planning public spheres rarely if ever endure the material, daily flux of animal injuries and suffering that the rehabilitator experiences.[60]

In the face of treating so many injured and dying animals, a sense of melancholy and pessimism sometimes permeates the rehabilitator community. One begins to ask oneself if all this effort, all this emotional strain when you see the life drain out of so many animals—is this all really worth it?

Do you think east San Diego County will have breeding golden eagles twenty years from now?
No [*Pause, and he takes a deep breath as he gazes out to the raptor cage, which is filled to over-capacity with injured birds of prey*], I don't think we will have breeding populations of goldens, I don't think we will have breeding populations of mountain lions. With the habitat requirement of each of those species, I don't think there will be enough habitat left for them. Will there be some around? Maybe. I see goldens gone in the not-too-distant future; golden eagles will be gone in ten to fifteen years, mountain lions less

than ten years. Part of the thing with mountain lions, more of them will be
shot than golden eagles will.

Nancy Scheper-Hughes utilizes the metaphor of triage to explicate how
human families prioritize who eats, and subsequently who lives, in the face
of extreme human poverty, hunger, and death.[61] In many ways, the concept
of triage, prioritizing who may live or die, not only explains how people
react in situations of widespread human misery, but also explains the actions
of rehabilitators who face daily a never-ending stream of injured and dying
wildlife caused by a seemingly unstoppable urban economic development
and expansion. As in human life and death situations, it is triage, a constant
choice of which animal must live and which animal must be put down, of
taking the risk by exposing a wild animal to humans so that it may be
rehabilitated and later released, that the rehabilitators experience each day.
It is this sense of grief and triage, or trying to save lives in a perceived
desperate situation, that for the most part pushes them to extend their ethic
of care to the public via environmental education.

Care as a political act of resistance
As I have stated before, traversing the borderlands is not simply about
recognizing our kinship with animals, but also entails political opposition to
current trends to dominate and destroy species such as the golden eagles.
Rehabilitators act upon their empathy with animals through a politics of
care that entails first caring for and then releasing injured animals, and
second conducting environmental educational outreach programs to spread
their notions of kinship with and their ethic of care for all species.
     In the above section on care, I stated that the historical roots of care stem
from the actions of women within the household or private sphere. Such
perceptions of care as maternal continue to prevail within the rehabilitator
community. One rehabilitator I interviewed said the animal care she
provides is very similar to the care a mother must give to an infant or a sick
child. In fact, she, along with the other female rehabilitators I interviewed,
insists that women are inherently qualified for this type of work:

> You get an injured animal in, you have got to give it fluids, you have got to
> get it on heat, you have got to run back and forth to the vet if it needs
> surgery, and then there is physical therapy afterwards. There is a lot of time
> on each animal.
> *Is it like taking care of a child?* Correct.
> *Have a lot of the women who get involved been mothers?*
> I think there are a lot of women who were never mothers and maybe they
> are fulfilling the mothering role with this. Like Helen, she has no children,
> did not want any. It is innate in us to *care and to give*. It is more difficult for
> the average man. The men enjoy the wildlife, but I don't think the men are
> into the nurturing, the same as women.
> *I go to the raptor rehabilitation workshop and there are women all over the place. And*

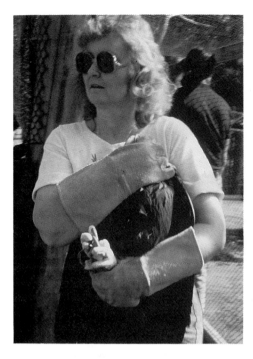

Figure 8.4 Donna Barron-Kiefer, a rehabilitator, transfers a golden eagle to a larger flight cage

I ask myself, what is drawing women to this? With rehab, it seems like there is a lot of everyday work— that is, work associated with raising children.
Right. Women have a need. My kids are grown and gone, and I get to do what I want in my life right now. And this is what I choose to take care of, something that can't speak.
There is a lot of dirty work associated with rehabilitation. I saw that in the raptor class. And it is not a very thankful job. They don't want to be near you.
They don't even tip a wing when you let them go. They are predators; they don't like us. Our release rate is about sixty percent total in 1992. This is a rate we are real proud of. If a raptor comes in with a broken joint, there is no surgery. He goes down. Which is not a pleasant task; a lot of people can't do that.

According to this rehabilitator (who is a woman), women rehabilitators are inherently more nurturing or caring than men due to a woman's desire to fulfill the traditional role of a mother. Indeed, as the quote above indicates, caring for injured animals requires similar attention and care to that a mother would give a sick child. Therefore it is no surprise that wildlife rehabilitation or care for animals is linked with culturally assigned maternal roles such as mother, caretaker, or giver and preserver of life (Figure 8.4).[62]

Recently, feminists such as Joan Tronto, Kathy Jones, and Joni Seager, insist that care can no longer be considered as synonymous with women.[63] Instead, these authors state that care is *defined culturally* as feminine or maternal, regardless of the sex of the person engaging in care.[64] In addition, since care has been culturally constructed as maternal and thus considered an emotive action relegated to the private sphere of the family, it has been delegitimatized within, even banished from, the realm of politics or the public sphere.[65] Given these assertions, Tronto argues that if we let go of the above gendered stereotypes associated with care, it is possible to interpret care as a political action which "includes everything we do to maintain, continue and repair our world, so that we can live in it as well as possible. The world includes our bodies, our selves, and our environments all of which we seek to interweave in the complex life sustaining web."[66] Tronto's redefinition of care as political is significant for environmentalists because she first denounces essentialist sexual, apolitical stereotypes associated with the action of care. In her eyes, care is redefined: it is not composed simply of moral actions associated with the maternal or private sphere, but of political actions that resist the devaluation and destruction of animals.

In this light, the rehabilitator's daily actions of care for young or injured wildlife are an example of borderland thinking and politics. First, rehabilitation demonstrates an example of Plumwood's relational self, in which the rehabilitator's identity is formed by and bound to their interactions with the animals or eagles on a daily basis. In addition, their politics of care are individualized acts of resistance, which rescue wildlife from urbanization's callous disregard for nonhuman life.[67] Hence, care is a distinct political process that differs dramatically from the dualistic thinking we witness in wildlife preservation literature and the hostile realpolitiks of planning public spheres.

In addition, rehabilitators not only have deep empathy for animal life but also recognize the importance of the subjectivity of the animals with whom they work. This animal subjectivity is recognized first by the rehabilitator's observations of individual animal intelligence and even preferences, then by involving the eagles themselves in political activities such as environmental education—a process I address in the next section. Hence, for rehabilitators, natural entities or the eagles they work with are not passive identities dominated or created by culture, but instead are important actors in the process of preserving all species.

Environmental Education as Environmental Politics

How do you voice your position? How are you an advocate?
When I appear on TV. When I go to wives' clubs, Lions Clubs, Kwanis Clubs, Rotary Clubs requesting these talks. When I go to the schools and talk to the kids. Even talking to individual visitors, whether they ask or whether I just say it. I try to always stress that the biggest threat to wildlife in this country today is loss of habitat. Every opportunity I have.

Probably the most successful borderland political activity orchestrated by environmental activists beyond planning public spheres is environmental education. Environmental education usually occurs in informal family settings such as campgrounds, zoos, nature stores, and even classrooms. This subtle form of political activism is a very successful coping strategy that allows pro-wildlife coalitions to voice their ethic of care for all species in presumed "unpoliticized" and receptive public spaces. In addition, through environmental education activists attempt to foster not only borderland identity construction (kinship with species) but borderland politics, or the political resistance to the destruction of eagles. This resistance is manifested by their encouragement of political activism for the rights of wildlife in both the local community and the household. Hence, environmental education is borderland activism that blurs the boundaries not only between nature and culture, but also between public and private political actions. The pro-wildlife political actions that occur at various sites are multiple, linked projects that in my opinion have the potential to foster political resistance at scales from the household to local to global (and back again).

Environmental coalitions at all levels, from community grassroots movements to national organizations, encourage some form of environmental education. Because eagles are such a charismatic species and are easy to transport, many raptor teams travel throughout the United States displaying live eagles and educating the general public about the species and their respective habitats. In fact, there is a national organization, known as the Raptor Education Foundation, whose sole mission is to provide "knowledge for preservation of ecological diversity." This organization publishes a quarterly newsletter, *The Talon*, and organizes raptor shows throughout the United States. In addition, the organization plans to build a raptor theme park, Raptor World.

In Ramona, raptor education is widespread. A grassroots organization, the Iron Mountain Conservancy, conducts raptor watches on a weekly basis and publishes educational hand-outs about the golden eagle. Additionally, the Ramona raptor rehabilitation center serves as a center of local wildlife education for children throughout San Diego County. In these raptor education-events, volunteers will bring birds of prey (who have sustained injuries that preclude them from surviving in the wild) to schools, community meetings, campgrounds, and, if invited, public government proceedings. Eagles then participate in wildlife politics by becoming charismatic and political identities who often take the center stage. Such involvement of live birds of prey subsequently encourages the public to become tangibly and materially connected to the plight of birds of prey. A live golden eagle, with an 8-foot wing span, piercing eyes, and huge talons that can shred your hands makes a lasting impression and consequently can become a source of political motivation to save these species from local extirpation.

It is within these presumed apolitical and educational settings that volunteers relay their experiences of the interrelatedness between themselves and the eagles. These relayed experiences, and the presence of the eagle itself, are demonstrations to encourage community residents to respect local wildlife rights to coexist with humans:

How do you voice your position concerning this conflict outside this planning process?
I spend at least three to four times a week going out to educate children at schools and adults at business ladies' meetings, gentlemen's meetings. Trying to let them know we have got these animals, what to do if you find an injured one. Our organization believes in a second chance, getting them back out there. And preserving them, respecting them, leaving them alone.

Hence, for the wildlife activist/educator (who may or may not be a rehabilitator), environmental education is definitely not just a "feel-good activity" for educators, rehabilitators, and families. It is a deliberate political attempt to "educate," or alter peoples' perceptions concerning local wildlife, and subsequently to encourage political actions to support wildlife preservation. Most "educational" brochures and magazines I examined relay not only educational messages (eagle wingspan, weight, food preferences, breeding behavior, habitat requirements, and so on) but also underlying political messages. The first message is the belief that animals have equal rights to coexist with humans. The second, and most important political message, is to "make a difference" and get involved. The involvement desired by environmental groups ranges from financial support to spreading the word about wildlife rights to active advocacy such as wildlife rehabilitation.

The community activists' pro-wildlife ideologies are rarely contested in the "innocent, powerless, and apolitical" children's learning environment as they are in state-sponsored public planning spheres. Even in the most normative children's learning environment, the public-school classroom, rehabilitators are encouraged to voice their advocacy for wildlife rights. A schoolteacher who incorporates rehabilitated, live birds of prey in her science curriculum comments upon the need to educate children about how the children's actions impact upon local raptor populations and habitat:

> *You have asked a rehabilitator to visit your classroom. What does he do?*
> He has come for our science club. Sometimes he comes during the course of our classroom day. Depending on what we are doing, we'll do predator-prey situations and talk about why there are more or less of what we are seeing now. For example, the owls and why children may not be able to see them in the future, and what we need to do to protect them. And why the animals that he has are his now, and why, for example, the owl he had brought in was not capable of living on its own. Its wing had been shot off. So kids start to realize that what they are doing may have impacted upon the animals in their area.

I stated before that the child's learning environment is presumed innocent, not political. Yet, due to environmental-education programs, children practice borderland politics by becoming political advocates for the rights of eagles. After one raptor "education" session, a group of students wrote letters of concern to the local newspaper. This letter-writing campaign indicates how the education programs which use live birds of prey can invoke public sentiment for the plight of wildlife. In fact, one community activist relayed to me that the children's letter-writing campaign gained much more recognition for the golden eagle than he/she had been able to do with conventional political strategies. But children's environmental education does more than bring environmental issues into the public light; this strategy also brings environmental politics into the private sphere or the household.

As one rehabilitator pointed out to me, educating children is a successful political effort utilized by activists to encourage care for the environment within the household:

> [A]t this point almost everyone I have talked to, especially in rehab or education, agrees that *education is the only place where we are going to make a change.* The adults have already got some sort of mind-set, and their habits are already formed. The only way that they [adults] are going to change is because the kids are changing. Even in our own group, the wildlife rehab, it is great. We get a good kick out of it, and we are doing something great. But it is not the answer. We can rehab 9 million animals a year, and that is not going to solve the problem. It is the kids that are coming home and saying "Dad, don't throw that out, put it in the recycle bin," or the kids that are gathering up trash, cutting up plastic rings. . . . The parents are following suit in what the kids are doing. If we can educationally get to the kids, that is the only place where change is going to be. (emphasis added)

By educationally getting to the kids, or targeting children, environmental education is a strategy that seeks to encourage borderland awareness and political activity within the household, the family unit. Community activists understand all too well that through environmental education, children are not the only ones who become environmentally aware: the parents in turn are educated by the children.

As demonstrated by this case study, environmental-education programs, much like Haraway's vision of cyborg (or borderland) politics, are simultaneously blurring the boundaries between nature and culture, public and private, by empowering persons at various sites (family, local community, schools, regional festivals) and scales (household, local community, region [watershed], nation, or even global) to encompass an ethic of care for local nature resources and wildlife. I believe the simultaneous public/private political activism exemplified by environmental education successfully resists the trends of destruction (which also occur at multiple sites and scales) of natural entities because it helps people identify with caring for their local natural environments, and it has the capacity to empower people to act in both the public spheres (through letter-writing campaigns or promoting watershed festivals), and the private spheres (through actions of recycling or even care of injured birds of prey).

Engendering Connections between Humans and Animals

The journey into the realm of the nature-culture borderlands sketched above first detailed the social histories of the nature-culture dualism and the pitfalls associated with dualistic thinking. Within the modernist perspective, humanity becomes externalized from an alien nature. This dualistic thinking fosters an unequal, hyperseparated relationship between nature and culture, in which the dominant side—humans—emphasize, even magnify, differences from the inferior other (nature). On the other hand, the Marxist social constructionist view of nature is also problematic in that it incorporates nature as purely a cultural product. This incorporation again reifies the power of culture over nature. Both the modernist and Marxist visions deny the subjectivity of natural entities, resulting in instrumentalist and/or unequal relationships between nature and culture.

Baker asserts that to see animals differently would require humans to see themselves differently.[68] Indeed, one way to overcome the problems associated with dualistic thinking and domination of nature is to reconstruct our own human identities in terms of mutuality with natural entities. Such a reconstruction entails a destabilized human self containing identities that simultaneously remain distinct from and share kinship with natural entities. This reconceptualization of self is not, as ecofeminist Val Plumwood would argue, a purely philosophical exercise.[69] Shared kinship with nature also entails a resistance to the domination of nature. It is this concomitant empathy with nature, and political resistance to destruction of natural entities such as wildlife, that I label as borderland politics.

In actions such as wildlife rehabilitation and environmental education, we find configurations of political activism rooted in an ethic of respect and care for local native wildlife communities. These alternative politics embody a sense of interconnectedness between humans and animals through daily care for animals being destroyed by an ever-growing urban expansion process. Care is reconceptualized herein as a political action that includes everything we do to maintain, continue, and repair our world.[70] This reconceptualization of care as political shatters the modernistic myth that political identities and actions that matter occur only in public spheres such as the land-use planning process. Instead, the politics of care demonstrate that the political negotiation for nature is a complex process, which occurs at multiple sites to include the household and involves the most unexpected agents (children and animals, for example).

Can this ethic of care influence land-use planning decisions? At present, I think not, due to the fact that our governmental political processes are so entrenched in the model of oppositional, competitive power politics. However, I assert that rehabilitators are successfully instilling this ethic of care (and concern) for disappearing wildlife through environmental education of children, who are the future voters and decision-makers in state-level politics. In fact, it is because people perceive family educational events as innocent and insignificant that activists are so successful in propagating their beliefs in wildlife rights to the public. For most participants, environmental education is a fun family event that allows kids to get up close to the present-day raptors.[71] On the other hand, for wildlife rehabilitators and environmental activists, environmental education is a very serious politics that may, over time, transform our perceptions, actions, even political processes concerning the coexistence of wildlife and human communities.

As demonstrated by the above case study, traversing the nature-culture borderlands is a complex process that involves nested social networks, institutions, and personal experiences. One future project for research would be to map these multiple sites, agents, and scales of borderland politics.[72] Such maps would reflect the destabilization of human and animal identities, the social histories of human-animal relations, and theories of agency-structure interactions.[73] In this mapping process, we must not forget that borderland travels entail the presence and actions of agents who are not human—natural entities such as animal species.

Within the field of ethnology, the presumed static boundary between human and animal is rapidly blurring as evidenced by recent research on "social" animals (such as primates, elephants, cetaceans, wolves, and even raptors) who exhibit characteristics associated with humanness (culture, complex group-hunting strategies, language, familial ties).[74] It is my hope that this essay traversed the borderlands between humans and animals from the social-science perspective by recognizing how human-animal relationships incite different styles of political action. In fact, I pose that the human connections with natural entities could become the most powerful and effective political messages anyone can send concerning the rights of natural entities such as wildlife. As exemplified by the work of rehabilitators, these interactions incorporate a sense of care for animal species. And as demonstrated by the children of Ramona, human-animal interactions foster political action and mobilization at not only the household but the community or even regional scales. This activism due to human-animal interactions (material or via the media) can even generate political action at the national or global scale, as evidenced by the political activism of families that followed the release of the movie *Free Willy*.[75] If any social scientist believes that he/she cannot be influenced by such interactions, I invite him/her to spend some quality time with the native fauna of his/her home region, or even observe how children react around live animals at the zoo. For myself, the greatest enjoyment of my research was spending time with golden eagles—just watching them preen, fly, or even scuffle over a dead rabbit. One afternoon, after I felt like I was being sized up by an eagle's piercing gaze (try to out-stare an eagle sometime), I wondered if it was really *she* who was studying *me* and trying to decide whether humans should be allowed to coexist with her species.

Acknowledgements

This chapter is dedicated in loving memory to Donna Barron-Kiefer.

Thanks to the San Diego County Fish and Wildlife Advisory Commission for funding this research. I extend my gratitude to Dr Kathleen B. Jones for her helpful and challenging comments and her mentorship. I am especially grateful to Dr Jennifer Wolch for her invaluable advice and inspiration concerning animal geographic research.

Notes

1. Jennifer R. Wolch and Jacque Emel, "Bringing the Animals Back in," *Environment and Planning D: Society and Space* 13, 1995, p. 632.

2. Helen Ingram, Nancy K. Laney, and David M. Gilliam, *Divided Waters Bridging the US-Mexico Border*, Tucson: University of Arizona Press, 1995.

3. Donna J. Haraway, *Simians, Cyborgs, and Women: The Reinvention of Nature*, New York: Routledge, 1991; Val Plumwood, *Feminism and the Mastery of Nature*, New York: Routledge, 1993.

4. Thomas A. Scott, "Human Impacts on the Golden Eagle Population of San Diego County from 1928 to 1981," Master's thesis, Department of Biology, San Diego State University, 1985.

5. Fred A. Ryser, *Birds of the Great Basin: A Natural History*, Reno: University of Nevada Press, 1985; Thomas C. Durstan, "The Golden Eagle," in William J. Chandler and Lillian Labate, eds, *Audubon Wildlife Report 1989/90*, New York: Academic Press, Inc., 1990, pp. 332–64; Thomas A. Scott, Telephone conversations with author, 19 April 1993 and 7 October 1993.

6. Margaret FitzSimmons, "The Matter of Nature," *Antipode* 21, 1989, p. 106; Sarah Whatmore and Susan Boucher, "Bargaining with Nature: The Discourse and Practice of 'Environmental Planning Gain,'" *Transactions of the Institute of British Geography* 18, 1993, pp. 166–78.

7. Cindi Katz and Andrew Kirby, "In the Nature of Things: The Environment and Everyday Life," *Transactions of the Institute of British Geography* 16, 1991, pp. 259–71.

8. Margaret FitzSimmons, "Reconstructing Nature," *Environment and Planning D: Society and Space* 7, 1989, p. 1.

9. FitzSimmons, "Reconstructing Nature," p. 1; Denis Cosgrove, "Environmental Thought and Action: Pre-modern and Post-modern," *Transactions of the Institute of British Geography* 15, 1990, pp. 344–58.

10. Neil Smith, *Uneven Development: Nature, Capital and the Production of Space*, Oxford: Basil Blackwell, 1984; Cosgrove, "Environmental Thought and Action."

11. Cosgrove, "Environmental Thought and Action."

12. Derek Gregory, *Geographical Imaginations*, Cambridge, Mass: Blackwell Publishers, 1994, p. 12.

13. Smith, *Uneven Development*, p. 4.

14. Paul Hirst and Penny Woolley, "Nature and Culture in Social Science: The Demarcation of Domains of Being in Eighteenth Century and Modern Discourses," *Geoforum* 16, 1985, p. 152. Not all early biologists accepted the uniqueness of humans. In *Systema naturae* (Known World) of 1758, Linnaeus, a Swedish biologist, classified humans in the same taxonomic order with animals: see Donna J. Haraway, *Primate Visions: Gender, Race and Nature in the Modern World of Science*, New York: Routledge, 1989.

15. Plumwood, *Feminism and the Mastery of Nature*, p. 49.

16. Haraway, *Simians, Cyborgs, and Women*, p. 184.

17. Haraway, *Simians, Cyborgs, and Women*, p. 184.

18. John A. Passmore, *Man's Responsibility for Nature: Ecological Problems and Western Traditions*, New York: Charles Scribner's Sons, 1974.

19. FitzSimmons, "Reconstructing Nature," p. 1; Katz and Kirby, "In the Nature of Things."

20. Katz and Kirby, "In the Nature of Things."

21. For modern natural-resource management, see Martin W. Lewis, *Green Delusions: An Environmental Critique of Radical Environmentalism*, Durham, N.C.: Duke University Press,

1992; for land-use planning proceedings, see Whatmore and Boucher, "Bargaining with Nature."

22. Smith, *Uneven Development*, p. 17.

23. Kathleen B. Jones, "The Marxian Concept of Community," Ph.D. diss., Department of Political Science, City University of New York, 1978.

24. Smith, *Uneven Development*; Jones, "The Marxian Concept of Community."

25. Smith, *Uneven Development*.

26. Smith, *Uneven Development*, p. 37.

27. Plumwood, *Feminism and the Mastery of Nature*.

28. Whatmore and Boucher, "Bargaining with Nature," p. 67.

29. Plumwood, *Feminism and the Mastery of Nature*.

30. Plumwood, *Feminism and the Mastery of Nature*, p. 155; Jennifer R. Wolch, Kathleen West, and Thomas E. Gaines, "Transspecies Urban Theory," *Environment and Planning D: Society and Space* 13, 1995, p. 735–60.

31. Plumwood, *Feminism and the Mastery of Nature*, p. 155.

32. Plumwood, *Feminism and the Mastery of Nature*.

33. Plumwood, *Feminism and the Mastery of Nature*, p. 156.

34. Haraway, *Simians, Cyborgs, and Women*, p. 154.

35. Haraway, *Simians, Cyborgs, and Women*, p. 151.

36. Haraway, *Simians, Cyborgs, and Women*, p. 154.

37. Haraway, *Simians, Cyborgs, and Women*, p. 151.

38. I am commenting upon wildlife preservation in Western societies, more specifically in the United States. Certain scholars have found borderland thinking present in non-Western culture: see Steve Baker, "Review Essay," *Society and Animals* 4, 1996, pp. 75–88; and James L. Wescoat, "The 'Right of Thirst' for Animals in Islamic Law: A Comparative Approach," *Environment and Planning D: Society and Space* 13, 1995, pp. 637–54.

39. See, for example, David Western and Mary C. Pearl, eds, *Conservation for the Twenty-first Century*, Oxford: Oxford University Press, 1989, or any issue of *Transactions of the North American Wildlife and Natural Resources Conference*.

40. Michael J. Dear and Adam I. Moos, "Structuration Theory in Urban Analysis," in David Wilson and James O. Huff, eds, *Marginalized Places and Populations: A Structuralist Agenda*, London: Praeger, 1994, p. 3.

41. Joan Tronto, *Moral Boundaries: A Political Argument for an Ethic of Care*, London: Routledge, 1993.

42. Plumwood, *Feminism and the Mastery of Nature*, p. 155.

43. Charles LeMenager, *Ramona & Roundabout*, Ramona, Calif.: Eagle Peak Publishing Co., 1989.

44. *Ramona Sentinel*, 22 July 1993.

45. Scott, "Human Impacts on the Golden Eagle Population." Because the golden eagle is populous in remote areas of California and other western states, California state and federal laws do not list the golden eagle as an endangered species. In 1963, the Bald Eagle Act of 1940 was extended to golden eagles. This act made it illegal to "take, possess, sell, purchase, barter, offer to sell, transport, export, at any time, in any manner any bald/golden eagle commonly known" (Durstan, "The Golden Eagle," p. 506). Today, golden eagles no longer suffer from direct persecution, but from the intrusion of human activity upon their nesting and foraging space.

46. Leslie Brown, *Birds of Prey: Their Biology and Ecology*, London: Hamlyn, 1976; Jon Rodliek and Richard M. Degraff, "An Overview of Special Session II: Wildlife Habitat in Managed Landscapes," in *Transactions of the Fifty-fourth North American Wildlife and Natural Resources Conference*, Washington, D.C.: Wildlife Management Institute, 1989, pp. 83–88.

47. Durstan, "The Golden Eagle."

48. Ryser, *Birds of the Great Basin*; Scott, "Human Impacts on the Golden Eagle Population."

49. Scott, telephone conversations with the author; Pete Bloom, telephone conversation with the author, 4 May 1993.

50. My study of golden eagle habitat preservation is the result of an ethnographic study of wildlife management, conducted in three phases: analysis of pertinent documents (articles in local newspapers and magazines, government planning documents, brochures, maps and newsletters [provided by governments and nongovernmental organizations]); observation of public planning meetings, nongovernmental organization meetings, and environmental community outreach/educational programs, and participation in a raptor rehabilitation program (which instructs activists on the physiology of raptors, trains activists how to triage and handle birds of prey, and finally trains activists how to handle birds of prey in family-oriented educational environments); and in-depth interviews of thirty informants (including technical experts, developers, landowners, environmentalists, members of citizen advisory boards, wildlife rehabilitators, and local environmental educators). Unlike most analyses of wildlife politics, my analysis does not ignore an ever-present actor in this case study—the golden eagle. I addressed the issue of agency of the golden eagle by examining how human-animal interactions influence people's perceptions and politics concerning the plight of the golden eagle.

51. Iris Marion Young, "Impartiality and Civic Public. Some Implications of Feminist Critiques of Moral and Political Theory," in Seyla Benhabib and Drucilla Cornell, eds, *Feminism as a Critique: On Politics of Gender*, Minneapolis: University of Minnesota Press, 1987; Craig Calhoun, "Introduction: Habermas and the Public Sphere," in Craig Calhoun, ed., *Habermas and the Public Sphere*, Cambridge, Mass.: MIT Press, 1992, pp. 1–40.

52. John Friedmann, *Empowerment: The Politics of Alternative Development*, Cambridge, Mass: Blackwell, 1992.

53. Timothy Beatley, *Habitat Conservation Planning: Endangered Species and Urban Growth*, Austin: University of Texas Press, 1994; Peter S. Menell and Richard B. Stewart, *Environmental Law and Policy*, Boston: Little, Brown and Company, 1994.

54. John O'Loughlin, "World Powers and Local Conflicts," in R. J. Johnston and Peter J. Taylor, eds, *A World in Crises? Geographical Perspectives*, Oxford: Basil Blackwell, 1989, pp. 289–332.

55. John Forester, *Planning in the Face of Power*, Berkeley: University of California Press, 1989.

56. Patsy Healey, "Planning through Debate: The Communicative Turn in Planning Theory," *Town Planning Research* 63, 1992, pp. 143–62.

57. Bettina Aptheker, *Tapestries of Life: Women's Work, Women's Consciousness and the Meaning of Daily Experience*, Amherst: University of Massachusetts Press, 1989.

58. Elizabeth Atwood Lawrence, "Conflicting Ideologies: Views of Animal Rights Advocates and Their Opponents," *Society and Animals* 2, 1994, pp. 175–89.

59. Lawrence, "Conflicting Ideologies."

60. Although rehabilitators must go through a very stringent federally mandated certification process, their knowledge of eagles is often considered biased, affective, and not expert, and therefore is not valid, not credible within the official, public sphere of scientific research and land-use planning.

61. Nancy Scheper-Hughes, *Death without Weeping: The Violence of Everyday Life in Brazil*, Berkeley: University of California Press, 1992.

62. Joni Seager, *Earth Follies: Coming to Feminist Terms with the Global Environmental Crisis*, New York: Routledge, 1993.

63. Tronto, *Moral Boundaries*; Kathleen B. Jones, *Compassionate Authority: Democracy and the Representation of Women*, New York: Routledge, 1993, p. 234; Seager, *Earth Follies*.

64. Certain visions of maternal or maternal or motherist politics which insist, for example, that women are inherently better nurturers than men (as asserted in the rehabilitator's above statement) reinforce essentializing roles culturally assigned to women. What is dangerous about this type of maternal politics is that it builds political action on the basis of sexual stereotypes: see Seager, *Earth Follies*.

65. Seyla Benhabib, "The Generalized and Concrete Other," in Benhabib and Cornell, eds, *Feminism as a Critique*, pp. 77–95.

66. Tronto, *Moral Boundaries*, p. 132.

67. Wolch, Emel, West, and Gaines, "Transspecies Urban Theory," p. 735.

68. Baker, "Review Essay," p. 79.

69. Plumwood, *Feminism and the Mastery of Nature*.

70. Tronto, *Moral Boundaries*.

71. Presently, raptors are a very "in" species with children, because pro-wildlife factions and author Michael Crichton portray raptors as the present-day version of the popular veloci*raptor* dinosaur character in the book/movie *Jurassic Park*.

72. Fredric Jameson, *The Geopolitical Aesthetic: Cinema and Space in the World System*, Bloomington: Indiana University Press, 1992.

73. Dear and Moos, "Structuration Theory in Urban Analysis"; Wolch, West, and Gaines, "Transspecies Urban Theory."

74. Haraway, *Primate Visions*; T. X. Barber, *The Human Nature of Birds*, New York: St Martin's Press, 1993; Lawrence, "Conflicting Ideologies."

75. The 1993 hit movie, *Free Willy*, is one example. Ironically, Keiko, the 21-foot killer whale who portrayed Willy in the movie, lived and performed tricks in an undersized 15-foot-deep tank in Mexico City, and suffered skin lesions because the tank water was too warm. Outraged at the plight of Keiko, a "Free Keiko" movement erupted, leading to his January 1996 transfer to a 2 million-gallon tank at the Oregon Coast Aquarium. Ultimately, activists plan to release Keiko in the wild (*San Diego Union-Tribune*, 1 and 8 January 1996).

PART III

THE POLITICAL ECONOMY
OF ANIMAL BODIES

9

The Spotted Owl and the Contested Moral
Landscape of the Pacific Northwest
James D. Proctor

It is my opinion that this bird should be listed on the threatened list. There are numerous scattered birds but each with its habitat is threatened. Even if there were no direct threats to wipe it out, the habits of our society to convert everything into the almighty dollar is threat enough.

I feel this spotted owl thing is a hoax. This bird has been seen nesting in secondary growth. I feel in order for me to provide an income for myself and my family, and to also stay living in Oregon these unrealist actions by these preservationists must stop.[1]

Theory: Habitat as Moral Landscape

Introduction

As little as a decade ago, few if any Americans would have heard of the northern spotted owl, *Strix occidentalis caurina*. This elusive, round-eyed creature is now one of the most prominent icons of the environmental movement, in large part due to its 1990 listing as a threatened species pursuant to the US Endangered Species Act.[2] The 1990 listing of the spotted owl led to a series of far-reaching policy actions affecting public forests of the Pacific Northwest, including the recently adopted Clinton Forest Plan.[3] Yet, as the epigraphs above suggest, the people of the Pacific Northwest were bitterly divided over whether the spotted owl was worthy of such protection. The question I would like to begin to answer in this essay is, why?

There are, of course, some rather straightforward explanations, the most common being that people who stood to benefit from the owl listing favored it, and people who stood to lose opposed it. If you were an environmentalist and worried about species extinction, the owl listing certainly would make you feel better. If you were an avid hiker and desired vast tracts of untrammelled wilderness, the owl listing would be a great way to lock up lots of prime forestland from logging. If, on the other hand, you were a logger, or if you owned a restaurant frequented by loggers, or if you sent your children to school in a district that depended upon federal timber revenues, the listing would look like nothing short of a threat to the very economic and social fabric of your life.

This interests theory is, however, a shallow explanation when taken alone, because it flattens people into knee-jerk reactive machines. The spotted owl debate was about interests, but not only interests; when we look deeper, we see a clash of meanings as well. It is these divergent meanings that I wish to explore in this essay, as they play an increasingly important role in the ways late industrial societies think and act toward animals.

The Ambivalent Symbolism of the Owl

People have long attributed divergent meanings to owls. Owl symbolism in Western civilization can be traced back as far as Lilith, the Mesopotamian goddess of death, who had wings and talons as well as owls at her side.[4] Lilith may have been the inspiration for Pallas Athene, the Greek goddess of wisdom and warfare. Lilith and Pallas Athene exemplify the multiple meanings owls have had over the last several millennia: owls as wise, owls as humanlike, owls as killers, owls as signs of death.

In some ways, this contradictory symbolic import has arisen due to characteristics of the bird itself.[5] Owls, of course, are predators, and could easily be linked to death and warfare. Yet their hunting occurs at night, largely behind a cloak of darkness. The owl of daytime that people encounter seems a far different creature altogether, with visual character-istics that more resemble humans than any other bird. The large eyes, designed so well for night vision and set in front of the head, which itself is broad and flat like that of humans, set on a vertical body posture, all contribute to a strong affinity between the human and the owl. Though there is little biological evidence that the owl is necessarily smarter than other avian species, its appearance lends itself to that ascription, and in fact since medieval times the quiet, all-seeing owl has been strongly associated with wisdom and learning. And so the owl has long had what Sparks referred to quite accurately as a "Jekyll and Hyde" duality in Western culture.

These differing meanings are clearly apparent in the Pacific Northwest.[6] Indeed, were the animal in contention a salamander or a mountain lion or even a songbird, there is little doubt that its symbolic role would have changed, as the valence of innocence, of humanness, of danger in these animals is far different than that of the spotted owl. The longstanding polarities of good and evil surrounding owls probably played a major role in providing a distinct focus to the pro- and anti-listing forces; following this line of reasoning, it was inevitable that some people would have found spotted owl protection to be more of a threat than others.

This mode of explanation has merit, but it narrowly circumscribes the contested meanings that arose in the Pacific Northwest, as the bulk of the debate over the last decade has focused on management of the region's old-growth coniferous forests,[7] which happen to be the northern spotted owl's preferred nesting and roosting habitat. Indeed, both supporters and opponents of spotted owl protection focused their attention primarily on forests: the environmental movement cautiously downplayed its interest in the spotted owl relative to protection of old-growth forests; and the pro-timber coalition preferred to discuss tree replanting over maintenance of spotted owl habitat. This avoidance is probably due in part to the longstanding potency and ambivalence of the owl as a symbol, making it a questionably faithful partner in either the environmentalist or pro-timber cause. Yet in many respects the owl became a living symbol for primeval nature as typified in Pacific Northwest old-growth forests. In fact, the spotted owl has played an official symbolic role in this regard as a management indicator species, or MIS.[8] The MIS is the proverbial canary in the coal mine, whose viability or downfall implies the trajectory of an entire ecosystem. In this sense, the health of the spotted owl population in the region has been taken as suggesting the status of old-growth forests.

What is thus necessary, I believe, to make meaningful sense of the spotted owl debate is to reconnect the animal with one of its most fundamental geographical elements: its home or habitat. This is true with other charismatic animals as well. Think of the wide-eyed giant panda staring down at us from a bamboo tree somewhere in the Szechwan province of China; the majestic elephant lumbering across the East African savanna; the tail fin of a humpback whale disappearing into the ocean as it continues its long migratory journey up and down the eastern Pacific Ocean. In each case we cannot help but see the animal as bound to its habitat. This habitat is both literal and symbolic; the threads of biology and ideology are intertwined in the ways we make sense of an animal's home. My interest here primarily concerns how ideology transforms habitat into a moral landscape, a geographical embodiment of the good.

The Moral Landscape

The term "landscape" as employed in everyday usage typically refers to a particular kind of place, often one with prominent biophysical features: a pastoral landscape, a wooded landscape, a wilderness landscape, a desert landscape. Yet, in contrast to more scientifically prevalent terms that describe nature (for example, ecosystem) landscape carries with it a sense of place; it is as much the appearance and feeling of a location as the location itself. Landscapes can, among other things, be pleasant, soothing, mysterious, frightening, and beautiful.[9] And this is the potent irony of the term: that the psychological and cultural constructedness of this sense of place is seamlessly embedded in the place itself, to the point that differentiated human meanings become embodied in apparently objective features of nature.[10] We all know that to one person the desert is a harsh landscape; to another it is simple and uncluttered; to yet another it is exceedingly lonely. All these attributes are, in reality, a complex interweaving of the subject and the object, yet they are rarely understood as such.

The common-sense usage of the term landscape thus reveals the tension inherent in the word: land (something objective, separate from the subject) as a view (something subjective by definition). This tension is evident in its use by geographers as well. James Duncan defines landscape as "a polysemic term referring to the appearance of an area, the assemblage of objects used to produce that appearance, and to the area itself."[11] The word has been used by geographers in both the objective sense, as a particular stretch of land, and in the subjective sense, as a way of seeing the land.[12] For example, Carl Sauer, the founder of the influential Berkeley school of geography, imported the term *Landschaft* from late-nineteenth-century German geography as a way to examine the human transformation of nature in specific places.[13] In a very different sense, the word landscape has been used more recently by cultural geographers as a social construction, one that reveals much more about the viewer than the land viewed.[14] Nature becomes, in this sense, an *ideological* landscape,[15] a meaningful representation of human values and interests, of social and human-environment relations, embodied geographically in the land, which nonetheless is generally apprehended not as ideology but as "reality."

I am using the term *moral* landscape in order to place attention on matters of value that permeate the spotted owl and old-growth debate;[16] yet of course this is a bit redundant. If landscape is a meaningful description of place, then landscape is inherently moral, since these meanings are never purely descriptive ("this is a forest") but normative as well ("this is a forest that has been imperiled by logging," or "this is a forest that has been improved by scientific management"). The modifier "moral" does remind us, though, that the landscape carries tremendous normative weight by geographically embodying an idea of the good.

Like biophysical landscapes, which are shaped by and respond to tremendous geological, climatic, and other forces, moral landscapes are a result of (and a particular moment in) a process of creating and interpreting meaning. One example of this dynamic approach to meaning is the work of Jacquelin Burgess, in the context of environmental values and mass communications media.[17] In Burgess's account, there are four phases in the social cycle of meaning: the context and act of production; the produced text; the interpretive readings or consumption of the text by people; and its subsequent interweaving with lived culture.[18] Meaning then is both process (production/consumption) and product (text/culture). The actual situation is even more complex, for not only is consumption of meaning highly variable, but there are commonly multiple, conflicting meanings being produced and distributed as well. The circuit of culture thus becomes an overlapping and contradictory set of paths of meaning creation and consumption.

I will adopt an analytical framework similar to that of Burgess, yet primarily informed by theory on the analysis of ideology.[19] This approach addresses three phases in the production and consumption of ideologically based meanings: intent, mechanism, and outcome. As production and consumption are separate, so are intent and outcome: what a particular social group intends to accomplish by means of propagating a set of meanings may or may not occur, both because these meanings are differentially interpreted by people and because these meanings may or may not result in the desired material effects (for example, impacts on popular opinion or policy enactment). The mechanism is the rhetorical content and form by which a particular meaning is propagated. Several common mechanisms have been identified in the literature: examples include universalization (making the interests of the few appear to be the interests of the many), rationalization ("defending the indefensible," in Terry Eagleton's account), and naturalization or reification (portraying a transitory, constructed state of affairs as fixed and eternal).[20] Critical analysis of mechanisms is particularly crucial in unmasking ideology, as these are the means by which the decidedly partial becomes all-encompassing and apparently irrefutable.

The three components of intent, mechanism, and outcome suggest a process of building moral landscapes in the Pacific Northwest in the context of the spotted owl and old-growth debate that focuses primarily on the production of meaning by environmentalists and the timber industry and supporters, and then on its popular consumption by the region's inhabitants. The process is not entirely linear, however; as in the framework adopted by Burgess, anticipated and historical patterns of ideological consumption guide ideological production. Nonetheless, I will adopt this sequence for clarity below, considering first the outreach by environmentalists and the timber industry and its supporters, and then the popular response by the region's inhabitants.

The empirical basis of this essay is the period immediately preceding the spotted owl listing, a particularly critical moment in the old-growth debate. I will focus on the case of Oregon, a state where the spotted owl debate was particularly pronounced, due to the prominence of the wood-products industry, as well as the relative prevalence of remaining old-growth forests. Primary sources for outreach include both published and unpublished materials, as well as interviews with interest-group representatives conducted in July 1991; sources for popular response include written comments submitted to the US Fish and Wildlife Service (FWS) regarding the proposed spotted owl listing, and a number of public opinion polls that were conducted during 1989 and 1990.[21]

What emerges is an animal geography shaped as much by the ideological production and consumption of moral landscapes as by the biology of the spotted owl and its habitat. The former element wove seamlessly through the latter, so that the debate over whether or not the spotted owl should receive special protection under the Endangered Species Act became in the last analysis a political struggle over whose moral landscape was to prevail.

Figure 9.1 Range of the northern
spotted owl in the Pacific Northwest
(shaded area)

Yet moral landscapes are not innocent entities; as we will see, there is a
decided partiality to all geographical embodiments of the good, as suggested
in the spotted owl debate.

Background: The Listing Debate

The Northern Spotted Owl Listing

The northern spotted owl is found only in the coniferous forests of the
Pacific Northwest, with habitat stretching from Northern California to lower
British Columbia (Figure 9.1). Though concern over the status of the
spotted owl dates from the early 1970s, there was no formal proposal to list
it until the late 1980s.[22] During this period old-growth stands dwindled
rapidly on federal lands in spite of federal programs designed explicitly to
protect owls.[23] In January 1987, a petition was submitted by the conservation
group Greenworld of Cambridge, Massachusetts, to the FWS proposing that
the species be listed as endangered under the Endangered Species Act
(ESA). The proposal to list the owl was ultimately rejected: in December
1987, the FWS concluded that the owl was not endangered and dismissed
the petition. A federal audit, however, found that the process used by the
FWS in arriving at its decision was "beset by many problems" which

ultimately "raise serious questions about whether FWS maintained its scientific objectivity during the spotted owl petition process."[24]

Subsequent to the rejection, the Sierra Club Legal Defense Fund sued relevant federal agencies on the grounds that biological evidence did in fact substantiate the proposed listing of the owl. The case, *Northern Spotted Owl* v. *Hodel*, eventually resulted in the status review being reopened, which led to the decision by the FWS in April 1989 that scientific evidence indicated the northern spotted owl was threatened throughout its range. The proposed listing of the owl as threatened was published in the *Federal Register* on 23 June 1989,[25] initiating what became nearly a year of controversy in the Pacific Northwest.

During the 1989–90 proposed listing, the FWS gathered additional biological information on the owl. One pivotal report, known as the ISC report after the Interagency Scientific Committee that produced it, concluded that the owl is "imperiled over significant portions of its range because of continuing losses of habitat from logging and natural disturbances," and argued that current protection, such as the Forest Service network of SOHAs (spotted owl habitat areas for individual pairs) totaling over 700,000 acres in Oregon, Washington, and northern California, was inadequate.[26] The ISC report recommended that a total of 8.3 million acres of forest should initially be set aside as blocks of habitat conservation areas (HCAs) to support multiple pairs of owls; timber management in between HCAs would be such that owls could safely migrate from one HCA to another; and ongoing research and monitoring would establish whether the strategy was effective and whether timber production could increase without endangering the owl. The areal extent of the ISC proposal was considerably more than environmentalists had ever dreamed of, and far worse than the timber industry had feared. The ISC report generated a flurry of controversy in the spring of 1990 over the biological necessity and human ramifications of such a large-scale effort to protect the owl.

The 1990 FWS status review of the owl, while conducted independently of the ISC report, arrived at essentially the same conclusion: the owl was in trouble due to loss of old-growth habitat from logging, and existing regulatory mechanisms were inadequate to protect remaining spotted owl habitat. The status review committee recommended that the owl be listed as threatened (a designation meaning that an organism, while not immediately in danger of extinction, would likely become so in the foreseeable future) throughout its range. The status review became the principal basis for the final decision, published in the *Federal Register* on 26 June 1990, listing the owl as threatened throughout its range effective 23 July.[27] Pursuant to Endangered Species Act provisions, the spotted owl listing decision was based solely on biological and management evidence; projected adverse economic impacts of owl protection, for example, were not considered as evidence. Following the listing decision, the FWS designated 6.9 million acres of public forest as critical habitat for the spotted owl and produced a draft recovery plan.[28] Ultimately, the environmentalists' dream, and the timber industry's nightmare, had come true.

Principal Organizations and Formal Response

The proposed spotted owl listing largely pitted national and regional environmental organizations against the timber industry and related interests, many of which were regionally based. One of the most prominent environmental groups involved in the spotted owl and old-growth debate is the Wilderness Society, which opened a Portland office in 1989 specifically in response to the magnitude and popularity of these issues. The Wilderness Society attracted considerable support for its role in the debate: membership doubled between 1989 and 1991 to 410,000 nationally. Examples of regional environmental groups include the Oregon Natural Resources Council (ONRC), a coalition of over ninety conservation, recreational, and other organizations, and the Native Forest Council (NFC), headquartered in Eugene, Oregon (though drawing nearly half of its membership from supporters outside of the region), whose policy bottom line was to halt logging of all "native" (that is, old-growth) forests on public lands.

Environmental groups unsurprisingly offered strong support for the proposed owl listing. The Wilderness Society, for example, cited as reasons drastic reduction and fragmentation of suitable old-growth habitat, demographic data suggesting that the owl is in danger of extinction, and the refusal of the Forest Service and Bureau of Land Management, who control over 80 percent of remaining habitat, to protect sufficient habitat to protect the owl. It strove in its formal comments to discredit timber industry claims that spotted owls do not require old-growth forests as habitat, arguing for instance that reported sightings of owls residing in second-growth stands in California are "inconsequential," since these are coastal redwood forests, where old-growth structural characteristics are attained far sooner than in other forests in the owl's range.[29]

On the other side, the diverse interests comprising the wood-products industry presented a relatively unified front in response to the threat posed by the spotted owl listing. One organization in particular, the Northwest Forest Resource Council (NFRC), acted as the chief industry voice in opposition to the listing. Based in Portland, Oregon, the NFRC is a coalition of industry associations in Oregon and Washington whose members are to some degree dependent on federal timber. Timber industry representatives were joined in their opposition to the proposed owl listing by a number of so-called grassroots groups, in which local community participation was emphasized. In Oregon, these grassroots groups joined forces as the Oregon Lands Coalition (OLC), a consortium of resource-use interests such as Associated Oregon Loggers, Oregon Fur Takers, and the Oregon Off-Highway Vehicle Association, devoted to "heightening community awareness and knowledge about the importance of the wise, multiple-use of public lands, balancing resource protection and resource production."[30]

The NFRC and the OLC were extremely active in responding to the Fish and Wildlife Service regarding the proposed owl listing, trying every conceivable means to forestall or prevent ESA protection of the spotted owl.

The OLC, for instance, petitioned the FWS to withdraw their proposed listing of the owl, arguing that "inadequate scientific data exists to justify the proposal to list," and giving examples such as taxonomic errors, unreliable population estimates, and sampling bias in estimating owl demography.[31] These groups also lobbied strongly for the Fish and Wildlife Service to extend the formal comment period and alter the method by which the FWS would consider evidence submitted regarding the listing. In September 1989, the NFRC formally requested that the owl listing comment period, slated to close on 21 September, be extended to 20 December. They argued that new and insufficiently-analyzed data were available, and requested that formal evidentiary hearings be held to "allow the Service to distinguish valid scientific information from mere speculation and hypothesis."[32]

The NFRC also attempted to discredit rival scientific opinions on the owl listing. For example, a review of the Wilderness Society's old-growth status report dated February 1989 charges that it utilized Forest Service timber inventory data not amenable to ecological classification, and applied a severely restrictive definition of old growth.[33] The NFRC concluded that it would be premature and misleading to use the Wilderness Society's figures. In another submittal, the NFRC charged that the ISC report was little more than a "theory" for preserving the owl, and would certainly entail major human implications if carried out.

The Owl, the Forest, and the Trees: Landscape Production/Counterproduction

During the proposed listing period, pro-timber and environmental groups waged a public outreach campaign to secure popular support for their positions. This outreach was the primary ideological vehicle by which the production of moral landscapes occurred and as such merits our close attention here. I will begin by considering the outreach of the environmental groups, and then turn to the very different geographical embodiment of the good as suggested in pro-timber outreach.

Environmentalist Outreach

Public outreach by environmental organizations during the period of the proposed spotted owl listing took a variety of forms, including television productions and radio commercials, pamphlets, and mass letter mailings to members. The following are representative selections.

The owl and its dwindling habitat
The spotted owl became an icon of the environmental movement during the Pacific Northwest battle, providing mute testimony to the plight of old-growth forests. It was frequently found in environmentalist outreach to its members (Figure 9.2) or publications intended for a sympathetic

Figure 9.2 Title graphic for a 1989 edition of the Sierra Club newsletter

audience. Yet in these and other cases, primary stress was placed on old-growth (what the environmentalists generally called "ancient") forests, not the owl.

Nonetheless, the owl became a familiar symbol to environmentalists of nonhuman nature threatened by humans. In one Wilderness Society video on old-growth forests, for instance, the spotted owl is heard at the outset, hooting in a peaceful old-growth forest.[34] This setting is shattered with the snarl of chain saws in the background, a sound that grows to a deafening roar and then ends with crashing trees and a ground-level scene of a clearcut. The sound of birds is now replaced with the buzzing of flies; the towering old trees are now stumps. The commentator says, "An age-old forest lost in a day." Old-growth forests are again the ultimate focus of attention when the discussion later moves more directly to the northern spotted owl, "A measure of the health of the entire old-growth ecosystem," and a wildlife biologist talks of owl habitat diminishing against background scenes of clearcuts.

The besieged old-growth forest

Old-growth forest management far overshadowed the status of the spotted owl as the focus of environmentalist outreach during this period. The Wilderness Society, for instance, produced a vast portfolio of communications with the public in the form of mass mailings, videos, special reports, and sponsored monographs.[35] One letter sent from then Wilderness Society president George Frampton to its members in 1989 states, "Our nation's last ancient forests have only one defense against the timber industry's bulldozers and chainsaws: An outraged American public." The undated letter highlights the timber industry's $12 million advertising campaign, designed "to mislead American citizens and Congress into believing that existing management practices pose no threat to the continued survival of our last pristine ancient forests!" The letter asks for financial support for the Wilderness Society's efforts to launch the "National

Forests Campaign," designed to counter the timber industry's public relations initiative.

The Oregon Natural Resources Council was chiefly involved in court
litigation and congressional lobbying on issues involving old-growth forest
protection, and did not mount a massive public outreach campaign, citing
insufficient funds. Nonetheless, the ONRC achieved a level of notoriety in
Oregon, chiefly because their spokespersons were frequently called on to
provide the environmentalist position regarding news items. One ONRC
mailing included an "Ancient Forests Action Packet," which encouraged
members to write letters to politicians in support of old-growth forest
protection, and offered ecological and economic facts members could cite,
such as "Less than 10 percent of Oregon's original forest remains in an old
growth condition."

The main public outlet of the Native Forest Council was its newsletter
publication, *Forest Voice*; NFC claims that nearly a million copies have been
distributed. The first *Forest Voice* was published in September 1989; its
headline reads, "An urgent appeal to citizens and Congress: Stop the
destruction of the last remnants of the public's native forests." The next
page shows another aerial view of a vast clearcut occupying hill after hill of
the Olympic National Forest. The caption reads, "Brazil? No, this is an
American National Forest!" The newsletter includes figures supporting the
NFC claim that plenty of private timberland exists to support the nation's
timber demand, cartoons depicting the greedy interests of the timber
industry, picture after picture of clearcuts in national forests of the Pacific
Northwest and elsewhere, an article attempting to debunk the "Myths,
deceptions and lies" of the timber industry, a critique of log exports, and a
ground-level scene of a clearcut forest with a quote from William Shake-
speare: "O pardon me thou bleeding piece of Earth, That I am meek and
gentle with these butchers."

Pro-Timber Outreach

Timber outreach dwarfed that of environmentalists during the period
preceding the spotted owl listing; I will accordingly devote more space to
this prodigious effort. This outreach generally increased as the stakes were
raised in the old-growth debate, culminating in a $12 million, three-year
long, nationwide public relations campaign launched in August 1989 by the
timber industry. The industry portrayed their campaign as a response to the
increased public interest in forest management issues, and the apparent
effectiveness of environmental outreach over the last decade. One organizer
said of the environmentalists, "I've never met anyone with less scientific
information and fact that's been able to dominate the media and program
the masses." This organizer cited the hard economic times of the early
1980s, which caused many timber companies to lay off their public relations
personnel, resulting in a critical communication gap in the latter part of
the decade.[36] Of the $12 million spent nationwide, some $300,000 was

earmarked for Oregon, and $150,000 went to the Green Triangle Project. In addition to the national outreach program, other existing public relations campaigns cited include Weyerhaeuser's "The tree growing company" ads and the Caterpillar Corporation's film "The Continuing Forest," for which they spent an estimated $500,000. These projects are reviewed below.

The public outreach of the timber industry and its supporters was done in forms similar to those of the environmentalists, although at the time its national network was insufficient to generate mass mailings to potentially sympathetic recipients, so mail campaigns were used far less.[37] Its regional network, in comparison, was far better geographically distributed, especially in smaller communities of the Northwest, so newsletters and pamphlets were a prime medium. Regional outreach also included a number of newspaper, radio, and television advertising campaigns.

The spotted owl hoax
The pro-timber coalition clearly differed with environmentalists on the status of the spotted owl and emphasized their differences in public outreach. For example, the North West Forestry Association, a sister organization of the non-profit NFRC, produced a nine-minute video entitled "The Northern Spotted Owl: A View from the Forest," which looks in detail at the proposed owl listing.[38] Over five hundred copies of this video were made, with one delivered to every member of the US Congress, and additional copies were sent to the heads of grassroots timber groups. The videos were used extensively on a local basis for talks on the owl listing in Chamber of Commerce and similar meetings. Its theme was the failure of the US Fish and Wildlife Service to look at all credible biological research on the owl.

The video features timber and wildlife specialists who argue that spotted owls can thrive in forests managed for timber production. Following discussion of "new" scientific evidence, the commentator concludes:

> It can no longer be said that the spotted owl is limited to preserved tracts of old-growth for its survival. By maintaining existing preserved areas, and by slightly modifying cutting practices in future harvesting units, spotted owl habitat areas can be maintained indefinitely, allowing forests to provide the Nation with a sustainable level of timber production.

In addition to providing an alternative biological reading of the status of the owl, the pro-timber movement argued that the owl listing was being promoted by environmentalists as a tactic to "lock up" forests. They frequently quoted a remark in support of this position made in 1988 by Andy Stahl of the Sierra Club Legal Defense Fund in 1988: "[t]he spotted owl is the wildlife species of choice to act as a surrogate for old-growth protection, and I've often thought that thank goodness the spotted owl evolved in the Northwest, for if it hadn't, we'd have to genetically engineer it."[39]

The spotted owl understandably generated a great deal of negative sentiment from people who opposed the listing and felt it constituted a threat to their livelihoods, resulting in a number of deliberate killings. In response, the pro-timber coalition distributed several flyers in rural communities urging people to restrain themselves and focus their attention instead on the environmentalists as the enemy.

Timber industry supporters also cited the owl as a means to express their desire for a "balanced" solution to the old-growth battle, as well as their distaste for the form of balance environmentalists proposed. Figure 9.3, for example, is the cover of an NFRC folder; Figure 9.4 reproduces the back page of a booklet distributed by the NFRC to timber-dependent communities, entitled "I'm Mad as Hell and I'm Not Going to Take It Anymore: A Resource Book for People Affected by Log Shortages." The booklet is filled with facts about the spotted owl, such as "[e]stimated numbers of spotted owls [have] increased from a few hundred ten years ago to almost 5000 today," and "[t]he spotted owl does not rely on 'old growth per se' but on a particular vegetative structure" that can be created by careful management of second-growth forests.

The renewable forest

Similar to environmentalist outreach, pro-timber outreach focused chiefly on forest management during the period between the proposed and final spotted owl listing decision. Yet their discussion rarely mentioned old-growth forests; instead, they unsurprisingly stressed the sound forest management practices of the timber industry. An example is the Green Triangle Project, a set of roughly fifty public relations television commercials aired statewide and particularly in metropolitan areas focusing on two themes: "Oregon will never grow out of trees" and "Oregon is timber country."[40] Another series was produced in 1990 by the Oregon Forest Industries Council (OFIC), which represents private timberland owners who have some dependency on federal forests. One OFIC ad spotlights replanting, a recurring theme in pro-timber outreach. It begins with a tree planter picking up his gear, then an aerial panorama of forests, with a whispered background saying, "Hundreds, thousands, millions." This background continues as the scene shifts to a tree planting crew walking through a lush forest to begin their work. As seedling after seedling is placed into the soil, replanting statistics are shown on the screen: "1960: 60 million trees," "1970: 81 million trees," "1987: 97 million trees." The ad closes with the caption "Oregon's Forest Industries" and an aerial view of a dense young forest.

The Weyerhaeuser Company launched a series of print advertisements in 1990 that ran in newspapers throughout westside Oregon. One ad in the Weyerhaeuser series features a forester from Coos Bay, who argues, "We aren't running out of trees in my corner of the Oregon forest. Not now. Not ever." He defends this position with figures: 7.5 million trees planted in Oregon every year; 1 million planted annually in the Coos Bay district; 300 to 450 seedlings planted per acre; only 2 percent of total forest holdings

Figure 9.3 Cover page of a folder produced by Northwest Forest Resource
Council in the late 1980s

Used by permission of Northwest Forest Resource Council

harvested every year. He also notes, "I've seen an increasing emphasis on
management for a broad spectrum of resources including watershed
protection, wildlife habitat, and scenic and aesthetic values."

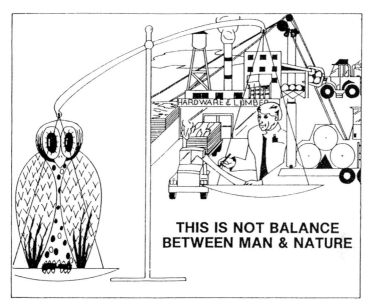

Figure 9.4   Back page of a Northwest Forest Resource Council
booklet distributed to timber-dependent communities

Used by permission of Northwest Forest Resource Council

Other forms of pro-timber outreach ranged from a series of radio ads
developed by the Associated Oregon Loggers to "put a human face on
loggers" and aired on thirty-five stations across the state during baseball and
football games, to a monthly magazine produced by the Evergreen Foundation, a strongly timber industry-sympathetic "advocate for science" in forest
management disputes, and distributed primarily through the Interstate 5
corridor running from Ashland to Portland, to a coloring book and tape for
children narrated by Timbear, "a big friendly old bear whose job it is to
guard the forest for all his human and animal friends."

Landscape Consumption: Popular Response to the Proposed Owl Listing

Public Comments

Over 20,000 comments were received on the proposed owl listing.[41] An
overwhelming majority of these—over 80 percent—opposed the proposed
listing (of which nearly four out of five were form letters prepared by

pro-timber interest groups and simply signed or copied by individuals). This response, while not statistically representative of Pacific Northwest residents, suggests that many people from the region were strongly opposed to the proposed action, and also implies high reliance on facts provided by the interest groups who distributed form letters. All public comments are found in the administrative record to the FWS spotted owl listing decision.[42]

Several dozen different form letters opposing the proposed listing were received by the FWS, most of which took issue with its biological basis. One form letter opposing the listing was sent by over 5300 people; it begins "[a] personal [*sic*] note to the US Fish and Wildlife Service" and gives eight reasons not to list the spotted owl.[43] Examples include, "[t]here is no conclusive scientific evidence that spotted owl populations are increasing or decreasing," "[a]n increasing number of spotted owls are being found in second growth timber stands, raising serious questions about the owl's level of dependence on old growth," and "[m]assive, natural disturbances—wind, fire and disease—are common. . . . There is no conclusive scientific evidence that timber harvesting, which mimics natural disturbances, has adverse impacts on owl populations." Another was sent by roughly 3250 opponents to the spotted owl listing.[44] It requested an extension of the public comment period, arguing that "[a]n independent survey is being conducted by wildlife biologists working on private, managed forestlands. . . . Hundreds of owls are being found in a variety of habitats, which will prove the 'threatened' listing is unnecessary."

Though fewer in number, personal letters were received by the US Fish and Wildlife Service as well, most again opposing the listing of the northern spotted owl, though largely based on projected personal impacts. One letter reads:

> I've spent all of my working life in the wood products industry. . . . Now, after around half of the forests have been set aside, groups of environmentalists are trying to stop or make it impossible for me to make a living in my owl homeland. This is not a fair or equable situation and its balance must be changed.[45]

Far fewer form and personal letters supported the owl listing. They offer a very different reading of the status of the owl and related economic issues. One form letter was sent by nearly 2200 supporters, and reads in part:

> I support listing of the northern spotted owl. . . . It is important to note that the protection of the spotted owl protects an entire ecosystem. . . . Rural economies should be maintained by prohibiting timber exports and managing forests for a sustained yield of various forest uses. . . . Sacrificing the few remaining groves of ancient trees and the many species dependent upon them (including the spotted owl) will not provide permanent stability to the

timber industry. Timber jobs have been lost to mill modernization and exports, not to environmental protection.[46]

Similar to comments opposing the listing, relatively few personal letters were mailed in by supporters; these also focus on the precarious condition of the spotted owl.

Public Opinion Polls

A number of public opinion polls were conducted in Oregon and the Pacific Northwest in the late 1980s and early 1990s on the topics of spotted owl and old-growth forest protection.[47] A majority of Oregonians (over 60 percent, with two out of five agreeing strongly) felt that spotted owl protection was excessive, even before the owl was listed. Opinion was more divided over whether timber jobs should be protected at the expense of owl habitat: in 1990, 46 percent agreed and 48 percent disagreed. Nearly two out of three Oregonians in 1990, however, opposed a halt on logging old-growth forests. Responses thus point to a strong public base of timber industry support in Oregon during this period.

Polls during this period also suggest that the public generally believed the timber industry position on ecological issues. For example, a majority disagreed that "cutting trees ruins the habitat of the spotted owl," with the largest subgroup disagreeing strongly. Also, a clear majority (nearly three out of four respondents) agreed that "cutting trees is necessary to keep forests healthy and productive." Their economic stance was generally pro-timber as well, in that a majority agreed that cutting trees is necessary for the region's economic health. The influence of pro-timber outreach is strongly evident in these results; additional polls confirm the degree of confidence people placed in the timber industry relative to environmentalists.[48]

Another statement in the 1990 survey elicited the opinion of Oregonians on the ISC owl-conservation report. The statement reads, "[t]o protect the spotted owl, we would stop logging on large tracts of federal timber land as recommended by the recent federal study, even if it means a loss of jobs." The ISC report received strong overall public disapproval in Oregon, with only one out of three supporting it, and 44 percent strongly disagreeing with the statement.

Disaggregation of these results provides us with some idea of the social basis of support for and opposition to spotted owl protection. The results suggest that the spotted owl issue divided longtime rural working-class residents from younger and better-educated recent urban immigrants to the state—those who, according to Carlos Schwantes, have increasingly flocked to the Pacific Northwest from California and elsewhere, lured by its environmental amenities.[49] At the same time, the social base of the pro-timber coalition is quite striking: not only was there solid support in rural, timber-dependent regions such as southern Oregon, where nearly two out of three strongly disagreed with implementing the ISC recommendations, but the divided opinion on the ISC strategy among urban residents, new residents, and college graduates suggests some measure of support in these sectors as well.

In short, people from timber-dependent households did not necessarily respond according to their immediate interests, and some people whose interests were apparently less tied to the timber industry nonetheless supported it. It would thus be a mistake to divide supporters and opponents of owl protection based on some simple interests theory. The influence of timber industry outreach is unmistakable in these polls. What is more, their message on the spotted owl and its forests—and ultimately the moral landscape they produced to provide a meaningful interpretation of the owl, the forest, and the trees—reached a broad audience. Though the pro-timber campaign lost the battle over the owl, it won the ideological war among a surprisingly large fraction of the Pacific Northwest.

Implications: The Meaning of Habitat

A distinct set of meaningful patterns emerges from the pro-timber and environmentalist outreach and popular response reviewed above. I wish to discuss these patterns in terms of the kinds of meanings that were produced and consumed in the spotted owl and old-growth controversy and the nature of the moral landscape that resulted, ending with a few more general observations.

Meanings Produced and Consumed

Intent

In a broad sense, the environmentalist and pro-timber ideological outreach in the Pacific Northwest during 1989 and 1990 was fashioned in order to serve their overarching interests. For the environmentalists, the objective was to secure old-growth forest protection. The timber industry groups sought to protect their corporate interests, which were clearly threatened by the environmentalist campaign. Pro-timber grassroots groups had a different set of interests: their concern was primarily for the economic stability of their local communities through continued timber jobs and revenues. The challenge to timber industry and grassroots groups was to maintain some sort of unified front against the environmentalists, as if their trajectories were parallel.[50]

One critical component of these broad objectives was to secure public support, which necessitated persuading laypeople not only to favor a particular set of policies but more fundamentally to favor a particular moral perspective on the spotted owl and old-growth problem. The greater activity of the pro-timber coalition in the Pacific Northwest is understandable in that people of the region were seen to be more likely candidates for political and moral suasion; the environmentalist emphasis on a national audience reflects a differing appraisal.

Mechanism

The ideological means by which environmentalists and the pro-timber coalition achieved these ends were complex. Some chief features included (a) legitimation/delegitimation of the messenger; (b) prioritization/marginalization of issues, and appeals to science and other sources of justification; and (c) universalization of the message to the audience.

The opposing bearers of news spent a great deal of energy legitimating themselves and delegitimating the other side. There was a particular dynamic between the two sides: the environmentalists seized the offensive in the spotted owl controversy, with a series of political victories most notably represented in the listing of the owl as a threatened species. Their challenge put the pro-timber coalition on the defensive, forcing it to justify recent timber practices in light of mounting evidence of their devastating impacts.

Curiously, this environmentalist challenge and the timber industry response hearkens back to debates of nearly a century ago, when timber barons were widely attacked for their cut-and-run policies, and is not directly applicable to the controversy over the spotted owl and old-growth forests, as forests have not been replanted to produce more old-growth habitat, but to produce more timber. This points to the overriding mechanism of legitimation and delegitimation at work, as opposed to some simple difference of fact about the relationship between timber practices and old-growth forests.

The pro-timber coalition also worked to discredit its attackers. This was a key function of the grassroots groups; for example, when the OLC charged that "preservationists" were intent not on protecting the environment, but on driving people off public lands, the implication was that the timber industry was more concerned about people than the environmentalists. The grassroots campaign to delegitimate the greens was assisted by the timber industry, which took a more scientific approach in its counterattack. Examples include the NFRC's critical review of the Wilderness Society's old-growth status report, and its charge that the Interagency Scientific Committee report was little more than a "theory" for preserving the owl.

Ecological and economic dimensions of human relations with Pacific Northwest forests dominated the spotted owl controversy and were framed in quite different ways by the two major players in the debate. Environmentalists prioritized the destructive impact of humans on nature, while the timber coalition prioritized the reliance of humans on nature, resulting in often incommensurate messages marked by absences as well as presences. Environmentalists, for instance, showed timber-dependent communities in their ads as seldom as the timber coalition discussed reduction of old-growth forests. Language played a key role in prioritization and marginalization of issues. The environmentalists described forests of the Pacific Northwest using terms such as "old-growth forests," "ancient forests," and "virgin forests." Their emphasis was clearly on that segment of the forest that had not yet been subject to human modification. The timber coalition, on the other hand, portrayed the forest as, for example, "the continuing forest" and "the working forest," and emphasized the word "timber." "Oregon," the series of timber industry ads read, "is Timber Country," certainly not "the Land of Ancient Forests." Weyerhaeuser, according to its motto, is "the Tree Growing Company." Pro-timber magazines and newsletters were called "The Seedling," "Evergreen," and "Timber."

The issues presented by both sides were justified by similar means, often involving science as an authority. The use of science was particularly evident in pro-timber outreach, which was put on the defensive by the science-based claims underlying the proposed spotted owl listing. For instance, the Northwest Forestry Association's spotted owl video is replete with claims of wildlife biologists that spotted owls are in fact thriving in Pacific Northwest forests. "The Continuing Forest" emphasizes the application of "modern science" in contemporary forest management, and the use of "experts" to help make critical management decisions. A Green Triangle Project ad also emphasizes sound scientific management of forests, noting that "before a single tree is harvested, studies are conducted by fish and wildlife biologists, forest engineers, soil specialists, botanists, and others."

Science was by no means the only authority, however. Environmentalists relied heavily on the aesthetic contrast between old-growth forests and clearcuts in their outreach. In this same sense, the spotted owl, with its large eyes and fluffy feathers, was a perfect charismatic species to support old-growth protection. The pro-timber campaign, especially that of the grass-roots groups, relied on appeals to common sense. Examples are numerous, ranging from the timber-affiliated organization called Common Sense, Inc. that distributes the "Timbear Unibearsity" tape and coloring book to the Associated Oregon Loggers ad that says the environmentalist message "makes the most noise, not the most sense."

Pro-timber and environmentalist outreach aimed to universalize their messages so as to appeal to as broad an audience as possible. For instance, environmentalists emphasized that national forests belong to all Americans, not just the timber industry; one ONRC brochure reads "This land is your land. . . . Help us save it.' The pro-timber coalition used a vast array of universalizing techniques to gain support. At the regional level, it portrayed its economic interests as not only central to the Pacific Northwest, but a part of its cultural identity. When a Green Triangle ad begins "for generations, people have worked hard on the farms and in the forests of Oregon," the very identity of Oregonians is bound up in logging. The "I'm Mad as Hell" booklet distributed to timber-dependent communities by the NFRC similarly says "[w]e can stand together and save our industry, our communities, and our rural lifestyle." The pro-timber coalition appealed to a national audience by noting how all Americans consume wood and paper products, including such identity-laden items as the home (the "American Dream," as commonly referred to in pro-timber outreach), and using wildlife symbols of national significance such as the bald eagle and grizzly bear in place of the spotted owl. The coalition often endorsed the multiple-use concept of resource management as a way to portray itself as sensitive to the full spectrum of human interests with respect to forests. An Associated Oregon Loggers ad says "[u]sing our forests doesn't have to be an all-or-nothing proposition. With a careful, balanced approach, we can have a healthy environment, *and* a healthy economy."

Outcome

The outcome of the pro-timber and environmentalist outreach was mixed for both parties. The environmentalists secured a good deal of national support for their ancient forest cause, as evidenced in their swelling ranks during the late 1980s and early 1990s, but their message was far less effective in the Pacific Northwest, where a substantial proportion of people felt that the environmentalist position was extremist. The pro-timber coalition lost the battle of the spotted owl, but won the sympathy of many of the region's residents, who trusted the industry position on forest issues far more than that of environmentalist groups.

What is striking in this pro-timber sentiment among the region's residents is that it was by no means confined to people from rural, timber-dependent communities. The timber coalition did an effective job of universalizing their message so that it made sense to a broad base of Oregonians, as suggested in public opinion poll results. Their responses in these polls indicates support both for the policies the timber coalition preferred and for the underlying ecological and economic assessment the coalition provided. Many people did not believe the scientific evidence that compelled the Fish and Wildlife Service to list the owl. They had heard that more and more owls were being found as surveys continued; they largely believed that the timber industry was taking adequate precautions to manage for wildlife in the forests. They knew that much of the state was critically dependent on the timber industry for jobs and revenues, and found little need to threaten the industry's economic role if little real scientific evidence existed that the spotted owl was imperiled. In these ways, production and consumption of the timber coalition message proved remarkably symmetrical.

The Moral Landscape of the Pacific Northwest

The outreach and response characterizing the period of the proposed northern spotted owl listing in the Pacific Northwest was clearly concerned with far more than the owl itself. The ultimate focus of the debate was the moral landscape of the region, as suggested in the relations between people and coniferous forests. The differentiated meanings people attached to this moral landscape emerged long before the spotted owl hit the news, providing a sedimented ideology that interest groups invoked in their owl outreach and people drew upon in their attempts to make sense of all the contradictory facts and figures these groups promulgated.

This moral landscape was more than a view of nature. Environmentalists did focus on the ancient forest a great deal, yet implicit in their argument is a sense of the proper boundary between humans and nature, of ancient forests as predating and existing apart from humans, and thus a view of people is equally involved. The bulk of interest group outreach and popular response addressed the morality of interactions between people and conifer-ous forests, whether typified by destructive impacts or economic and social reliance. To the environmentalists, the plight of the spotted owl was suggestive of the tremendous destructive impact of logging on old-growth coniferous forests and associated wildlife. To the pro-timber coalition, a very different interpretation emerged, one in which logging mimicked nature's own extreme events that served to renew forests and allow vigorous young trees to grow where decadent old ones had once reigned. The two sides also differed as to whether the economic and material reliance of people on the region's timber resources justified continued logging of old-growth forests. Though the various moral landscapes that emerged con-trasted markedly with each other, they were quite consistent internally. Few if any pro-timber supporters, for instance, argued that the heavy economic reliance of people on timber justified continued logging of old-growth forests in spite of recognized ecological impacts. Their position was a consistent one: people depended on logging, which at any rate did not do harm to the forest.

As a moral landscape, Pacific Northwest forests provided a sense not only of the good in nature and human relations with nature, but of the ways that the good is to be valued. There was, in other words, a decidedly axiological discourse at work in the spotted owl and old-growth debate, in which both intrinsic- and instrumental-value arguments were employed. For example, the timber coalition's emphasis on the human benefits of sustainable timber production follows an anthropocentric argument in which these practices are good as measured by their instrumental value to people. It was, of course, no coincidence that the timber industry provided strong support for anthropocentrism during the spotted owl controversy. The idea of nature as resource inherent in anthropocentrism is critical to the profitability of the timber enterprise; it has allowed an onslaught of the forests of the United States that has transformed the ecology of millions of acres for the extraction of wealth.

The axiological basis of the moral landscape defended by environmental-ists was more complex than that of the timber coalition, and in some ways contradictory. Their condemnation of old-growth destruction has been read by many commentators as a condemnation of anthropocentrism itself, as this ethic has effectively justified the conversion of old-growth forests to far more instrumentally valuable managed-timber stands.[51] Indeed, it seems to be hard to justify spotted owl protection on instrumental-value grounds, as there would likely be few negative human consequences if owls were to go extinct. Yet environmentalists would be incorrectly characterized as wholly non-anthropocentric. For instance, their frequent point that national forests belong to everyone, not just timber interests, and their stress on recreational, aesthetic, and other instrumental values of forests to Americans all follow broadly anthropocentric logic. In many ways, the axiological position of environmentalists was subservient to their overarching interest of achieving public support for old-growth protection; they deployed whatever intrinsic- and instrumental-value arguments were necessary to attract public sympathy.

The moral landscape, however, cannot speak for itself; people must represent it. In this sense, environmentalists portrayed themselves as the defenders of the old-growth forests, setting up the timber industry as the enemy. Conversely, the pro-timber coalition exhorted people to trust it as the proper representative and caretaker of the forest. The messenger was thus closely allied to the message in the construction of moral landscapes during the spotted owl controversy; people accepted or rejected many of the characteristics of particular moral landscapes based on their assessment of the messenger.[52]

When a particular messenger and message resonated in the social identities, the meaningful lives of people, they would tend to listen and believe. The moral landscape as they interpreted it also gave them a context to interpret the plight of the spotted owl, which then acted as a symbol of this moral landscape. To environmental sympathists, the owl provided sage though silent testimony to the beauty and goodness of nature, and the human foolishness and depravity of destroying the ancient forest. To timber industry-dependent communities, the spotted owl was a very different symbol, one suggestive of all the tree-hugging extremism and lack of human concern displayed by the environmental movement in its campaign to save old-growth forests.

Beyond the Moral Landscape: Larger Implications

In many ways, the spotted owls of the Pacific Northwest were caught in a struggle for ideological control between environmentalists and the timber industry. This struggle spills far beyond the region: indeed, American forests have long been a contested moral landscape.[53] Yet, for all the magnitude of the questions under scrutiny in the context of the spotted owl debate, both sides offered a decidedly shallow response.

The environmentalist critique indeed constituted a resistance against many of the taken-for-granted timber practices that had long been deployed in the Pacific Northwest; yet it generally avoided challenging the commodifying social and human-environment relations embedded in industrial capitalism, which must ultimately be held responsible for the precarious position of forests, forest-dependent animals, and forest-dependent communities alike in the region.[54] The kinds of meanings that pervade the kind of environmentalism that prevailed in the Pacific Northwest are decidedly partial meanings. They are primarily effective as means to arouse anger among sympathists against the desecrators of the ancient forests and their nonhuman inhabitants; they are less effective in answering the question of how humans fit into the moral landscape, and decidedly ineffective in explicating the political terrain in which this landscape has been cast.

For its part, the timber industry's ideological defense of completing the commodification of nature in the Pacific Northwest, cloaked under the guise of sound timber management, community well-being, and even environmental concern (in spite of its shaky biology and ecology), is understandable from an interests perspective, but clearly offers even less in the way of vision than the environmentalist critique. The massive resonance the pro-timber outreach campaign received among Pacific Northwesterners betrays ideological vulnerability as much as an affinity of interests. In fact, this groundswell of anti-listing sentiment against the spotted owl suggests the extent to which the democratic ideal of open public discourse on such important decisions as the owl listing can be quite readily twisted by distorted, convincing meanings.

The environmentalist resistance to wholesale habitat alteration and the pro-industry defense of resource extraction are now bundled into much larger ideological campaigns, such as the pro-extraction "wise use" movement, which has gained great strength in the US West and the halls of Congress,[55] and the global-scale biodiversity conservation movement, which aims to protect species and their habitats worldwide from destruction by human hands.[56] Similar to the spotted owl case, these larger movements will in many cases use animals to represent their agendas of meaning.

Indeed, discourse has already shifted in the Pacific Northwest. Though the old-growth forest debate has continued, the spotted owl has effectively receded from the forefront—in part because people are tired of talking about the owl after such sustained prominence during the late 1980s and early 1990s. The animals that now occupy the spotlight include the marbled murrelet, an elusive bird that inhabits coastal forests, and salmon and other anadromous fish species that periodically migrate up forested streams to spawn.[57] These animals too will become bound up with the contested moral landscape of the region's forests, once more intermingling habitat and ideology in the geography of the Pacific Northwest. But the divergent meanings people of this region attached to the spotted owl and its habitat will probably persist in the American consciousness for some time. Like the diminutive snail darter—that mid-1970s symbol of endangered species protection in the extreme—the spotted owl, threatening and wise, will long remind people of the wisdom and the threat of preserving nature.

Notes

1. Comments I.249 and I.54 from USDI "Administrative Record to Proposed Determination of Threatened Status for the Northern Spotted Owl," US Fish and Wildlife Service 1990.

2. USDI, "Determination of Threatened Status for the Northern Spotted Owl; Final Rule," *Federal Register*, vol. 55, 26 June 1990, pp. 26114–94.

3. USDA-USDI, *Forest Ecosystem Management: An Ecological, Economic, and Social Assessment*. Washington, D.C. 1993; USDA-USDI., *Final Supplemental Environmental Impact Statement on Management of Habitat for Late-successional and Old-growth Forest Species within the Range of the Northern Spotted Owl*, Portland, Oreg. 1994.

4. Paul A. Johnsgard, *North American Owls: Biology and Natural History*, Washington, D.C. 1988.

5. John Sparks, "Owls and Men," in John A. Burton, eds., *Owls of the World: Their Evolution, Structure and Ecology*, Glasgow 1984, pp. 18–26.

6. Steven L. Yaffee, *The Wisdom of the Spotted Owl: Policy Lessons for a New Century*, Washington, D.C. 1994.

7. William Dietrich, *The Final Forest: The Battle for the Last Great Trees of the Pacific Northwest*, New York 1992; Keith Ervin, *Fragile Majesty: The Battle for North America's Last Great Forest*, Seattle, Wash. 1989; David Seideman, *Showdown at Opal Creek*, New York 1993.

8. The management indicator species concept was introduced via the National Forest Management Act (NFMA) of 1976, which was the first piece of legislation that mandated concern for protection of biodiversity on US national forests; see R. Edward Grumbine, *Ghost Bears: Exploring the Biodiversity Crisis*. Washington, D.C. 1992. The idea was that NFMA regulations could be adhered to more efficiently by measuring success in terms of the status of MIS populations. This designation of the spotted owl follows 1982 revisions to the NFMA.

9. Yi-Fu Tuan, *Topophilia: A Study of Environmental Perception, Attitudes, and Values*, Englewood Cliffs, N.J. 1974; Yi-Fu Tuan, *Space and Place: The Perspective of Experience*, Minneapolis, Minn. 1977.

10. Compare Raymond Williams, "Ideas of nature," in *Problems in Materialism and Culture*, London 1980, pp. 67–85.

11. James Duncan, "Landscape," in R. J. Johnston, Derek Gregory, and David M. Smith, eds, *The Dictionary of Human Geography*, Oxford 1994, pp. 316–17.

12. James Duncan, "Landscape geography, 1993–94," *Progress in Human Geography*, vol. 19, no. 3, 1995, pp. 414–22.

13. Carl O. Sauer, "The morphology of landscape," in J. Leighly, ed., *Land and Life: Selections from the Writings of Carl Ortwin Sauer*, Berkeley, Calif. 1963, pp. 315–50.

14. Alan R. H. Baker and Gideon Biger, eds, *Ideology and Landscape in Historical Perspective*, Cambridge 1992; Denis E. Cosgrove, *Social Formation and Symbolic Landscape*, London 1984; Stephen Daniels, "Marxism, Culture, and the Duplicity of Landscape," in Richard Peet and Nigel Thrift, eds, *New Models in Geography: Volume II*, London 1989, pp. 196–220; Stephen Daniels, *Fields of Vision: Landscape Imagery and National Identity in England and the United States*, Princeton, N.J. 1993; James S. Duncan, *The City as Text: The Politics of Landscape Interpretation in the Kandyan Kingdom*, Cambridge 1990.

15. Kenneth Olwig, *Nature's Ideological Landscape: A Literary and Geographical Perspective on Its Development and Preservation on Denmark's Jutland Heath*, London 1984.

16. The term is not original; other geographers have employed it as a means of focusing on the complex values embedded in particular landscapes. This is, for instance, the general intent of David Ley in his discussion of co-operative housing and postmodern landscapes in Vancouver, B.C.; see David Ley, "Co-operative Housing as a Moral Landscape: Re-examining 'the Postmodern City,'" in James Duncan and David Ley, eds, *Place/Culture/Representation*, London 1993, pp. 128–48. Ley's conceptualization of moral landscapes is, however, unfortunately brief. Recent use of the related term "moral geographies" has been reviewed in David Matless, "Culture Run Riot? Work in Social and Cultural Geography, 1994," *Progress in Human Geography*, vol. 19, no. 3, 1995, pp. 395–403.

17. Jacquelin Burgess, "The Production and Consumption of Environmental Meanings in the Mass Media: A Research Agenda for the 1990s," *Transactions of the Institute of British Geographers*, vol. 15, 1990, pp. 139–61; Jacquelin Burgess and John R. Gold, eds, *Geography, the Media and Popular Culture*, London 1985.

18. Burgess, "Environmental Meanings," pp. 145ff.; compare R. Johnson, "The Story So

Far: And Further Transformations?" in D. Punter, ed., *Introduction to Contemporary Cultural Studies*, London 1986, pp. 277–313.

19. Mike Cormack, *Ideology*, Ann Arbor, Mich. 1992; David McLellan, *Ideology*, Minneapolis, Minn. 1986; John B. Thompson, *Studies in the Theory of Ideology*, Berkeley, Calif. 1984.

20. Terry Eagleton, *Ideology: An Introduction*, London 1991; John B. Thompson, *Ideology and Modern Culture: Critical Social Theory in the Era of Mass Communication*, Stanford, Calif. 1990.

21. A more detailed presentation and analysis of these materials is found in James D. Proctor, "The Owl, the Forest, and the Trees: Eco-ideological Conflict in the Pacific Northwest," Ph.D. diss., University of California, Berkeley, 1992. All unpublished materials and interview notes are in possession of the author.

22. An excellent discussion of the spotted owl listing is presented by Yaffee, *The Wisdom of the Spotted Owl*.

23. USDA, *Final Supplement to the Environmental Impact Statement for an Amendment to the Pacific Northwest Regional Guide*, Portland, Oreg. 1988.

24. GAO, "Spotted Owl Petition Evaluation Beset by Problems." US General Accounting Office, 21 February 1989, Report B–226076.2. In essence, it appears that the fate of the owl in 1987 was apparently determined in advance by top interior and FWS officials; biological evidence supporting a listing decision was ignored and in fact deliberately omitted.

25. USDI, "Proposed Threatened Status for the Northern Spotted Owl," *Federal Register*, vol. 54, no. 120, 1989, pp. 26666–77.

26. J. W. Thomas, E. D. Forsman, J. B. Lint, E. C. Meslow, B. R. Noon, and J. Verner, *A Conservation Strategy for the Northern Spotted Owl*, Portland, Oreg. 1990.

27. USDI, "Final rule."

28. USDI, "Determination of Critical Habitat for the Northern Spotted Owl; Final Rule," *Federal Register*, vol. 57, 15 January 1992, pp. 1796–1838; USDI, "Draft Recovery Plan for the Northern Spotted Owl," USDI Fish and Wildlife Service Spotted Owl Recovery Team, April 1992.

29. Wilderness Society, "Comments to Fish and Wildlife Service on Proposed Listing of Northern Spotted Owl." FWS administrative record letter #19443, 19 December 1989.

30. Unpublished OLC pamphlet, 1990.

31. Oregon Lands Coalition, "Comments of the Oregon Lands Coalition to the Proposed Rule by the U.S. Fish and Wildlife Service to List the Northern Spotted Owl as Threatened Pursuant to the Endangered Species Act," Report prepared by the Oregon Lands Coalition, Salem, Oreg., 19 December 1989.

32. Northwest Forest Resource Council, "Comments to Fish and Wildlife Service on Proposed Listing of Northern Spotted Owl," FWS administrative record letter #II.c.7, 8 September 1989. A formal evidentiary hearing would take much longer (up to a year or longer) than a hearing under the informal structure already used by the FWS. It would involve using a judge as hearing officer, with all parties represented by attorneys and all testimony subject to cross-examination. In addition, witnesses may be subpoenaed under this structure.

33. USDI, "Administrative record," index IV.E.3.a.

34. Wilderness Society, "Ancient Forests: Vanishing Legacy of the Pacific Northwest," Produced by Kathy Kilmer of The Wilderness Society, Videocassette, 1989.

35. Wilderness Society, "Ancient Forests: Vanishing Legacy"; Elliot A. Norse, *Ancient Forests of the Pacific Northwest*, Washington, D.C. 1990; Jeffrey T. Olson, *Pacific Northwest Lumber and Wood Products: An Industry in Transition*, Washington, D.C. 1988; Wilderness Society, *End of the Ancient Forests: A Report on National Forest Management Plans in the Pacific Northwest*, Washington, D.C. 1988.

36. Timber industry public relations campaigns, however, far predate their media blitz of the late 1980s. For a critical review of outreach spanning the last several decades, see Robert Pyle, *Wintergreen: Listening to the Land's Heart*, Boston 1986.

37. This in all likelihood is no longer the case, as the wise-use coalition has taken on a national stature and developed a strong organizational alliance in order to address policy

concerns such as the Endangered Species Act reauthorization and private property rights: see John Echeverria and Raymond B. Eby, eds, *Let the People Judge: Wise Use and the Private Property Rights Movement*, Washington, D.C. 1994.

38. North West Forestry Association, "The Northern Spotted Owl: A View from the Forest," Videocassette, 1989.

39. Cited in Yaffee, *The Wisdom of the Spotted Owl*, p. 215.

40. KVAL Television, "Oregon Is Timber Country: Molalla Timber Association," Produced by KVAL Television, Eugene, Oreg., Videocassette, 1989; KVAL Television, "Oregon Will Never Grow Out of Trees," Produced by KVAL Television, Eugene, Oreg., Videocassette, 1989.

41. USDI, "Determination of Threatened Status for the Northern Spotted Owl; Final Rule," *Federal Register*, vol. 55, no. 26, June 1990, pp. 26114–94.

42. USDI, "Administrative Record."

43. Spotted Owl Administrative Record Supplemental Index heading #K.11 (no record number is available for this letter).

44. USDI, "Administrative Record," #3618.

45. USDI, "Administrative Record," #I.6.

46. USDI, "Administrative Record," #22663.

47. See Griggs-Anderson Research, "Forest Product Industry and the Spotted Owl Controversy Statewide Survey I," Survey conducted for *The Oregonian*, 21 June 1989; Griggs-Anderson Research, "Forest Product Industry and the Spotted Owl Controversy Statewide Survey II," Survey conducted for *The Oregonian*, 1 May 1990. The surveys involved random telephone interviews of 400 (1989) and 600 (1990) Oregon adult residents, with sampling errors of 4.9 percent (1989) and 4 percent (1990). For a survey commissioned by the timber industry, see the Nelson Report, "Oregon Timber Industry and the Northern Spotted Owl," Prepared by the Nelson Report, 18 June 1990.

48. See the Nelson Report, "Oregon Timber Industry"; Cambridge Reports/Research International, "Survey Results on Forest Management Policy," conducted for the Timber Industry Labor Management Committee 20–23 June 1991, Report #CR 3203; KOIN Television, "Summary of Green Triangle Project Survey Results," Portland Stowell Data 1991.

49. Carlos A. Schwantes, *The Pacific Northwest: An Interpretive History*. Lincoln, Nebr. 1989.

50. As William Robbins has argued, these interests have rarely coincided in the region's history: when logging proved unprofitable, the timber industry would move on, leaving independent communities behind. See William G. Robbins, "The Social Context of Forestry: The Pacific Northwest in the Twentieth Century," *The Western Historical Quarterly*, vol. 16, no. 4, 1985, pp. 413–27; William G. Robbins, "Lumber Production and Community Stability; A View from the Pacific Northwest," *Journal of Forest History*, vol. 31, no. 3, 1987, pp. 187–96.

51. See, for example, Douglas E. Booth, *Valuing Nature: The Decline and Preservation of Old-Growth Forests*, Lanham, Md. 1994.

52. This is a common argument made in science-studies literature; for a discussion, see Brian Wynne, "Misunderstood Misunderstandings: Social Identities and Public Uptake of Science," *Public Understanding of Science*, vol. 1, no. 3, 1991, pp. 281–304.

53. Samuel P. Hays, *Conservation and the Gospel of Efficiency: The Progressive Conservation Movement, 1890–1920*, Cambridge 1959; Roderick Nash, *The Rights of Nature: A History of Environmental Ethics*, Madison, Wis. 1989; Michael Williams, *Americans and Their Forests: A Historical Geography*, Cambridge 1989.

54. John Bellamy Foster, "Capitalism and the Ancient Forest," *Monthly Review*, vol. 43, no. 5, 1991, pp. 1–16.

55. See Echeverria and Eby, *Let the People Judge*.

56. See, for instance, Robert Burton, ed., *Nature's Last Strongholds*, New York 1991; Vernon H. Heywood, *Global Biodiversity Assessment*, Cambridge 1995; Anatole F. Krattiger, Jeffrey A. McNeely, William H. Lesser, Kenton R. Miller, Yvonne St Hill, and Ranil Senanayake, eds, *Widening Perspectives on Biodiversity*, Gland, Switzerland 1994.

57. USDA-USDI, *Forest Ecosystem Management*.

Shrines and Butchers:
Animals as Deities, Capital, and
Meat in Contemporary North India

Paul Robbins

Establishing a normative ethical position around the question of animals and animal rights is confounded by the diversity of positions that animals occupy in the economic, cultural, and social worlds of people. Leaving aside the complexities of a "wild" animal ethic, even the animals who live and work alongside people seem to elude simple definition or identification owing to their contradictory status as personalities, sources of capital, and objects of consumption. Attempts to reconcile these distinct positions run the risk of sterilizing the rich and difficult nature of these relationships.

In India, these complexities are written large on an ideological landscape with well-known elements of nonviolence and an economic landscape of animal-oriented agropastoralism. Such complexities are not readily understood by American and European travelers, who are often upset and confused by the sight of a man whipping a cow or a goat on the streets of an Indian city; in the land of nonviolence, cruelty is seen as hypocrisy by the indignant outsider. Similarly, amongst foreign scholars the several positions of animals in social and economic life have long been viewed as contradictory and received the attention of Orientalist scholars mystified by the "sacred cow" and other apparently "irrational" ideas about animals.

In fact, domesticated animals in India have long occupied a wide range of economic and cultural positions. Within the subcontinent, the roles of animals as vessels for the divine, high-interest capital, and food have been allied in an evolving and complex ecological, economic, and spiritual triad. The last hundred years has brought accelerated change and pressure to the relationship between people (as worshipers, producers, and consumers) and animals (as divine vehicles, locations of stored value, and dinner), while all of the complex elements of this relationship have been retained. These ongoing pressures and tensions are increasingly evident in the political and social events of the country.

The controversy around and closure of Delhi's largest slaughterhouse,

Idgah, in 1994 reflects an unprecedented growth of India's national meat industry and a transformation in the relationships between domesticated animals and people in Indian culture and economics. Culturally, the traditional sense of ritual significance invested in many domestic animal species, especially in cattle, sheep, and goats, has declined. Economically, changes in the rural market and village ecology have resulted in a dramatic increase in small stock and the decline of the revered cattle population; at the same time, the ongoing and growing demand for meat, both in the cities and for export, has pushed the limits of infrastructure. These changes have collided against deeply held ideologies of nonviolence and ritual purity. The Idgah slaughterhouse debate reveals the contradictions in these simultaneous upheavals in economy and culture.

In an effort to elucidate the complex and contradictory roles of animals in Indian life, this review examines the events surrounding the slaughter-house closure and goes on to describe the multiple roles of domesticated animals in India as sacred vessels, sites of production, and objects of consumption. The review concludes that the ongoing interaction of emer-gent and residual cultural forms is often manipulated by politically powerful groups and individuals through the imposition of a strategic "moral monism." The complexity of human relations with animals, and indeed all of the rest of the natural world, may require a "plural ethic" capable of encompassing the multiple sites and positions of natural subject/objects like domesticated animals in a social world. Practical solutions to the problems that simultaneously drive human and animal suffering demand the ethics of a humane economy.

A Slaughterhouse in Delhi

On 18 March 1994 the Idgah slaughterhouse in Delhi was closed by court order.[1] One of only two slaughterhouses in the city, the facility had been constructed for a capacity of 2500 animals a day at the turn of the century. At the time it was closed, the abattoir was handling approximately 14,000 animals (goats, sheep, and buffaloes) daily under conditions that were described by the court as cruel and unsanitary. Some 13,000 liters of blood and offal were being discharged daily into the nearby Yamuna river.[2] Consumption and export of meat from the capital had outpaced the capacity of the ninety-year-old structure.

At first appraisal, this appears to be little more than another indicator of unprecedented growth in one of India's largest cities, outstripping its ancient infrastructure and design. The furor, debate, and politics that ensued in the months following the closure, however, suggest a deeper tension in Delhi's supply and consumption of meat than in its overused roads and crowded parks. Despite long-standing complaints concerning the sanitation and efficiency of slaughtering facilities in many major cities in northern India,[3] politicians have done little to adapt public infrastructure to the recent growth in the animal economy. Slaughterhouses are not good political capital in Delhi; modernization, investment, and attention to the meat industry in an ostensibly vegetarian culture do not garner votes for a ruling party. The growth of the meat industry and the closure of the abattoir are not, therefore, simply a product of unchecked growth in population and economy. Tied to religious beliefs and ethics, these developments must instead be seen as part of a larger set of political and ethical debates.

In particular, the politics and meaning of the Idgah controversy touch on complex relationships between producers and consumers, Hindus and Muslims, and people and animals. A *Times of India* poll taken a few months into the closure revealed a number of popular interpretations of the situation.[4] Few of those interviewed were aware of the court order, and while most acknowledged the official hygienic reasons for the closure, many responded that it was intended to reduce the slaughter of animals and nonvegetarianism.[5] Many activists warned that continued meat-eating would lead to an explosion in the goat population and the environmental destruction of northern India.[6] Conversely, others argued that rampant meat-eating would lead to the total demise of India's goats and so to a loss of milk and fuel sources for the poor.[7]

Political responses varied greatly. The conservative Hindu fundamentalist Bharatiya Janata Party (BJP), governing in Delhi, celebrated the closure. BJP satisfaction with the reduction of slaughter fit well with their conservative Hindu ideology.[8] A separate BJP action had restricted cow slaughter only a year earlier,[9] and vegetarianism as public policy linked the party to its conservative and religious base. Emerging from the electoral turmoil of the late 1980s, the BJP offered a program of Indian renewal through fundamental Hindu values. Along with several other communal political organizations, the party relied on a discourse of "return" to a purer and more stable past.[10] While claiming to represent many low-caste and Muslim minorities, the party's constituency is built overwhelmingly of Hindu upper-castes, businessmen, and underemployed professionals.[11] The party continues to enjoy national success in the wake of its incitement of the 1992 destruction of the Babri mosque in Ayodhya. There, the rioting crowds of militants tore down a Muslim holy building rumored to be built on the site of an ancient Hindu temple. The riots and retributive bombings that followed the incident represent the ebb of secular rule in modern India and the high-water mark in the power of the BJP.[12] For the BJP, a public and ideologically conservative reaction to the explosion of meat production in the city reflects the careful deployment of the kind of sectarian symbolism that made them so popular. As Parikh noted,[13] it is this kind of "manipulation of cultural symbols" that has "engendered nationalistic pride in the upwardly mobile Indian elite," a voting population that has prospered greatly in recent decades. A public celebration of the Idgah closure in the name of Hindu vegetarianism and conservative values in this way turned a public-health issue into a divisive religious and ethical question that cements BJP power.

At the same time, Delhi's predominantly Muslim butchers perceived the closure as a sectarian attack on their livelihood. In protest to the closure, the butchers went on strike.[14] Subsequently taking their case to the supreme court, they complained that their livelihood was being threatened and refused to work.[15] Meat became nearly unavailable throughout the city and public debate on the closure began with heavy coverage in the capital's newspapers. The events of those weeks were commonly discussed and debated on the streets throughout the city. Fears of riots resulting from the closure and the strike rose in the months immediately following the court order.[16] While not nearly as inflammatory as the events at Ayodhya, the closure and strike brushed closely against many strongly held convictions in Delhi and throughout northern India.

While the conservative idealist conceptions of Indian economy and society espoused by the BJP describe a fundamental collapse in the position and treatment of animals in India and call for a return to a consistent ethic, they disguise the complexity of the history of human-animal relations on the subcontinent. In their call for a pure and singular Indian culture, conservatives are responding to genuine changes in the culture and economy. Westernization of closely held cultural beliefs and loss of economic security are driving many Indians to the BJP and other conservative parties.[17] At the same time, these groups are capitalizing on inflamed emotions and eliding the complexity of India's cultural past. The pace of cultural and economic changes may be accelerating, but recent changes continue to follow complex, traditional patterns.

This is reflected in the changing position and meaning of animals and meat. Conservative interpretations of history notwithstanding, meat production and consumption have lived alongside the Hindu concern for sacred animals for millennia. Though constantly evolving, people's relationships to domesticated animals have been richly contradictory since even early Vedic times.[18] Domesticated animals have long played the simultaneous roles of locations of the *sacred*, sites of *production*, and articles of *consumption*. Using the case of recent changes in culture, production, and consumption in Delhi and the neighboring desert state of Rajasthan, it is increasingly clear that transformations in international demand and agricultural production, along with changes in caste identity and power, have fundamentally altered the bargain between people and domesticated animal species. By examining these changes, the complex threads of tradition and the implications of cultural transformation become evident. In marked and apparent contradiction, livestock are dirty and clean animals, capital on the hoof, and meat imbued with meaning.

Dirty and Clean Animals

It was early recognized by Western Orientalist scholars that animals, including and especially domesticated ones, held special and sacred meanings in South Asia. It was often mystifying to them, however, that these same "sacred" animals would be treated harshly, slaughtered, and sacrificed. One colonial officer saw in the symbolic acts of Hindu reverence for animals "the topsy-turvy morality of the east" and noted that the rough treatment of India's working animals demonstrates that "Oriental tender mercy has always been liable to the taint of grotesque exaggeration."[19] Apparent contradictions in the Indian animal ethic were happily seen as hypocritical weaknesses by colonial agents. "Sacred cow" continues to be common currency in the West for irrational and counter-productive beliefs.

More recent versions of Orientalist scholarship have also been befuddled by the "sacred cow" of India. Following Harris, cultural materialists championed a rational if unconscious structural explanation for apparently irrational beliefs.[20] In contrast, other scholars claim the power of an autonomous culture to form economic and material conditions.[21] All of these scholars have developed a long publication history over the debate concerning a rational explanation for sacred animals in India. Like their predecessors, these recent Orientalist scholars share an urge to identify a *single* Eastern logic in order to reveal its workings.[22] Their project turns on the explication of an exotic order of human/environment relations, relying alternately on ecological or religious factors in creating a monolithic and identifiable South Asian psyche.

The notion that domesticated animals are sacred would seem worthy of complex psycho-social explanation only to an observer from the industrialized West. The linkage between everyday life and interaction with the sacred that marks much of South Asian religious experience makes domesticated animals particularly likely vehicles for the sacred. Animals have shared environmental niches with humans and have been at the core of daily practice, the center of production, and the focus of religious tradition throughout the region's history.[23]

Even from the earliest recorded periods of South Asian history, domesticated animals have been depicted in art and scripture. The Indus Valley Civilization left images of cows, bulls, and goats in terracotta, steatite, and stone.[24] While it is difficult to verify these as "sacred" objects, they reflect the prominence of domesticated animals in the minds and material of these artists.[25] Later traditions revolve around sacred domesticated animals and animal sacrifice. Early Hindu law establishes animal protection and reverence. Throughout the Manu-*smriti* (Laws of Manu), cows are associated with gurus and Brahmans (ritual specialists), and rules are laid out which govern a wide range of behaviors towards cattle.[26] While many of these notions of nonviolence and cow-worship appear quite late in the Vedic tradition, the prominence and ritual associations of animals is present from the earliest record.[27]

In cultural and ritual practice developing out of this Vedic tradition, cattle, especially cows, continue to be held sacred throughout the subcontinent. They are adorned and worshipped directly as the vehicle of the mother goddess. In Rajasthan, legends of saint-like *bhomiaji* heroes grow around figures who die in battle protecting cows.[28] The bull also occupies a sacred position in this tradition, on its own as a symbol of reproductive power, or in association with deities, especially *Shiva.*[29]

Figure 10.1 A cow in festival garb, Ramdevra, Rajasthan

Other practices related to the meaning of animals, including the ritual use of dung and some forms of sacrifice, also persist today. Hindu sanctuaries known as *Goshalas* provide another example of the ritual status of animals. These are refuges for sick, old, or useless cattle similar to the Jain institution of *pinjrapoles*, which are for the protection of all animal life. These sanctuaries are perhaps the most prominent examples of a regional ethic for the reverence of animal life, especially for domesticated animals.[30] In sum, cattle who work, produce milk, dung, and leather are simultaneously held sacred. They are revered in themselves and as abodes of the souls of past generations.[31]

Other domesticated animals, including buffaloes, chickens, goats, and sheep, are generally not regarded as sacred in the same sense as cattle, but are filled with complex meaning and association nevertheless. Following Simoons and Debysingh,[32] the cultural meaning of animals in both traditional and contemporary Indian tradition is based around two distinct principles: the concept of *ahimsa* (nonviolence) and the notion of ritual purity. The concept of *ahimsa* holds that life itself is sacred and that the taking of life leaves deep karmic wounds. Those who would aspire to the highest level of proper living under this conservative ethic must remain vegetarian.[33] The notion of ritual purity, on the other hand, attributes to certain behaviors, contact, and actions a status of purity or pollution. In this case, the handling or managing of certain animals or animal products, especially fowl, pigs, and small stock, is a polluting and spiritually damaging act. Between these two concepts, vegetarianism and animal handling are given meaning and sacred significance.

Although there are wide geographic variations in ritual practice and the sacred significance of animals, some regionwide tendencies are discernible. Ideally, a perfect vegetarian never comes into contact with animal flesh and thereby lives a ritually pure life. Similarly, these ritual elites do not manage or herd small stock or fowl. Generally, this kind of practice is associated most prominently in the region with elite, high-caste (*jati*) groups, especially Brahmans.

In practice, some combination of meat eating and animal keeping is a reality for the vast majority of rural Indians. The particular form this combination takes, however, is often a complex one. Caste rank and meat eating often do not conform; low-caste vegetarians and high-caste meat eaters often live alongside one another.[34] In Rajasthan, elite *rajputs*, a caste of former rulers and landlords, eat goat, sheep, chicken, and buffalo meat without regard to concepts of *ahimsa*-related vegetarianism. They historically avoid the polluting profession of animal handling, however, seeing it as an ignoble, dirty, and impure practice. It is more polluting, in this case, to manage animals than to eat them.

In apparent contradiction, many members of the *raika* caste, sheep herders by trade, are practicing vegetarians who will not eat mutton or goats. Nor would they themselves slaughter the animals or tan the skins, leaving those tasks to yet other groups. While many of these herders describe sheep as clean, lucky, and *shub* (auspicious), their neighbors may describe them as dirty and impure. Tensions between animal production and impurity often develop in the actual application of the principles of *ahimsa* and ritual pollution. Animals are sacred, but in no simple or monolithic way. The notions of ritual purity and *ahimsa* are not always mutually reinforcing: livestock are sacred and profane, clean and unclean, revered and shunned in complex ways.

Nor are these traditions simply stagnant codes, inherited solely from abstract and cemented notions out of ritual tradition. Rather, they continue to evolve in the face of ongoing cultural and economic change. The primary sources of this cultural change are the forces of "Sanskritization" and Westernization. Sanskritization is the tendency among lower-caste groups to live by increasingly stringent Brahmanical codes of behavior and, over time, assume a higher social status as a group. In this way, many lower-caste groups eschew carnivorism as part of a purifying path to better status.[35] Simultaneously, modernization tends to ease the ritual obligations of even the highest castes. The fast-food restaurants of Delhi are daily filled with Hindus from a variety of backgrounds.

In Rajasthan, the trends of Sanskritization and modernization are realized in dramatic changes in herding and animal keeping. Where small stock were considered too dirty in the past to be herded by higher-caste

Figure 10.2   A herder with his
sheep flock (*chang*) on the daily
trip to water

households, now *rajputs* find great pride and public power in large herds of
sheep and goats. Where traditionally the camel-herding *raika* subcaste (*Maru
Raika*) was considered distinct and superior to sheep-herding *raika* (*Chalkia
Raika*),[36] now these groups are indistinguishable and keep large mixed
herds of small stock. Many marginal groups, who would not have been
associated with animal raising in the past, keep herds. The taint of animal
management has lessened considerably.

Similarly, the meaning of animal species themselves, along with their
associated taboos, is changing. Having lost ritual significance, goats are no
longer the traditional subject of sacrifice and enjoy some of the protections
of *ahimsa*. In many Rajasthani villages where sheep were historically considered unclean and allowed only on the outskirts, they now enjoy free
movement within the village center. Camel milk, never before sold or
consumed, is appearing in isolated markets.[37]

Some of this change is the result of an increased disassociation of caste
status from profession. Most desert producers, while still separated through
the ranked endogamous groupings of caste, practice nearly identical forms
of mixed agropastoralism. Some of the change is the result of increased
cash market value for animal products. The value of goat and sheep
products makes them a source of capital and pride in a way that could never
have been possible in previous centuries. The market is exerting a stronger
influence on cultural and ethical configurations than ever before. Even so,

a "market logic" of animal value has not emerged to remove cultural meaning and sacred value from domesticated animals. The associations of animals with the sacred, the profane, the dirty, and the clean persist but are reconfigured under these changing conditions.

These changes in vegetarianism and animal status have made pastoral adaptations more common and important in the rural economy. More groups are keeping animals. The decline of pollution rules associated with small stock in particular is linked to changes in the second role of domesticated animals; these sacred objects are also productive resources, providing fuel, power, protein, and a location of stored value. They fill both sacred and productive roles simultaneously and changes in one sphere are closely connected to changes in the other.

Capital on the Hoof

If domestic livestock in India, from the bull to the goat, are imbued with ritual significance, they have never been exempted from the demands of the lively agropastoral economy in the region. As a supply of energy and fertilizer, draught power, high-interest stored value, and protein, the sacred domesticated animals of India are the central source of capital in the largely rural country. Like the meaning of animals as sacred vehicles, the position of these animals in production has changed greatly in recent years, causing a large-scale reordering of agropastoralism in the region.

For most villagers in the arid northern and northwestern parts of the country, wood is a highly unusual source of calories for cooking and heating. More commonly, animal waste, especially *gober* (cow dung), pro-vides the bulk of fuel for most households.[38] By cycling fast-burning silica into solid, slow-burning cakes of fuel, the cow solves one of the most fundamental energy problems in a region perennially plagued with short-ages of wood. This is especially true for the bulk of economically marginal households.

Similarly, *gober* and *mingni* (sheep and goat dung) provide the main nutrient inputs into dry-field farming in the region. By cycling off-field nutrients from village forests, wastes, and grazing lands into agricultural holdings, animals retain the yields of farms increasingly pressured by the soil nutrient loss resulting from shorter fallow periods.[39]

Also, in support of India's agropastoral lifestyle, many cattle breeds of the region are carefully selected for draught power. Even among the strong milk breeds of northern and northwestern India, many breeds are capable and powerful plow pullers.[40] The sacred bull of India is most commonly seen yoked to a large plow or cart.

Animals also serve as the traditionally most reliable, high-interest location for stored value. The cattle wealth of the Indian herder is not simply a status symbol, it is one of the few places where capital can be safely stored. This is especially true under conditions where land can only be owned by the feudal elite, as was traditionally the case, or where scarcity makes land accumulation difficult, as is presently the case. The reproductive return per unit is very high in livestock, especially in small stock, who may reproduce one or two times a year. In this case, animals, rather than animal products, are themselves the productive capital in a rural society. The sacred attribute of fertility associated with domesticated animals is, in this case, realized in a very practical way.

Finally, livestock serve as key sources of protein in India. Not only are there a myriad of vernacular preservative methods for the milk products of goats and cows, but the meat of animals, though rarely acknowledged, has long been a source of protein for its consumers and capital for its producers. The remarkable variety of dairy products and processes, from creaming and condensing to making yogurts, butters, and cheese, is a testimony to the importance of animal proteins in regional subsistence. Methods of heating, souring, and agitating dairy foods are cornerstones of almost all household food preparation techniques.[41] Even the diets of the most elite vegetarians rely upon *ghee* (clarified butter) in food preparation. Meat production and consumption does not enjoy as public a position in the Indian diet, yet meat traditionally has been eaten by many groups and, as will be shown, meat eating has had a role in Hindu culture since the Vedic period.

Yet like the sacred tradition of animals in India, the tradition of production has not gone unchanged through recent history. The extensification and intensification of agricultural production has led to increased demands for animal fertilizer. As timber falls to the plow, the value of animal fuel also increases. Simultaneously, the loss of pasture has made the management of some livestock more difficult, with herding being pushed to more marginal locations. Local institutional change in the village sites of production has further led to the fragmentation of pasture land and to seasonal bottlenecks in production. By surviving on the margin, sheep and especially goats can fill increasingly critical roles in production and are in great demand.[42]

Growth in urban consumer capital has also led to an increased demand for dairy products, while prices for milk and cheese have remained relatively low through the institution of dairy co-operatives. The demand from urban areas for meat continues to rise as well. This demand is felt quite directly by herders in rural areas. Families raising goats in rural Rajasthan report that they make regular sales within their village or to traveling *beopari* (animal merchants) throughout the year. The market has been described as favoring the herder. Good prices are usually available for most small-holders.[43] This is excepted during drought years, when desperation sales allow the buyer to set the price. Even then, goat sales are one of the few avenues for access to capital. In good years and bad, goats provide the steadiest, most reliable source of income for many households. As the market grows, agropastoral households continue to produce animals to meet the demand.

Taken as a whole, this increase in demand, coupled with the decrease in supporting resources for livestock raising, has led to a demographic shift in animal species, with small stock coming to dominate larger stock. At the same time, changes in agricultural practices and crop choices have led to the decline of quality crop residues upon which large stock are heavily dependent. This has made small stock, who browse armed and toxic plants, a cheaper and more viable investment choice. Similarly, the relatively low maintenance costs of goats as the source of family dung and protein have caused many of the poorest families to substitute goats for cattle in the household. In villages throughout northern India, the decline in the quality and availability of resources has made the turn away from cattle and into goats and sheep a ubiquitous strategy for people in a wide range of caste communities.[44] Maneka Gandhi, the former environmental minister who initiated the investigation and closure of Idgah, ironically reported that the burgeoning goat population which feeds this meat industry constituted a severe environmental threat.[45] It is likely that the growth of the goat population is as much a response to ecological changes as it is a cause of them.

Between agriculturally and ecologically driven pressures, the character of India's livestock population has changed dramatically in the last fifteen years. In Rajasthan state, where much of north Indian cattle, goat, and sheep production is bred, the populations of sheep and goats each climbed 22 percent between 1977 and 1992, while the cattle population fell by just over 10 percent.[46] Under shifting subsistence and market systems, the sacred cow is giving way to the goat and the sheep.

In the process, the increased capital demands of the livestock economy have pushed some members of traditional herding groups out of the practice of pastoralism and invited many new groups in. In some cases, wealthy landholders have come to dominate livestock production in place of their poorer, semi-itinerant, pastoralist neighbors. More commonly though, the demands of marginal household production have led to the most dramatic changes in livestock production. The poorest rural households are turning to small-stock production to meet deficits in household fuel, fertilizer, protein, and cash and, in the process, are contributing greatly to changes in regional animal demography.

Even while the role of animals in Indian life is being dramatically reconfigured, livestock and traditional herding groups and techniques continue to thrive. The productive system of livestock capital is changing, but the features of the agropastoral economy continue to center around animal resources. Like the sacred realm of animals, the productive realm continues to evolve. Specifically, an increasing population of small stock has emerged in recent years, facilitating changes in the third role of domesticated animals, that of meat. The outpouring of the meat supply, in the form of small stock, has driven prices down and fed an increasingly hungry market. Animals are sacred vessels and stored value but they are also meat, and this third, terminal role also has a long history and tradition in north India. Livestock have been consumed goods since long before Vedic times. This largely unacknowledged role exists alongside the others in increasing tension.

The Meaning of Meat

The changes in the cultural and productive roles of animals have increased the supply of meat animals in the economy and produced an increased tolerance for meat consumption. Despite the prominence of these recent events, animals have actually been products of consumption in Indian society for a long time; meat eating has a long and enduring, albeit silent, history. Looking back to earlier traditions of meat eating in South Asia, the implications of the development of a vegetarian ethic alongside an agro-pastoral system with meat consumption become more apparent.

Meat eating is a prehistoric Indian phenomenon. Domesticated meat animals, including goats and sheep, appear in the earliest sites known in the Indus Valley (from 4000 B.C.E.).[47] Meat was also an important part of the post-Harappan diet, with the charred animal bones marked by cutting evident in the sites of the Black and Red Ware cultures (1100–600 B.C.E.).[48] Meat eating was common and pervasive throughout the early culture of the region. The vast changes in culture brought about during the post-Indus period linked meat with the creation of social hierarchy and elite kingship.

The earliest Vedic traditions of animal slaying and meat eating are tied closely to ritual hierarchy, sacrifice, and, curiously, to *Ayurvedic* medical practice.[49] The traditions of hunting and meat eating sit uneasily alongside the early evolving concepts of *ahimsa* and ritual purity in this early tradition. Many *dharmasastric* (religious legal) texts, including the Laws of Manu, emphasize both the role of meat eating in the universal hierarchy and the limits on the consumption of meat in pursuit of purity. Manu explains:

> What is destitute of motion is the food of those endowed with locomotion; [animals] without fangs [are the food] of those with fangs, those without hands of those who possess hands, and the timid of the bold.[50]

A hierarchy of creatures is acknowledged here, and the eating of beings lower on the food chain is justified in Hindu religion and law. At first glance then, meat eating and animal slaughter seem to be a traditional practice. At the same time, however, the eating of meat is reserved especially for those animal foods prepared by ritual sacrifice by Brahmans. Meat eating became regulated by and for elite groups through sacred law:

> The consumption of meat [is befitting] for sacrifices, that is declared to be a rule made by the gods; but to persist [in using it] on other [occasions] is said to be a proceeding worthy of Rakshasas [demons].[51]

In his unique work on Ayurvedic medicine, Zimmerman emphasizes the role of meat in Vedic ritual practice and in defining kingship.[52] *Dhatu* (meat tissues) were to be fed to a patient to "fatten the body," "produce muscle," and "firm the flesh." In particular, the eating of deer and buffalo were known to have powerful effects. More exotic (and even fantastic) animals were also included as important remedies and fortifications. Blood itself was prescribed for its powerful biospiritual effect. As Zimmerman explains, in the very violence of the remedy are its healing properties. Meats are not only present in early Vedic tradition, they are acknowledged as potent medicines, given power through the violence of their source.

These medicinal foods were not intended as the course for the everyday person, however. In the Ayurvedic prescriptions are the remedies and techniques intended for the health and maintenance of the ruler and the elite through whom the health of society is realized. Meats were the privilege of kingship and reserved for the *kshatriya* (warrior kings), who functioned, at least ideally, as hunters and rulers.

Even so, these meats could be eaten only under Brahmanic ritual conditions, through sacrifice. By sacrificial purification, these meats became free of the taint of violence. It is likely that the goat, always a good source of meat and free from the ritual status of cattle, therefore took on its now-traditional role as the sacrificial animal at this time. Sacrifice balanced out violence in the ritual equation. In this way, Zimmerman observes that the "therapeutic system of force" represented in the eating of meat and even raw blood is here reconciled with the "therapeutic system of purity" represented by the Brahmanic order of *ahimsa*. In this early period, meat is both pure and polluted, clean and dirty, a source of high status and a source of impure, ritual taint.

The source of ritual purity and the association of certain species, many behaviors, and most castes with ritual pollution also emerges in the texts of this period. Some groups or communities, for example, were associated exclusively with small stock raising or leather tanning, others with the keeping of cattle. Like the concept of *ahimsa*, the concept of ritual purity created a set of metaphoric associations that established a dharmic order of people and behaviors, intimately linked to the world of animals. The role of the individual in life was tied to her/his role in animal production and consumption.

Through the imposition of Vedic order, meat, animals, and animal management were all invested with meaning, linked to concepts of nonviolence and tied to the sacred order of purity and pollution discussed previously. The punishment for violation of these Vedic laws was ostracism. Rules had been created and imposed which governed diet and, indirectly, agropastoral practice; by implication, the keeping of animals strictly for meat production and consumption was taboo. Manu is littered with references to cattle herding and management but never mentions smaller stock, fowl, or other meat animals. The proscriptions of *ahimsa* and ritual purity are, *in text*, an iron-bound ordering of people and animals in the world.

The Vedic system would seem to create a general vegetarian order. Such texts were constructed and interpreted by elites, however. In effect, rather than simply establishing an abstract order, these rules served to form communities and create hierarchies through religious practice, justifying domination through diet and control of ritual. Missing from the religious and legal documents of the period is any account of the actual practice of meat eating among the non-elite. Archeological data, however, suggests that meat eating continued through this period, unabated. Concurrent Painted Gray Ware sites (from 900–500 B.C.E.) in the Gangetic plain show evidence of mixed agropastoralism with farming supported through the keeping of domesticated animals. Remains of the contemporary range of livestock are found at these sites, including those of the pig, an animal with little value but meat consumption.[53] It can be safely assumed that despite elite proscriptions, meat eating continued throughout and following the Vedic period.

With the establishment and continuation of a complex agropastoral system that relied upon the handling and management of ritually impure animals and at least in small part upon meat-based extraction, the Vedic prescriptions of nonviolent behavior and ritual purity begin to seem less like a universal cosmology and more like an ideological system of status legitimation. Rather than creating a universal vegetarian order, the Vedic proscriptions for diet and behavior served to elevate the status of ritual and land-holding activity at the expense of the vast base of agropastoral practice over which the ritual order was built. In the early history of the region, therefore, we see the rise of a vegetarian ethic and the simultaneous emergence of an agropastoral tradition with its domesticated animals and dependence on animal proteins. The persistence of meat eating in the north and northwest parts of the country continues into recent centuries.

The arrival of Muslim invaders and settlers and the rise of the Sikh faith reinforced the ranks of meat eaters, although both faiths observed a general ban on the public consumption of beef. The Muslim diet is directed by the Koran's strictures against "carrion, blood, the flesh of swine."[54] In India, the actual practice of meat eating for Muslims is slightly more restricted. In mixed communities, beef is generally avoided out of deference to Hindu neighbors. Nonmeat foods dominate the diets of most western Rajasthani Muslims; their diets closely resemble those of other groups in the region.[55] Muslims make up most of the population of *beopari* (animal traders) and *kassai* (butchers), assuring slaughter of animals is *halal* (lawful) for consumption. Hindu *khatiks* (butchers) also exist but in far fewer numbers. The Muslim tradition across north India thus brought its own dietary concerns and practices, contributing to and compromising with the region's tradition and economy of meat.

In Rajasthan, where these faiths mixed in a conservative Hindu context, a balance of meat eating and vegetarianism thus emerged. Administrative documents from the turn of the century describe a well-founded trade in goat meat and sheep mutton.[56] A regional census of castes conducted during this period not only records the number of castes traditionally involved in the meat trade, it further documents the practice of meat eating amongst Muslims and many lower- and middle-caste Hindu groups.[57]

The colonial period transformed the economy and ecology even more profoundly. The influx of a Western colonial presence increased the demand for meat. Colonial surveys from the period extol the quality of

Figure 10.3 A butcher (*kassai*)
in Jodhpur, Rajasthan

Marwari (Rajasthani) sheep, which, "though small, fatten well and, if properly fed, yield mutton second to none."[58] Moreover, changes in the nature of land ownership, community property regulation, and the export market dramatically altered the livestock economy, driving small stock like sheep and goats onto the market in place of cattle and camels.[59] Moreover, the capitalization of the rural economy during this period, most clearly seen in the change from rents in kind to rents in cash, created an increased demand on the rural producer for cash. The sale of animals for meat was an important avenue to meet the increased demand under colonial rule. It was likely that during this period the large-scale national livestock market began to blossom, initiating the transport of goats and sheep in cramped trucks, carts, and trains to distant markets. The ground was laid here for the complex carnivore/vegetarian culture and economy that dominates contemporary north India.

In Rajasthan today, the traditional rules barring consumption do exclude meat from the diets of some higher caste groups, but meat production and consumption continue to thrive throughout the state. Today, animals are handled for slaughter by both Hindu *khatiks* and Muslim *kassai* butchers.[60] Muslims and upper-caste Rajputs continue to eat meat. Further, despite the process of Sanskritization, which historically turns lower castes away from inauspicious practices like the eating of meat, meat is eaten by many marginal Hindu caste groups throughout the region. Much of this meat eating is limited to urban or ex-urban areas, but the consumption of meat in rural areas is an increasing trend. While the proportion of meat-eating Hindus is not known, it is estimated at not much less than 30 percent of the population.[61] Butchers in the city of Jodhpur, Rajasthan, report an explosive increase in business during recent years. One explained that "it used to be our best sales were only to the [military] bases, now I sell one or two animals a day on the street."[62]

While traditional Rajasthani culture and economy always had meat and skin elements, recent trends have clearly accelerated the growth of trade. Demand within the state continues to grow rapidly: meat production increased from 17.29 thousand tons in 1986 to 25.50 thousand tons in 1993. This is a 47 percent increase in less than a decade.[63]

The demand for meat outside the state in the urban areas of Delhi, Ahmadabad, and Bombay is also rising dramatically. Meat production is not supported or subsidized by the government in any way and goes untaxed. Statistics on production and export are difficult to find. However, some 3,286,000 goats (from a state population total of approximately 14,000,000) were recorded to have been exported from the state of Rajasthan in 1990–91.[64] These were for export to the larger cities and to foreign markets, predominantly in the Gulf States.

This growth of meat- and animal-based industry has caught the attention of Hindu fundamentalists, represented politically by the BJP. They cite a decline in traditional moral authority in India, including a decline in vegetarianism. In an action prior to and separate from the Idgah closure, BJP politicians tightened the bans on the production and possession of beef in Delhi, where they are the governing party.[65] This despite the minimal proportion of meat consumption represented by beef. As part of their call to a purer Hindu past, they include a "return" to vegetarianism and a call for a renewal of the sacredness of animals, especially cattle.

As discussed previously, BJP claims about the sacredness of animals in Indian life are far more than a simple cry for a moral vegetarianism. The call is closely linked to the parallel tendencies in BJP politics towards economic isolationism and anti-Muslim rhetoric. In the first case, meat and hides represent a large and growing export market in India.[66] The long-standing Minimum Export Price on these as well as several other exports was recently dropped as part of Indian trade-liberalization policy.[67] These measures are strongly opposed by the BJP, who favor economic isolationism and close controls on imports and exports.

More significantly, this highly public anti-meat rhetoric follows a long history of anti-Muslim rhetoric. The Idgah slaughterhouse debate follows on the heels of the much more explosive and violent debate over the Babri mosque in Ayodhya, which ended in riots and retaliatory bombings. While nowhere near as violent or emotional as the Ayodhya controversy, the conservative treatment of the Idgah controversy shares a few key elements with the earlier event.

First, conservatives invoke a discursive "return" to a singular and ideal configuration of culture, one in which all of India was Hindu and all Hinduism was one tradition. Despite evidence to the contrary, a uniform vegetarian past is offered as a historic reality and a desirable goal. Constructed by social elites, this discourse mimics and mirrors the late Vedic appearance of vegetarianism as a wedge for elite status.

Second, the vegetarian cultural critique focuses on practices or symbols broadly associated with Islam in the popular mind. Few, if any, Muslims in India eat beef (generally out of respect for their Hindu neighbors) and many of India's meat eaters are, in fact, Hindus. The call for a beef ban and stiffer anti-beef laws, nevertheless, is directed as a clear, symbolic attack on the Muslim minority. The connection of meat eating with moral degradation and from there to Islam is a chain of association that legitimates the sectarian message of the BJP. This association reinforces a version of South Asian history in which, as Metcalf explains, Muslims are seen as oppressors "who ultimately ushered in a period of decline."[68] Such a logic is a significant and dangerous one.

Meat is a powerful ideological tool. Vegetarianism, though never a universal practice in Delhi, serves as a trope for group identification and sectarian struggle. As a deeply worn analogy, vegetarianism is easily appropriated in this way by powerful interests defending their own status. The familiar ethical configuration of *ahimsa* and ritual purity is appealing under conditions of uncertainty and change. The explosion of illicit slaughtering throughout the city that followed the closure, however, draws into question the possibility of a viable vegetarian future. The language of purity and singularity fits poorly into the complex social landscape of north India.

As in the case of Western interpretations of Indian cultural history, the nationalist and neoconservative conceptions of Hindu tradition are linked through discourse to larger agendas. Where the essentialism and determinism of colonial and Orientalist interpretations of Indian history and tradition served to justify domination,[69] the singular and monolithic interpretations of Vedic tradition and ethics offer a wide field of play for Hindu revivalism. The tropes are powerful and the remoteness of the time in question is no barrier to its powerful role in contemporary politics.[70]

Ethical Pluralism and a Humane Economy

The case of animals, meat, and politics in India is a complex one, but it elucidates several key points. First, it is clear that the ethics that govern the treatment of and interaction with nonhumans are tied to cultural forms. These cultural forms are in a state of constant flux; residual forms exist from previous social formations while emergent ones develop under new conditions. Second, the study demonstrates that the complexity and contradictions of these multiple cultural and ethical formations are easily manipulated by political players drawing on the urge for consistency and tradition.

Understanding this, an alternative and more plural approach to ethics is required if normative issues in human-environment relations are to be addressed in conversations beyond theoretical abstraction. To this end, a practical and plural regional ethic might be imagined to take the place of the divisive and distracting discourse that elides more fundamental ethical problems. Commonalities between the positions of animal and human communities are the launching points for such a program; animal and human rights may serve as the foundation of a humane economy.

To conclude, each of these issues will be addressed in turn.

The connection between culture and ethics is a complex one and is not restricted to Indian tradition. "Folk ethics" (along with their modern counterparts, "philosophical theories") are linked to cultural values and ascriptions of meaning.[71] The ethic of nonviolence to animals in India is tied closely to the meanings that grow from and affect the positions of animals in economic and social life. While not determined by a material "base," these ethical and cultural configurations are linked to contextual and material circumstances that are multiple, complex, and simultaneous. The multiplicity of the roles of animals has, for this reason, led to a complex and fractured ethical landscape.

This complexity is compounded by the intricate history of the social formations within which these ethics have evolved. As described by Williams,[72] some cultural forms are the residual traces of previous social formations while others are emergent under changing social conditions. The cultural and therefore the ethical configuration of contemporary Indian society is a mix of old and new forms, each tied to the varying and contradictory roles of animals in differing parts of the region's history. This is not simply indicative of a modern or postmodern condition; rather, it is the perpetual state of complexity within which culture emerges and is transformed. In India, as in other regions of the world, the cultural and ethical configurations remain complex and contradictory and are difficult to reduce to particular historical moments, traditions, or economic conditions.

Despite this, attempts to reduce ethical form to one or another economic or historical "fact" are a constant canard in the political and intellectual life of the region. Crude cultural materialism seeks such a reduction, as did earlier Orientalist conceptions of the "Indian mind." More recently, the BJP and other fundamentalist thinkers have worked to unify the apparent contradictions in the region's ethical landscape through a return to the past. Yet this past is itself littered with contradictory ethics and proscriptions. Recent fundamentalists, like their Vedic precursors, seek ethical order in the natural world as a bridge to hierarchical order in the social world. The Idgah controversy is made to be a reflection of a "moral breakdown" in such an order, rather than the collision of several ethical and material configurations. The political implications of this kind of "moral monism" are clear.[73] Ethics are postulated prior to the social world to justify relations within that social world. These kinds of calls for order are greeted with broad responses in a large number of cultural and geographical instances, as much now as in earlier historical periods. Such discursive political efforts to iron flat complexities and relationships and to establish either a firm "market logic" or a "vegetarian ethic" attack the rich relationships amongst people and between people and animals. In an ethnically and ethically plural society where animals enjoy complex status, such efforts are themselves a threat.

Yet the needs and desires for normative guidelines for human behavior remain. The models of *ahimsa* and other normative notions in Indian cosmology continue to be attractive to people within and outside the region because they point towards a richer and more humble position for humanity in a world of human material domination. In reclaiming ethics from divisive politicians, a way forward may emerge for an emancipatory project founded in a regional ethic. Consider the simultaneous impoverishment of local animal raisers and suffering of animals in transport and slaughter. The marginal rural household turns to small stock raising when other avenues to capital are closed, resources decline, and institutions for production stability collapse. They turn the animals over to a transport middleman at small marginal profits, earning little of the surplus these goats and sheep will turn on the urban market. The middleman crams dozens of small animals into trucks, transports them without food or water for several days, and turns them over for slaughter in the kinds of facilities previously described. The profit from human labor and animal suffering is consumed at a distant site; the animals and animal raisers are rendered invisible to the consumer.

Ahimsa, taken in its literal meaning (the absence of the desire to harm or kill), applies powerfully to both the situation of the rural producer and to that of the animals. The violence of the extractive economy is mirrored in the suffering of the animals. Guidelines for an ethical economy, grounded in *ahimsa* and related concepts, should speak simultaneously to the conditions of the producer and the animal. Indeed, turned away from the divisive, normative, and exclusionary chauvinism of an ahistorical vegetarianism, *ahimsa* and other ethical constructs from South Asian tradition may provide contemporary guidelines for a humane economy.

Such an economy may not be able to enforce or institute a universal vegetarianism. Conceiving and negotiating such a singular ethic, grounded in first principles, is impractical in a plural context. In more abstract ethical terms, Stone explains, "[w]e simply may not be able to devise a single system of morals, operative throughout, that is subject to closure, and in which the laws of non-contradiction ... are in vigilant command."[74] The task, therefore, may not be to enforce a vegetarianism that has no practical or historical precedent. It may instead be to build an ethical patchwork and form a practical program of ethical action. Such an ethic might be built from the Hindu notion of *ahimsa*, the Muslim obligation to redress economic and social injustice and assure social welfare embodied in the notion of *zakat*,[75] as well as from Western ethical concepts like the inherent value of subjects-of-a-life,[76] agreeable work,[77] and public value.[78]

This plural ethic may be formed to point towards a more humane form of production.

Producer-managed dairy and meat co-operatives founded on such ethical and economic standards provide such conditions for production. While not a pure vegetarian manifestation of *ahimsa*, this form of production may be humane and geared towards minimizing suffering. While designed for profit, it may follow a social obligation to a larger group or community (*umma*). It extends the concept of agreeable work from humans to their animal partners and puts the economic institution in a context of public value, to be evaluated in terms of human and social effects. Reasonable and fair practices may be institutionalized by mutual consent of producers to extend the notions of duty or piety (*dharma* and *taqwa*) to the environment.[79] Small producers, managing their own slaughtering and transportation facilities locally, are not driven by the same capital demands as middlemen and, equally importantly, have a far different relationship to the animals under their care. The quality of agricultural and animal products and the quality of animal lives is certainly better assured under such an arrangement than the present one. Far from romantic, such a configuration closely resembles the guiding model of the highly successful AMUL milk co-operatives of India's "white revolution."[80]

Rather than finding and enforcing a singular ethic, such a project would engender a pluralism that "welcomes diversified material out of which moral judgments can be fashioned."[81] Differing from a moral or ethical relativism, such an environmental "ethical pluralism" would not allow all ethics for all conditions but would allow multiple ethical proscriptions in guiding action and interaction, each with an appropriate context and range of subjects. This follows closely the ecofeminist notion of a contextualist ethic,[82] demanding a certain range of first principles but deploying them in practical social contexts.

The formation of such a constellation would be, no doubt, a difficult task. It is far less daunting, however, than the idea of implementing a single environmental ethic or moral framework. It is a task that seeks commonality within and between cultures and that may galvanize environmental action amongst disparate communities. While normative, it may not be reductive. While concerned for the world of humans, it need not be anthropocentric. Finally, it is practical and implementable, putting ethical considerations to work for animals and people.

In Indian tradition, animals inhabit several worlds simultaneously. Animals are understood to occupy the *lokas* (universal realms) of the natural world, the human world, the divine world, and the transcendent world;[83] domesticated animals are wild, human, divine, and more. They have occupied these positions, simultaneously within, outside, and beyond the human world, since earliest record. The challenge of animal and environmental ethics is to incorporate all of these positions and to situate ethics in practical and material contexts. This challenge remains for the future.

Notes

1. "HC Shuts Down Idgah Abattoir," *Times of India*, 21 March 1994, p. 8.

2. N. S. Ramaswamy, "Shocking Conditions in Abattoirs," *Times of India*, 4 April 1994, p. 12.

3. Government of India, Ministry of Food and Agriculture, "Report of the Ad-Hoc Committee on Slaughterhouses and Meat Inspection Practices," Delhi: Manager of Publications, India, 1958; M. S. Rathore, *Marketing of Goats in Rajasthan*, Jaipur: Institute of Development Studies, 1993; M. Usha, "Survey, Analysis, and Disposal System of Hazardous Waters: A Case Study of Jaipur City," Ph.D. diss., University of Rajasthan, Jaipur, 1992.

4. "The *Times* Question: Meat Matters," *Times of India*, 8 May 1994, p. 24.

5. The *Times of India* telephone poll (8 May 1994) reports that 87 percent of those interviewed perceived the problem, at least in part, as that of hygiene; 76 percent reported that the reduction of animal slaughter was a reason for closure; 40 percent thought that it was intended to reduce nonvegetarianism.

6. M. Gandhi, "Criminal Record of the Idgah Abattoir," *Times of India*, 25 April 1994, p. 12.

7. S. S. Saikia, "Meat and Dust," *Times of India*, 3 April 1994, pp. 15–16.

8. A. Kala, "The Vegetarian and Non-vegetarian Controversy," *Times of India*, 13 April 1994, p. 16.

9. H. McDonald, "Revivalist Rally: The BJP Launches Its Campaign for Power," *Far Eastern Economic Review*, 18 April 1991, p. 19.

10. T. Basu, P. Datta, S. Sarkar, T. Sarkar, and S. Sen, *Khaki Shorts and Saffron Flags*, New Delhi: Orient Longman, 1993.

11. McDonald, "Revivalist Rally."

12. Basu, Datta, Sarkar, Sarkar, and Sen, *Khaki Shorts*.

13. M. Parikh, "The Debacle at Ayodha: Why Militant Hinduism Met with a Weak Response," *Asian Survey*, vol. 33, no. 7, 1993, pp. 673–84.

14. "Butcher's Stir: Affected Parties May Move SC," *Times of India*, 11 April 1994, p. 7.

15. "Abattoir Issue Is Tough Meat for Delhi Government," *Times of India*, 6 April 1994, p. 11.

16. H. McDonald, "Where's the Beef? It's Banned in Delhi and Mutton Isn't Available," *Far Eastern Economic Review*, 5 May 1994.

17. Parikh, "The Debacle at Ayodha."

18. The Vedic period, named for the hymnal scriptural Vedas, followed the demise of the Indus Valley Civilization between 1500 B.C.E. and 500 B.C.E..

19. J. L. Kipling, *Beast and Man in India*, 1891; reprint, Lahore: Al-Biruni, 1978.

20. M. Harris, "The Cultural Ecology of India's Sacred Cattle," *Current Anthropology*, vol. 7, no. 1, 1966, pp. 51–66.

21. D. Western, "Cultural Materialism: Food for Thought or Bum Steer?" *Current Anthropology*, vol. 25, no. 5, 1984, pp. 639–45; J. W. Bennett, "On the Cultural Ecology of India's Sacred Cattle," *Current Anthropology*, vol. 8, no. 1, pp. 51–66.

22. Said, 1979.

23. D. O. Lodrick, "Man and Animals in India," in R. L. Singh and R. P. B. Singh, eds, *The Roots of Indian Geography: Search and Research*, Varanasi, Uttar Pradesh: National Geographic Society of India, 1992.

24. J. W. Bowers, "Animal Art and Mythology of India," M.A. thesis, San Francisco State University, 1989.

25. The Indus Valley Civilization spanned the period 2500 B.C.E. to approximately 1700 B.C.E. The seals found at Indus Valley sites, upon which many of these animals are represented, bear a yet untranslated script.

26. G. Bühler, *The Laws of Manu*, Delhi: Motilal Banarsidass Publishers, 1886.

27. F. J. Simoons, *Eat Not This Flesh: Food Avoidances from Prehistory to the Present*, 2nd edition, Madison: University of Wisconsin Press, 1994.

28. D. C. Shukla, *Spiritual Heritage of Rajasthan*, Jodhpur: Books Treasure, 1992.

29. T. C. Majupuria, *Sacred Animals of India and Nepal*, Gwarlior: M. Devi, 1991.

30. Lodrick, "Man and Animals in India."

31. In Buddhist lore, especially the traditional Jataka Tales, the Buddha spends many lives in the form of animals.

32. Simoons, *Eat Not This Flesh*; M. Debysingh, "The Cultural Geography of Poultry Keeping in India," in D. Sopher, ed., *An Exploration of India: Geographical Perspectives on Society and Culture*, Ithaca, N.Y.: Cornell University Press, 1980.

33. K. N. Jacobsen, "The Institutionalization of the Ethics of 'Non-injury' to All 'Beings' in Ancient India," *Environmental Ethics* 16, 1994, pp. 287–301.

34. Debysingh, "Cultural Geography of Poultry Keeping."

35. This process of Sanskritization, originally explored by M. N. Srinivas, *Social Change in Modern India*, Berkeley: University of California Press, 1966, overlaps into other animal-related tasks and practices.

36. M. H. Singh, *The Castes of Marwar*, Jodhpur, India: Books Treasure, 1993.

37. Kohler-Rollefson, personal communication, 1994.

38. B. Bowonder, N. Prakash Rao, B. Dasgupta, and S. S. R. Prasad, "Energy Use in Eight Rural Communities in India," *World Development*, vol. 13, no. 12, 1985, pp. 1263–86.

39. R. Cincotta and G. Pangare, "Population Growth, Agricultural Change, and Natural Resource Transition: Pastoralism amidst the Agricultural Economy of Gujarat," Overseas Development Institute, Pastoral Development Network Paper 36a, 1994, pp. 17–35.

40. S. George, "Agropastoral Equations in India: Intensification and Change of Mixed Farming Systems," in J. G. Galaty and D. L. Johnson, eds, *The World of Pastoralism*, New York: Guilford Press, 1990.

41. George, "Agropastoral Equations in India."

42. Cincotta and Pangare, "Population Growth."

43. Rathore, *Marketing of Goats*.

44. P. Robbins, "Goats and Grasses in Western Rajasthan: Interpreting Change," Overseas Development Institute, Pastoral Development Network Paper 36a, 1994, pp. 6–12.

45. Gandhi, "Criminal Record of the Idgah Abattoir."

46. Department of Animal Husbandry, Jaipur.

47. F. R. Allchin and B. Allchin, *The Birth of Indian Civilization*, New Delhi: Penguin, India, 1993.

48. R. P. Singh, *Agriculture in Protohistoric India*, Delhi: Pratibha Prakashan, 1990.

49. *Ayurveda* is a form of traditional and holistic physical and spiritual health practice dating from the early Vedic period. It remains a current part of Indian medicine.

50. Manu V:29, in Bühler, *Laws of Manu*, p. 173.

51. Manu V:31, in Bühler, *Laws of Manu*, p. 31.

52. F. Zimmerman, *The Jungle and the Aroma of Meats: An Ecological Theme in Hindu Medicine*, Berkeley: University of California Press, 1988.

53. F. R. Allchin and B. Allchin, *The Rise of Civilization in India and Pakistan*, Cambridge: Cambridge University Press, 1982, pp. 317–25; H. D. Sankalia, "Functional Significance of the OCP and PGW Shapes and Associated Objects," *Purattva* 7, 1974, pp. 47–52.

54. Koran V:1, in A. J. Arberry, *The Koran Interpreted*, New York: Macmillan Publishing, 1955.

55. Simoons, *Eat Not This Flesh*.

56. Rajputana Gazetteers, 1908.

57. Singh, *The Castes of Marwar*.

58. Rajputana Gazetteers, 1908, p. 106.

59. M. S. Jain, *Concise History of Modern Rajasthan*, New Delhi: Wishwa Prakashan, 1993.

60. Rathore, *Marketing of Goats*.

61. Kala, "Vegetarian and Non-vegetarian Controversy."

62. From interview, December 1993.

63. Figures from the Department of Animal Husbandry, Jaipur. This statistic reflects the total tonnage of all species produced for meat in the state. Of this, a considerable

majority of meat passing through slaughterhouses is goat meat (interviews with health officers, Jaipur).

64. Rathore, *Marketing of Goats.*

65. McDonald, "Where's the Beef?"

66. Saikia, "Meat and Dust"; I. Rajaraman, "OECD Imports of Leather: Indian Performance and Real Exchange Rates of the Indian Rupee," *Journal of Development Studies*, vol. 29, no. 3, 1993, pp. 541–60.

67. Government of India, Ministry of Food and Agriculture, "Report of the Ad-Hoc Committee on Slaughterhouses."

68. B. D. Metcalf, "Presidential Address: Too Little and Too Much: Reflections on Muslims in the History of India," *The Journal of Asian Studies*, vol. 54, no. 4, 1995, pp. 951–67.

69. R. Inden, *Imagining India*, Cambridge: Blackwell, 1990.

70. R. Thapar, *Interpreting Early India*, Delhi: Oxford University Press, 1992.

71. J. B. Callicott, "Traditional American Indian and Western European Attitudes towards Nature: An Overview," in M. Oelschleger, *Postmodern Environmental Ethics*, Albany: State University of New York Press, 1995.

72. R. Williams, *Problems of Materialism and Culture*, London: New Left Books, 1980.

73. Following C. D. Stone, "Moral Pluralism and the Course of Environmental Ethics," in M. Oelschleger, *Postmodern Environmental Ethics*, Albany: State University of New York Press, 1995.

74. Stone, "Moral Pluralism," p. 253.

75. N. N. Ayubi, *Political Islam: Religion and Politics in the Arab World*, London and New York: Routledge, 1991.

76. T. Regan, *The Case for Animal Rights*, Berkeley: University of California Press, 1983.

77. P. Kropotkin, *The Conquest of Bread*, Montreal: Black Rose Books, 1990.

78. J. R. Commons, *Institutional Economics: Its Place in Political Economy*, New Brunswick and London: Transaction Publishers, 1990.

79. Following I. H. Zaidi, "On the Ethics of Man's Interaction with the Environment: An Islamic Approach," *Environmental Ethics* 3, 1981, pp. 35–47.

80. A. Q. Alvi, "Cooperatives: Effective Tool of Development in India," Land Tenure Center, University of Wisconsin, 1989.

81. Stone, "Moral Pluralism," p. 253.

82. K. Warren, "The Power and Promise of Ecological Feminism," *Environmental Ethics*, vol. 12, no. 2, 1990, pp. 125–46.

83. S. Snead, W. Doniger, and G. Michell, *Animals in Four Worlds*, Chicago: University of Chicago Press, 1989.

Building a Better Pig:
Fat Profits in Lean Meat

Frances M. Ufkes

Introduction

Since 1976, US red-meat consumption has waned. Unleashed by this crisis of consumption was a competitive cannibalism among meatpackers, culminating in reverberations of restructuring throughout the beef and pork commodity chains, from the farm to the meat case. In the early post–World War II years, meatpacking was a dynamic sector, with changing lifestyles and rising real wages resulting in a national "meat and potatoes" diet and the intensification and concentration of US livestock production. Today's meatpackers see fat profits in lean meat, with this imperative facilitating a new wave of agro-industrialization in US beef and pork production.

The new accumulation strategies of US meatpackers hinge upon major transformations in livestock production and animal physiology. Changes are evident in the types of animals raised, how they are raised, where they are raised, and who is raising them. During the first wave of agro-industrialization, meat animals were designed to grow bigger faster. This made sense during a time of escalating domestic meat consumption. Now genetics are being altered in the hopes of developing animals with less fat and more lean muscle tissue. Animal physiology is clearly an arena of industrial accumulation as growth processes and immunological responses are manipulated to produce new animal "interior geographies" more befitting American dietary trends. Changes "down on the farm" are manifested in the reconfiguration of genetics, livestock buildings, veterinary inputs, labor and capital requirements, and feedstuffs; changes at the meat plant include the use of lasers, nuclear magnetic resonance, and other non-invasive probes to reliably measure animals and reward producers for raising animals with specific carcass leanness traits.

In this chapter, I focus on recent agro-industrialization in US pork production. I argue that there have been two main expressions of accumulation in the US pork complex in the post-1976 period, and that the lean-meat imperative is integral to both. There has been a movement toward greater scale and standardization of production to provide low-cost, lean pork for general consumption by major meatpackers: IBP, ConAgra, Smithfield Foods, and Cargill's Excel. There has been a concurrent shift toward the industrial production of boutique pork products, including organic pork and extra-lean pork. Both result in intra-industry alliances and farm-level changes linked to the production of leaner hogs.

Capitalist Accumulation and Agro-industrialization

The confluence of recent thought in geography and other fields calls for examining agriculture's linkages with both the international and the national political economy, and for explicating the ways in which agriculture, over time, has become more fully integrated into circuits of industrial and finance capital.[1] Understanding the shifting and uneven relations between farming, manufacturing, and consumption within specific commodity chains, and the role of the state and other political forces in this process, is key. The agro-industrialization model of Goodman, Sorj, and Wilkinson,[2] the food-regime approach of Harriet Friedmann,[3] aspects of regulation theory,[4] and elements of the "new" industrial geography have been employed to theorize capitalist industrialization in a rural and agricultural context.[5]

Goodman, Sorj, and Wilkinson elaborate trajectories of innovation, or processes of agro-industrialization, central to the rise of "agroindustrial complexes."[6] Agrarian change is anchored in the expansion and deepening of markets for industrial goods. This occurs via appropriationism, the construction of demand for an ever-increasing array of industrially-produced farm inputs, or by substitutionism, whereby advances in industrial-process technologies reduce agro-food capital's reliance upon specific raw materials, evident in the fractioning of crops into "intermediate" components—amino acids, carbohydrates, and fats—to be used interchangeably in downstream processing. Both processes are contingent upon the constraints of "nature" in agricultural production—"the historical and contemporary configuration of the agri-food system [being] attributed to industrial capital's inability to subordinate agriculture and food to direct industrial transformation."[7] Innovations allowing either a greater regulation of natural processes or an increase in agricultural productivity afford agro-food capital with new arenas of accumulation, thus enlarging agro-industrial complexes built around the initial source of appropriation.[8]

Complementary insights are provided by food-regime theorists who stress agriculture's role within the international political economy.[9] Friedmann and McMichael describe two world food regimes and the contours of a third emerging since the mid 1970s. The first food regime (circa. 1880–1914) was based upon extensification, with world food output rising via the cultivation of more land. The second food regime (circa 1945–73) was anchored in international and national regulatory structures (for example, the GATT and domestic farm supports), fostering world trade expansion, and the intensification of farming (that is, increasing output per acre). This was expressed in the proliferation of highly processed, "durable" foods, the convergence of rural and urban consumption norms, and specialization and concentration in farming.[10]

The energy-food-currency shocks of the mid 1970s destabilized the second food regime.[11] The 1972 US-Russia grain deal and the 1973 US soybean embargo triggered wide shifts in world food prices,[12] and by the mid 1970s Keynesian full-employment policies could no longer be sustained given rising oil costs, declining rates of productivity growth in manufacturing, and escalating food prices.[13] A neoliberal, national economic regime stimulating disinvestment and industrial restructuring emerged.[14]

This deregulatory ethos paved the way for new accumulation strategies in US food processing. A high-wage sector in the early postwar era,[15] food processing turned into one dependent upon low-wage, rural, female, and immigrant workers.[16] Competition among firms culminated in greater consolidation "across the food chain."[17]

Within this era of agro-food restructuring, two main expressions of accumulation in industrial food production are evident.[18] The mass production of food remains important as an array of mass-produced inputs— sugars, meats, cheeses, spices—is combined in assorted ways to create the seemingly endless variety of low-cost, durable foods found in US food stores, these products being aimed at major market segments like singles, time-centric, dual-income families, and the working poor.[19] There is another thrust towards the production of boutique foods, with changes in social life and health concerns representing opportunities for new product development as firms create markets within the exploitation of difference.[20]

Both expressions prompt attention to the farm production process. Standardized inputs are needed in making mass-market foods, especially in the case of branded items which requires that product quality be predictable. Making boutique foods often requires similar controls to document the presence or absence of discrete commodity attributes linked to consumer concerns. Such is the case of "BST-free" milk, chosen for its absence of growth hormones. Firms are experimenting with contracting and other production/marketing relations to manage risk, expand sourcing options, and gain a competitive edge. Control over production may also be attained by regulating biological and physiological processes to engineer crops and animals with the traits needed in downstream processing; innovations of this sort represent new appropriations within farming.[21]

The US Meat Industry in Transition

Rise of the meat complex

The US livestock/feed/meat complex is a product of the second food regime.[22] Within this era of rising real wages, many foods that had been luxuries were transformed into commodities for the masses. Such is the case with beef and pork.

From 1945 through the 1970s, intensification of US livestock production centered around getting more and larger animals to market faster.[23] Emphasis was on reducing feed conversion ratios (that is, the amount of feed needed per pound of weight gain), the length of the gestation cycle, the number of nonreproductive days (that is, times when females are not pregnant, nursing, or in heat), and the total number of days on feed. Reductions in feed conversion ratios and days on feed were evident in pork and beef production, as were steady increases in the average "dressed" weight of cattle and hog carcasses.[24]

The intensification of production further entangled livestock feeders in circuits of agro-industrial and finance capital. Industrially produced feedstuffs replaced farm-produced fodder,[25] land as a factor of production declined in importance,[26] and aggregate farm-level debt escalated.[27] Undergirding this rationalization of production was state action by the US Department of Agriculture (USDA) and the land-grant college system in the research, development, and public diffusion of production technologies,[28] and federal farm programs resulting in cheap feedgrains.[29]

The maturing of the livestock/feed/meat complex was mirrored in expanding aggregate US meat production and consumption. In 1945, Americans consumed 149 pounds of meat (that is, red meat, poultry, and fish) per capita; by 1976 and 1990, this climbed to 205 and 235 pounds, respectively (see Figure 11.1).[30] From 1945 to 1976, the composition of the average US meat marketbasket changed significantly, with the consumption of beef and poultry eclipsing that of pork, lamb, and fish.[31] Advances in intensive livestock feeding translated into rising meat supplies and falling real meat prices. Fueled by rising real wages, population gains, and remembrances of "meatless Tuesdays," more Americans eating more meat per capita fostered the development of a dynamic US meat sector. Beef became the "center of the plate" of the American diet, with production and consumption increasing 153 percent and 116 percent, respectively, from 1945 to 1976.[32]

Cattle feeders responded to this buoyant market by expanding the national cattle herd from 86 million head in 1945 to 132 million head in 1975.[33] Increases in beef production outpaced increases in the number of cattle slaughtered due to advances in genetics and feedstuffs enabling the production of larger-framed, "meatier" animals.[34]

In grain-fed beef production, intensive feeding was targeted towards animals with the greatest physiological potential for fast, efficient growth: steers (young, castrated males) and heifers (i.e. young females) from larger-framed breeds and genetic lines. From 1960 to 1978, the number of grain-fed, or fed, cattle marketed in the US doubled;[35] fed steers and heifers as a percentage of total US cattle slaughter volume rose from 54 to 71 percent. Intensive feeding regimes hastened processes of concentration in US beef cattle production, with the large feedlots of the Plains states overshadowing the market share of midwestern farmer-feeders by the early 1970s. (Figure 11.2).[36] Prior to intensification, beef cattle production, whether by grain or

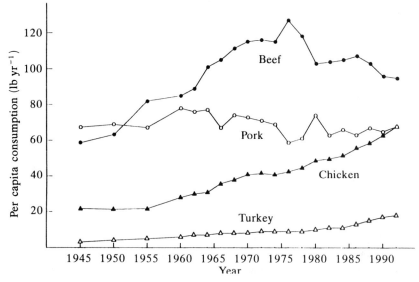

Figure 11.1   Per capita consumption of beef, pork, chicken, and turkey in the
US, 1945–92

Sources: B. Senaur, E. Asp, and J. Kinsey, *Food Trends and the Changing Consumer*, St Paul,
Minnesota: Eagen Press, 1991; "Food Marketing Review, 1989–90," Agricultural
Economic Report 639, US Department of Agriculture, Washington, DC, 1990

forage feeding, was largely the domain of diversified, midwestern farmer-
feeders, with midwestern marketings comprising 68 percent of US cattle
slaughter volume in 1955.[37] From 1970 to 1992, the number of feedlots in
the US fell over 70 percent, with the small, diversified midwestern lots being
hit the hardest.[38] By 1993, eighty-two companies raised a third of the
nation's cattle.[39]

From the 1950s to the mid 1970s, the beef complex incorporated high-
wage workers in meat processing and retailing with an increasingly specialized feedlot sector.[40] At a time when "it seemed that the whole United States
was alit coast to coast with backyard barbecues,"[41] meatpacking was a
competitive, high-wage, urban industry, with wages 15 percent higher than
the national manufacturing wage average.[42] Master wage contracts, whereby
the first company-union accord set the pattern for all subsequent contracts
in the sector, guaranteed uninterrupted output during this time of rising
demand.[43] At the retail level, the meat department was the hallmark of the
supermarket, comprising up to 25 percent of gross sales,[44] with skilled,
unionized butchers receiving top wages.[45]

Technological and regulatory developments in the 1950s, including the
emergence of refrigerated trucks, large supermarket chains with their own
cold-storage facilities, and a new federal meat-grading system, spurred

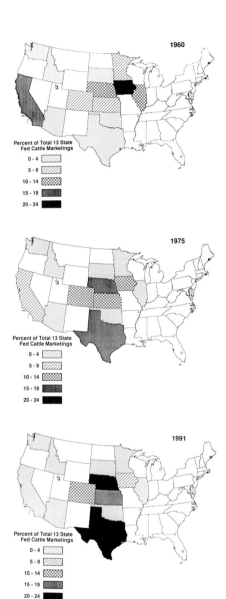

Figure 11.2   Leading US cattle-feeding states, 1960, 1975, and 1991

Source: *Cattle on Feed*, US Department of Agriculture, Washington, DC, various issues 1960–93

processes of decentralization in meatpacking, which had been centered in the major railheads of the Midwest. Formerly, livestock were hauled long distances to stockyards and auctions in midwestern cities, where marketing intermediaries purchased animals for resale to packers. At the plant, skilled "disassembly-line" workers processed carcasses for sale to retailers. By the late 1960s, old-line firms shut down beef plants in midwestern cities and built new ones in the Plains states near major feedlots. Increasing volumes of cattle were being sold "directly" from feeders to packers, marking declines both in the number of public markets (that is, auctions and stockyards) and in the volume of animals marketed through them.[46] In 1960, 39 percent of all US slaughter cattle were sold directly; by 1965, 1970, and 1980, this was 45 percent, 65 percent, and 77 percent, respectively. East of the Mississippi, packing-plant workers lost jobs and cattle-feeders lost markets; west of the Mississippi, beef processors found cheaper land, labor, and livestock costs.[47]

Single-species meat plants emerging as the norm, the center of beef packing shifted westward, whereas pork packing remained in the Midwest near ample hog supplies, with farmer-feeders raising 70 percent of the nation's hogs.[48] Firms that had been household names for decades continued to be the major players in the industry, but more hogs were being sold directly to packers.

Crisis, Competition, and Restructuring

In the mid 1970s, the US meat industry was in crisis due to health concerns and competition from the poultry sector.[49] US beef consumption fell 28 percent from 1976 to 1990 (see Figure 11.1). Meatpackers could not depend upon rising incomes, increases in consumption, falling input costs, or expanding export markets to pull them out of the hole.

Major change came when Iowa Beef Processors (now IBP Inc.) sold "boxed beef" to the food service trade in 1967 and to supermarkets shortly thereafter.[50] IBP broke from the master wage accord in the early 1970s,[51] setting in motion similar tactics by older firms, who cut costs via union busting, plant closures, and plant relocations.[52]

IBP, ConAgra, and Excel expanded their share of the US beef market by building large plants in rural areas near large cattle supplies, using non-union workers in meat fabrication, deskilling tasks, and increasing line speeds.[53] Wages in meatpacking fell and the production workforce shrunk 10 percent.[54] Cutthroat bargaining among states and localities for "economic development" subsidized these activities.[55]

Competition led to high concentration ratios, first in beef packing and later in pork packing, and to a declining number of meat plants (see Figure 11.3). A small number of very large plants owned by a handful of firms now comprise the majority of US meat production.[56] Until the early 1970s, beef plants with capacities to kill several hundred head of cattle per day were considered large, but are minuscule by today's standards, with firms now killing up to four thousand head per plant daily.

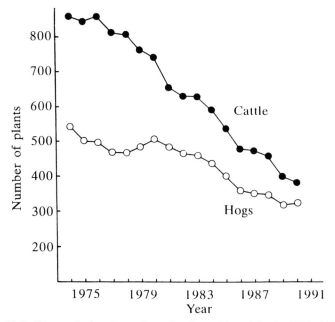

Figure 11.3   Livestock slaughter plants (cattle and hogs) in the USA, 1974–90

Source: *Packers and Stockyards' Statistical Report*, US Department of Agriculture,
Washington, DC, 1973–90

Since the mid 1980s, the US pork sector has undergone rapid change as the big three expand their hog slaughter capacities and as corporate integrators and poultry firms enter the scene. Four firms—IBP, ConAgra, Smithfield Foods, and Excel—control about 40 percent of US pork output volume.[57] These firms own plants capable of processing up to 14,000 hogs daily, nearly triple the size of the largest plants of the early 1980s.

Increased scale is only part of the picture. Another area of growth in the US meat sector relates to boutique meats, which are made in small batches for specialty consumption. Examples are organic, ethnic (for example, kosher meats), and micro-farmer "clean" foods (that is, foods raised under certain social/environmental conditions by small-scale producers).[58] Free-range chickens are an example of the latter. Overall, boutique packers, federally inspected or otherwise, account for less than 5 percent of US meat output.[59]

In the pork sector, some mid-sized packers focus on extra-lean, branded pork, these products being aimed at high-income, health-conscious adults in the US and abroad.[60] An example is Premium Standard Farms (PSF). Founded in 1988, PSF built nucleus herds from Pig Improvement Company (PIC) and Newsham Hybrids, British and Belgian swine seed-stock firms, respectively, both specializing in lean genetics. PSF now operates a 80,000-

sow operation in Missouri and is developing a similar facility near Hereford, Texas. Hogs are processed at PSF's 5500 head/day packing plant in Missouri, with the pork being marketed under PSF's branded label.

Lean and Mean: Marketing Relations and Livestock Production

The imperative for lean meat is evident within both mass-market and boutique production. Meatpackers can no longer base growth around aggregate increases in US meat consumption; thus, emphasis is on selling smaller volumes of higher value-added meats. Markets for industrial by-products—tallow and lard—are also waning. Given that a pound of animal fat is now worth about $.11 whereas a pound of lean pork is worth $1.20 or more, the industrial imperative for leaner animals rings clearly.

This imperative is mirrored in changes in the types of animals raised, how they are raised, and by whom they are raised. Rather than owning and raising meat animals themselves to any significant extent, the new leaders in meatpacking have established contractual relations with feeders to assure ample input supplies. By making production and marketing contracts with feeders, meatpackers are able to influence the timing, volume, and quality of input supplies, yet avoid the risks and costs associated with direct livestock production.[61]

Production contracting, whereby "independent" farmers raise company-owned pigs, has become an important accumulation strategy for major and specialty pork packers alike, with these relations often involving feed companies as intermediaries.[62] Marbery estimates that 25 percent of all US hogs are now raised on contract or sold via marketing agreements.[63] With the proliferation of contracting came new leaders in US hog production— Murphy Farms, Tyson Foods, Carroll Foods, and Goldkist.[64] Contracting took root in the Carolinas, but now flourishes in the remnants of the midwestern farm crisis and in parts of the central and southern Plains.[65]

First the Sow: Technology and Animal Physiology

Recent advances in carcass-evaluation and pricing technologies heighten packer control over the livestock-production process.[66] Using lasers, ultra-sound, nuclear magnetic resonance, and other probes, meatpackers are now able to reliably and non-invasively measure the relative fat, bone, and muscle tissue of each carcass. This shift to "value-based marketing" allows packers to reward producers for raising animals with specific leanness traits and to penalize others for their absence. Increasing numbers of US cattle and hogs are no longer priced "subjectively" (that is, via live appraisal) or via federal carcass grading standards (USDA "grade and yield"), but upon lean-value systems established by individual meatpacking firms.[67] Lean-value programs result in a wider range between the prices paid for the "best" and "worst" animals, with producers who are able to supply large volumes of consistent-quality, lean animals on a regular basis receiving prices signifi-

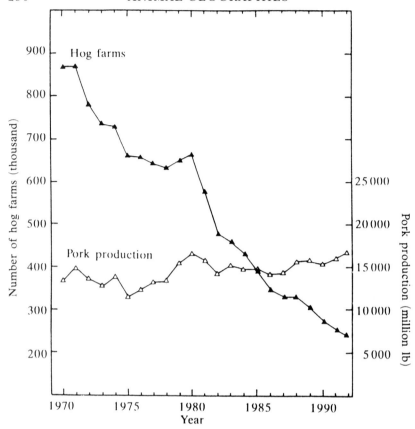

Figure 11.4   The number of hog operations and total pork production in the
USA, 1970–92

Sources: "Meat Facts," American Meat Institute, Washington, DC, 1992; *Hogs and Pigs*, US
Department of Agriculture, Washington, DC, various issues 1971–93

cantly higher than those quoted to the average producer.[68] In that premiums
have become the difference between profit and loss for even the better
producers, these programs intensify processes of consolidation and concentration in the US cattle and hog subsectors. These developments come at
the heels of extensive consolidation in US hog production stimulated by the
production efficiencies made possible by confinement feeding (the first
"wave" of agro-industrialization), the number of hog farms in the US falling
70 percent from 1970 to 1991 (Figure 11.4).[69]

   Lean-meat imperatives stimulate changes in swine-production inputs,
facilities, and management.[70] Until recently, most US hog farms were
"farrow-to-finish." That is, they were involved in the birth (farrowing) of
pigs through the fattening of hogs for market. Lean hog production involves

the spatial separation of farrowing and finishing.[71] Other inputs and production systems linked to lean growth are split-sex feeding, whereby female and male pigs are fed different rations in different locations,[72] artificial insemination,[73] segregated weaning,[74] and phase feeding.[75] Standardized two-phase weaning and finishing diets characteristic of the first wave of pork sector agro-industrialization are giving way to very specialized feeding regimes. Feedstuffs are increasingly customized to specific farm/ herd needs,[76] and markets are booming for new enzymes, amino acids, genetics, growth hormones, and medications,[77] as hog producers try to build a better pig to remain competitive within this industry evolution. As a result, advertisements for veterinary inputs dominate the pages of pork industry magazines.

Concurrent with this new wave of agro-industrialization is an increasingly globalized US pork sector. Manifestations of globalization include European and Australian seedstock firms establishing joint ventures and licensing agreements with US meatpackers and integrators. Japanese firms, including Mitsubishi Corporation, are buying US meat plants.[78] Lean-value, carcass probes have European origins, and foreign veterinary suppliers are increasing their US sales. This trend is driven home by an advertisement for swine vaccine, sporting a two-toned pig with spots matching the map of the world, which heralds "A product evolution of global proportion."

Conclusion

Current agro-industrialization in the US pork sector is propelled by changes in consumer demand and crisis and competition in US meatpacking. Competition, as always, is an eviscerative process whereby growth and restructuring hinge upon a massive reordering of social and economic relations. During this time of tremendous change in the US pork sector, family-farm operations find it difficult to afford lean-growth production inputs, which in turn reduces the market prices of their animals under the more discriminatory lean-value marketing system instituted by major meatpackers. The consequence is heightened consolidation of US pork production. Another result is that living and working conditions of meat-industry employees are now on unsure footing. The current existence of meatpacking workers is precarious, characterized by economic insecurity, hardship, and hazard, as they face union busting, low wages, and few benefits.

The successful resolution of the pork industry's crisis of consumption pivots on major changes in swine physiology to achieve leaner meat. Continued profitability in the US pork sector thus intimately binds human and animal fortunes. Producers who are unable to afford lean-growth inputs face marginalization and the sector becomes more consolidated, while future industry profits lie in the immune responses, muscle, bone, and fat of living pigs. For humans and pigs alike, this industry today is lean and mean.

Acknowledgements

This chapter is adapted from "Lean and Mean: US Meatpacking in an Era of Agro-industrial Restructuring," which first appeared in *Environment & Planning D: Society & Space*, vol. 13, no. 6, December 1995, pp. 683–705. Thanks are due to its publishers, Pion Limited, for permission to reprint. I am grateful also to Lacy Daniels and the University of Iowa Medgraphics for assistance with figures . This research was funded in part by a Dartmouth College Walter Burke Research Initiation Award.

Notes

1. D. Goodman, B. Sorj, and J. Wilkinson, *From Farming to Biotechnology: A Theory of Agro-industrial Development*, Oxford: Basil Blackwell, 1987; D. Goodman, "Some Recent Tendencies in the Industrial Reorganization of the Agri-food System," in W. H. Friedland, ed., *Towards a New Political Economy of Agriculture*, Boulder, Colo.: Westview, 1991, pp. 37–63; Margaret Fitzsimmons, "The New Industrial Agriculture: The Regional Integration of Specialty Crop Production," *Economic Geography* 62, 1986, pp. 334–53; J. Gilbert and R. Akor, "Increasing Structural Divergence in US Dairying," *Rural Sociology*, vol. 63, no. 1, 1988, pp. 56–72; T. Marsden, R. Munton, S. Whatmore, and J. Little, "The Restructuring Process and Economic Centrality in Capitalist Agriculture," *Journal of Rural Studies*, vol. 2, no. 4, 1986, pp. 271–80.

2. Goodman, Sorj, and Wilkinson, *From Farming to Biotechnology*.

3. H. Friedmann, "The Political Economy of Food: The Rise and Fall of the Post-war International Food Order," *American Journal of Sociology* 88, pp. S248–86; Friedmann, "Changes in the International Division of Labor: Agro-food Complexes and Export Agriculture," in Friedland, ed., *Towards a New Political Economy of Agriculture*, pp. 65–93.

4. For example, M. Aglietta, *A Theory of Capitalist Regulation: The US Experience*, London: New Left Books, 1979.

5. For example, M. Storper and R. Walker, *The Capitalist Imperative: Territory, Technology, and Industrial Growth*, Oxford: Basil Blackwell, 1989.

6. Goodman, Sorj, and Wilkinson, *From Farming to Biotechnology*.

7. Goodman, "Some Recent Tendencies," p. 41.

8. Case in point: the development of crop hybridization techniques in the 1930s (J. Kloppenburg, Jr, *First the Seed: The Political Economy of Plant Biotechnology*, Cambridge: Cambridge University Press, 1984). The use of hybrid seeds improved yields, yet required the purchase of farm chemicals and other commercial inputs, resulting in the rise of a plant genetics/farm chemicals agro-industrial complex.

9. Friedmann, "Political Economy of Food"; H. Friedmann and P. McMichael, "Agriculture and the State System: The Rise and Decline of National Agricultures, 1870 to the Present," *Sociologia Ruralis* 29, 1989, pp. 93–117; P. McMichael, "World Food System Restructuring under a GATT Regime," *Political Geography*, vol. 12, no. 3, 1993, pp. 198–214.

10. Friedmann, "Political Economy of Food"; F. H. Buttel, "The Land Grant System: A Sociological Perspective on Value Conflicts and Ethical Issues," *Agriculture and Human Values* 2, 1985, pp. 79–95.

11. McMichael, "World Food System Restructuring."

12. Buttel, "The Land Grant System," pp. 51–3.

13. Aglietta, *Theory of Capitalist Regulation*.

14. S. Christopherson, "Market Rules and Territorial Outcomes," Paper presented at the Annual Meetings of the Association of American Geographers, Atlanta. Available from the Department of City and Regional Planning, Cornell University, Ithaca, N.Y.

15. C. Glenn, C. E. Morris, P. M. Dillon, and L. Mancini, "State of the Food Industry: The Challenge of Restructuring," *Food Engineering*, vol. 65, no. 5, 1993, pp. 75–97.

16. Papademetriou et al., "The Effects of Immigration on the US Economy and Labor Markets," US Department of Labor, Washington, D.C., 1989.

17. Glenn, Morris, Dillon, and Mancini, "State of the Food Industry."

18. C-K Kim and J. Curry, "Fordism, Flexible Specialization and Agri-industrial Restructuring," *Sociologia Ruralis*, vol. 33, no. 1, 1993, pp. 72–4.

19. Senauer, Asp, and Kinsey, *Food Trends*.

20. F. Ufkes-Daniels, "Agrarian Ideology, Market Structure and the Reproduction of Consent: Producer-Packer Relations in an Era of US Meat Industry Restructuring," Ph.D. diss., Department of Geography, University of Iowa, 1994.

21. Goodman, Sorj, and Wilkinson, *From Farming to Biotechnology*.

22. Friedmann, "Political Economy of Food"; Friedmann, "Changes in the International Division of Labor."

23. Ufkes-Daniels, "Agrarian Ideology."

24. USDA, "Livestock and Meat Statistics, 1984–88," *Economic Research Service Bulletin* 784, Washington, D.C., 1989, p. 132.

25. D. Goodman and M. Redclift, *Refashioning Nature: Food, Ecology and Culture*, London and New York: Routledge, 1991, p. 107.

26. Goodman, Sorj, and Wilkinson, *From Farming to Biotechnology*, p. 178.

27. Ufkes-Daniels, "Agrarian Ideology," pp. 128–9.

28. Buttel, "The Land Grant System."

29. Friedmann, "Political Economy of Food."

30. Senauer, Asp, and Kinsey, *Food Trends*, p. 16.

31. Senauer, Asp, and Kinsey, *Food Trends*, p. 16.

32. Senauer, Asp, and Kinsey, *Food Trends*, p. 16; see various issues of USDA, *Livestock Slaughter*, Washington, D.C.

33. American Meat Institute, *Meat Facts*, Washington, D.C.: AMI, 1992.

34. K. Krause, "Cattle Feeding 1962–89: Location and Feedlot Size," USDA, Agricultural Economic Report 642, Washington, D.C., 1991.

35. Krause, "Cattle Feeding 1962–89," p. 1.

36. Also stimulating these spatial shifts was a matrix of federal irrigation, farm, energy, and tax policies (Krause, "Cattle Feeding 1962–89"). Many derivatives of southwestern oil refining were used in intensive cattle feeding (Ufkes-Daniels, "Agrarian Ideology," p. 129).

37. Krause, "Cattle Feeding 1962–89."

38. "Feedlot Consolidation Continues," *Cattle Buyers Weekly*, 30 August 1993, pp. 1–3; Krause, "Cattle Feeding 1962–89," pp. 1–11.

39. S. Kay, "A Tale of Tomorrow," *National Cattlemen*, vol. 8, no. 6, 1993, pp. 5–10.

40. K. Stanley, "Industrial and Labor Market Transformation in the US Meatpacking Industry," in P. McMichael, ed., *The Global Restructuring of Agro-food Systems*, Ithaca, NY.: Cornell University Press, 1994.

41. J. M. Skaggs, *Prime Cut: Livestock Raising and Meatpacking in the United States, 1607–1983*, College Station: Texas A and M University Press, 1986, p. 166.

42. Stanley, "Industrial and Labor Market Transformation," p. 133.

43. S. W. Heimstra, "Labor Relations, Technological and Structural Change in US Beef Packing and Retailing," Ph.D. diss., Department of Agricultural Economics, Michigan State University, 1985.

44. R. E. O'Neill, "Is the Meat Department Slipping?" *Progressive Grocer*, 1980, pp. 95–102.

45. J. P. Walsh, *Supermarkets Transformed: Understanding Organizational and Technological Innovations*, New Brunswick, N.J.: Rutgers University Press, 1993.

46. USDA, *Packers and Stockyards' Statistical Report*, Washington, D.C., 1990.

47. Ufkes-Daniels, "Agrarian Ideology."

48. Iowa State University Swine Task Force, *The Iowa Pork Industry*, Ames: Iowa State University Press, 1988, p. 38.

49. Kim and Curry, "Fordism, Flexible Specialization," p. 67.

50. L. Milkovics, "Iowa Beef Packers Aims to Be Nation's Top Packer," *Progressive Grocer*, 1970, pp. 118–24.

51. Heimstra, "Labor Relations."

52. Stanley, "Industrial and Labor Market Transformation."

53. Ufkes-Daniels, "Agrarian Ideology."

54. Stanley, "Industrial and Labor Market Transformation," p. 133.

55. L. Gouveia, "Global Strategies and Local Linkages: the Case of the US Meatpacking Industry," in A. Bonanu, ed., *From Columbus to Contigra: The Globalization of Food and Agriculture*, Lawrence, Kans. 1994, pp. 125–48.

56. American Meat Institute, *Meat Facts*, p. 21; USDA, *Packers and Stockyards' Statistical Report*.

57. "Top Packers Increase Capacity and Share," *Cattle Buyers Weekly*, 9 August 1993, pp. 2–4.

58. R. Smith, "Microfarmer 'Clean Foods' Could Reach 25% of Consumers," *Feedstuffs*, vol. 66, no. 28, 1994, p. 8.

59. Ufkes-Daniels, "Agrarian Ideology."

60. S. Marbery, "Pork Industry Gears-Up for Consumer Driven Market," *Feedstuffs*, vol. 65, no. 8, 1993, p. 17; Marbery, "Canadian Link Gives PSF Pipeline to Dutch Swine Genetics," *Feedstuffs*, vol. 66, no. 42, 1995, p. 19; Marbery, "IBP Bonds with Nippon," *Feedstuffs*, vol. 67, no. 20, 1995, pp. 18–19; Marbery, "Producers Weigh Alternatives to Packer-Directed Marketing," *Feedstuffs*, vol. 65, no. 23, 1995, pp. 20–21; A. Oppedal, "Productivity, Lean Value Define New Standards for Pork," *Feedstuffs*, vol. 65, no. 13, 1993, p. 14.

61. V. Baxter and S. Mann, "The Survival and Revival of Non-wage Labour in a Global Economy," *Sociologia Ruralis*, vol. 32, no. 2/3, 1992, pp. 231–47; Ufkes-Daniels, "Agrarian Ideology."

62. C. Hurt, K. Foster, J. Kadlec, and G. Patrick, "Industry Evolution," *Feedstuffs*, vol. 64, no. 35, 1992, pp. 18–19; Marbery, "Canadian Link"; Marbery, "Pork Industry Gears-Up"; S. Marbery, "Swine Industry Consolidation Spells Coordination of Suppliers," *Feedstuffs*, vol. 67, no. 17, 1995, pp. 47–52; R. Smith, "NPPC Chief Maintains Producers Can Choose to Succeed," *Feedstuffs*, vol. 67, no. 23, pp. 1, 22–3.

63. Marbery, "Swine Industry Consolidation."

64. D. Houghton and K. B. McMahon, "*Hogs Today*'s Top Ten Producers," *Hogs Today*, vol. 8, no. 9, 1992, p. 17.

65. B. Page, "Hogging the Market: The Origins and Consequences of Contract Farming in U.S. Livestock Production," available from the Department of Geography, University of Colorado-Denver.

66. Oppedal, "Productivity"; "Designer Cattle with Ultrasound," *Science* 263, 1994, p. 327.

67. Ufkes-Daniels, "Agrarian Ideology."

68. Although lean-value programs among firms vary, the range of prices paid for the best and worst hogs can be as high as $35/head (Oppedal, "Productivity," p. 23). Producers receiving premiums for large volumes of consistent-quality, lean hogs receive prices about 25 percent higher than those bid to farmers with low-grade hogs. Feeders differ also regarding costs of production, with the least-cost producers/best marketers making 20 percent return on investment, and high-cost producers barely breaking even (W. Marsh and G. Dial, "The PigCHAMP Grow-finish Herd: A Profile of North American Swine Production," *Pigense* 1, 1992, pp. 1–7). Feeders who can't change production methods lose marketing options: "there is a day coming very soon when [their] local hog buyer may not let [them] unload, much less give them any kind of a decent bid" (B. Fleming, "Opinion Page," *National Hog Farmer*, 15 January 1992, p. 42).

69. See various issues of USDA, *Hogs and Pigs*, Washington, D.C.

70. J. Lawrence, "The US Pork Industry in Transition," Department of Economics Staff Paper 240, Iowa State University, 1992.

71. J. Connor, "Medicated Early Weaning, Multiple-site Production," 1993, available from Carthage Veterinary Clinic, Carthage, Ill.

72. J. Goihl, "Research Examines Protein, Lysine Requirements for Barrows, Gilts," *Feedstuffs*, vol. 65, no. 31, p. 12.

73. S. Muirhead, "Swine Industry Realizing Benefits, Flexibility of Using AI Technology," *Feedstuffs*, vol. 66, no. 31, p. 10–11.

74. Smith, "NPPC Chief."

75. T. Stahly, "Impact of Nutritional Regime on Lean Growth and Carcass Composition," 1992, available from Department of Swine Nutrition, Iowa State University; Stahly, "Emerging Trends in the Pork Industry that Impact Swine Feeding Programs," 1993, available from Department of Swine Nutrition, Iowa State University.

76. S. Marbery, "Custom Diet Demand Drives Hog Feed Market," *Feedstuffs*, vol. 66, no. 9, 1994, p. 27.

77. R. Brown, "Enzymes: Interest in Products Continues to Rise," *Feedstuffs*, vol. 66, no. 17, 1994, pp. 1, 4; R. Fountain, "Animal Health Field Churning to Increase," *Feedstuffs*, vol. 66, no. 46, 1994, p. 12.

78. "Japanese Company Plans Panhandle Pork," *Feedstuffs*, vol. 64, no. 31, 1995, pp. 1, 5.

PART IV
ANIMALS AND THE
MORAL LANDSCAPE

The "Right of Thirst" for Animals in Islamic Law: A Comparative Approach

James L. Wescoat, Jr

It is the She-camel of Allah, So let her drink!
The Qur'an[1]

Introduction

Outrage at the abuse of animals and concern for their welfare encompasses many problems, but little comparative attention has been directed toward animals' access to water in different cultural and legal contexts. Substantial research has been done on the physiological consequences of water deprivation for various species. In dry environments, water deprivation can lead to dehydration, renal failure, brain seizure, and death.[2] Aquatic and amphibious species face depleted and degraded habitats. Terrestrial species are losing access to ponds and streams. Domesticated animals receive more consideration, but even they suffer too frequent neglect of basic water needs. The obligation to provide access to water, or water itself, has not been the subject of focused comparative research.

In this paper I examine the "right of thirst" (*haq-i shurb*) for animals in Islamic law and its relevance in two geographic contexts. Under Islamic law, animals have a right to drinking water similar to that of human beings.[3] Because this right appears to be stronger in Islamic than Western water law, it raises many important questions. Why do animals have a right of thirst in Islam? In what sense is the "right of thirst" a "right" (*haq*)?[4] Does it extend to all animals and situations, or just some? What makes it "Islamic"? Did it originate in the revelations of the prophet Muhammad or build upon pre-Islamic law from Arabia, Rome, and customary traditions? Is it an adjustment to arid environments and pastoral lifeways, or is it more general in scope and application?

In the first section of this paper, I address these questions by situating the right of thirst within the broader context of Islamic law. In Islam, there is a close relationship between ethics, law, politics, and religion. Thus, the first

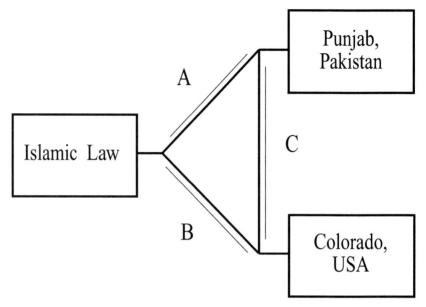

Figure 12.1   Comparative framework for the legal relevance of Islamic law in
Muslim and non-Muslim societies

task is to introduce the basic structure of Islamic law, which sets the stage
for examining the doctrinal bases of the right of thirst for animals.

In the second section of the paper, I ask, what relevance does the right of
thirst have for Muslim and non-Muslim societies? I focus on its legal
relevance in two countries I have studied—the Islamic Republic of Pakistan
and the US.[5] Assessing the legal relevance of Islamic law for Muslim and
non-Muslim societies entails three types of comparison, as shown diagram-
matically in Figure 12.1. First, one expects Islamic water law to have direct
relevance in Islamic states such as Pakistan (A in Figure 12.1). Islam is the
dominant religion in many arid regions of the world and a significant
minority religion in others (Figure 12.2). Islamic political movements are
seeking to instate religious law (*shari'a*) in many of these countries.[6] One
would expect these Islamic movements and states to recognize readily the
right of thirst for animals; or, if they do not presently recognize it, to find
strong legal evidence and arguments compelling. However, the Pakistan
case study reveals that the imperatives of religious law are complicated by
deeply rooted struggles among colonial, regional, and international legal
traditions.

The relevance of Islamic law for places with Muslim minorities, such as
the US, may seem far-fetched. US courts have difficulty with even basic
disputes involving Islamic law.[7] But a long-term perspective on the origins
and diffusion of water laws calls, in my view, for a more open perspective.[8]

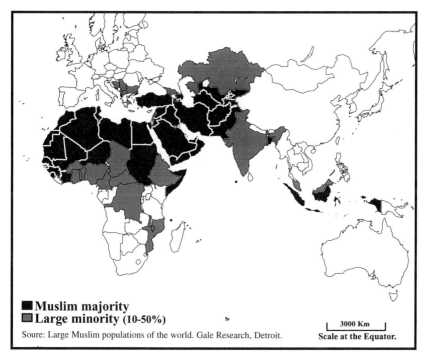

Muslim majority
Large minority (10-50%)

Soure: Large Muslim populations of the world. Gale Research, Detroit.

3000 Km

Scale at the Equator.

Figure 12.2 Large Muslim populations of the world

Source: *Countries of the World and their Leaders 1994*, Gale Research, Detroit

Water rights in the US fall far short of the Islamic right of thirst for animals. States in the American West have belatedly, awkwardly, and inadequately sought to provide "minimum" instream flows for aquatic species and wildlife. Anti-cruelty laws offer some protection against water deprivation for domesticated animals. Comparison with Colorado water law (B in Figure 12.1) highlights these differences between legal traditions.

If the initial aim of comparison in the Pakistan case study is to assess the applicability of Islamic law, the aim in the US case study is to stimulate questions about the adequacy of its water law. In the final section of the paper, I reassess these initial aims by comparing the Pakistan and US case studies (C in Figure 12.1). It asks, in the case of Pakistan, what lessons, if any, may be drawn from legal analogies in the US; and, in the US, what lessons, if any, may be drawn from the applicability of religious law in Pakistan?

These questions call for a type of cross-cultural comparative research that is underdeveloped in the field of "law and geography" in general and water resource geography in particular.[9] Comparative studies in related fields range from grand syntheses and thematic essays to formal classifications.[10] These examples, and others from comparative law, indicate that comparison has been variously regarded as a method of analysis; a survey of the "varieties" of legal systems; and a pursuit of practical transactions between places and cultures.[11]

This third, practical, perspective ultimately seeks to link comparison with legal reform.[12] But the relations between comparison and reform are complex. On the one hand, legal historians draw attention to the distortions that occur when comparison accompanies political and cultural influence.[13] On the other hand, comparison can help identify problems and alternatives. As John Dewey put it, "[a] sense of possibilities that are unrealized and that might be realized, when they are put in contrast with actual conditions, are the most penetrating 'criticism' of the latter that can be made."[14] This line of reasoning has been more widely recognized in human rights than in animal rights research.[15]

Comparative research on animal rights has tended to deal separately with religious and public law traditions on animal sacrifice, experimentation, and hunting.[16] Here, I present a body of religious law and assess its relevance for water law in two geographic areas.

The Structure of Islamic Law

For serious Muslims, Islam is a complete guide for life, including but not limited to religious belief and ritual. Islam does not acknowledge sharp distinctions (or rigid relations) among law, morality, politics, and religion. The source of laws for water use and animal welfare are the same as for other aspects of human understanding and conduct. Thus, the literature on "Islam and ecology," "Islam and human rights," and "Islam and animals" follows a general pattern that may be sketched out below.[17]

In Islam, this unified approach to thought and conduct is known as *shari'a* or "the way." Its original connotation as "the path to water" reflects the significance of access to water in the Arab landscape.[18] *Shari'a* provides the basis for all branches of jurisprudence (*fiqh*). Although the evolution of *shari'a* lies beyond the scope of this paper,[19] the four major "roots of the law" (*usul al-fiqh*) became focuses of jurisprudence during the eighth century C.E. to address an expanding range of situations and concerns.[20] In descending order of importance, the roots of the law are:

1. The *Qur'an* (the word of Allah as revealed to and uttered by the prophet Muhammad).
2. The *Sunna* (the example of the prophet, including *hadiths* [authenticated sayings and actions ascribed to him by his companions]).
3. *Ijma'* (consensus among the Muslim community [*umma*]).
4. *Qiyas* (analogies drawn from one precedent or situation to another by persons of religious authority).

Despite widespread agreement on these roots of the law, there are many divisions among Sunni and Shi'a schools of law, Sufi orders, local customs, as well as disputes about the role of personal inquiry (*ijtihad*), opinion, and innovation. Distinctions that bear upon the right of thirst for animals are taken up later in the paper.

The primary sources of guidance for Muslims are the *Qur'an* and *Sunna* (especially the *hadith*s). This paper concentrates on relevant passages in the *Qur'an*, identified with the concordance of Kassis,[21] and *hadith*s, compiled from Wensinck, Muslim, and secondary sources.[22] Although I write as a non-Muslim for a largely non-Muslim audience, I attempt to transmit the evidence in a manner consistent with its "external understanding and interpretation following [Muslim] tradition" (*tafsir bi'l-ma'thur*), both to avoid misrepresentation and to realistically appraise the relevance of Islamic water law.[23]

Doctrinal Bases for the Right of Thirst

Preliminary considerations
Many verses in the *Qur'an* make it clear that animals have a spiritual role in the creation and that sustenance is provided for them as well as for humans. A widely quoted verse states that "there is not an animal [*dabba*] in the earth, nor a creature flying on two wings, but they are peoples [or communities] like you" [6:38]. The word *dabba* refers to all beasts, creatures, and crawling things, and it is the most common word for animals in the *Qur'an*.[24] Communities of animals praise Allah through their existence and their actions.[25] They have souls (*nafs*) of various types and qualities that constitute essential parts of the world soul.[26] Some Muslim scholars infer from this passage that animals will be resurrected on the day of judgement along with humans.[27]

At the same time, the *Qur'an* states that animals are subordinate to human beings and that they are created to serve human beings. The use of animals for meat and hunting is permitted with certain limitations. A noted Muslim scholar on animal welfare, al-Masri provides a clear discussion of controversies concerning animal slaughter, ritual sacrifice, animal fights, experimentation with animals, and hunting for sport in Muslim societies.[28] Although neglected in the recent literature, the relative rights of thirst of humans and animals in different situations figure prominently in treatises on water law.[29]

Notwithstanding the dominance of human beings over animals, the *Qur'an* warns that humans are ultimately no different from animals in their dependence upon Allah for water and life. "And We send the winds fertilizing, and cause water to descend from the sky and give it to you to drink. It is not ye who are the holders of the store thereof" [15:22]. Allah, not human beings, controls the resources that support all life, for "there is not a beast in the earth but the sustenance thereof dependeth on Allah" [11:6]. Resources have been "measured out" in the creation, and human beings must not forget either the divine origin or measure of those resources [54:49]. Because animals exist as part of the divine creation, because they have souls, because they constitute communities like people, because they provide valuable services for humans, and because they depend upon resources sent for them by Allah—for all of these reasons, human beings must recognize animals' rights to water.

The qualities of these rights may be discerned from four main sources: (1) the *Qur'an*; (2) the *hadiths*; (3) jurisprudence regarding pollution and purity; and (4) early historical incidents that reinforce or complicate legal doctrine. These sources help us answer the questions posed early in the paper about what makes the "right of thirst" a "right" and what makes it "Islamic."

The Qur'an on water for animals

Four groups of verses in the *Qur'an* deal with water for animals. The first group speaks of the rains sent down for animals and human beings on an equal basis: "and We send down purifying water from the sky, That we may give life thereby to a dead land, and We give many beasts and men that We had created to drink thereof" [25:48–9]. In this passage, as al-Masri notes, animals are mentioned first, and they have a "fair share" in the waters that are sent.[30]

A second set of verses refers to water for domesticated animals, especially livestock. Water is provided by Allah, "for you and your cattle" [79:33; 80:32]. Here, special recognition is given to the animals that serve human beings. In addition to being a wonderful fact of life, the provision of water is a sign (*ayah*) for human beings: "[h]ave they not seen how we lead the water to the barren land and therewith bring forth crops whereof their cattle eat, and they themselves? Will they not then see?" [32:27] In contrast with the first set of verses, this group underscores the instrumental value of water for fodder and crops for animals, which ultimately provide food for humans [10:24]. But the overarching themes are the beneficence of Allah; the dependence of all creatures on Allah; and, above all, the imperative for human beings to recognize these signs of beneficence and dependence.

A third set of verses deals with the story of Moses and the shepherdesses of Midian. When Moses came to the oasis at Midian, he saw two women with their flocks, standing at a distance from the water, prevented from approaching by male shepherds [28:22–28]. Moses took their flocks and watered them, and for righting this wrong, he was rewarded by their father. This incident has its origins in the book of Exodus [2:15–22]. It is a complex story involving tribal relations, gender relations, and Jewish history (compare Genesis 24:11ff and 29:10ff). But some Muslim commentators suggest that the denial of water was a wrong against the animals as well as the owners.

A final example is exemplified by the verse at the beginning of this paper, which commands that water be provided for a camel [91:13]. It involves another pre-Islamic tribe known as the Thamud, who demanded a sign from Allah. The sign was a female camel whom the Thamud were commanded to treat properly and give a fair share of food and water. Instead, they denied her those things and hamstrung her, for which they were destroyed by an earthquake.

> He said: (Behold) this she-camel: she hath the right to drink (at the well), And ye have the right to drink (each) on an appointed day. And touch her not with ill. . . .

> But they ham-strung her, and then were penitent. So the retribution came on them [26:155–8]

And in another passage:

> Lo! We are sending the she-camel as a test for them; so watch them and have patience; And inform them that the water is to be shared between (her and) them. Every drinking will be witnessed [54:27–8]

Although the central point of this passage is obedience to clear signs (*bayinnah*), the point is illustrated with a story about the denial of water to a camel, an animal of great worth and affection in Arab culture.

It is appropriate to conclude with a brief discussion of water in the imagery of spiritual rewards and punishments. The *Qur'an* has many beautiful passages describing the paradise gardens that await the faithful on the day of judgement. For believers, there will be shady gardens "underneath which rivers flow" where fruits and cooling beverages will be served by beautiful companions.[31] The wicked, by contrast, will experience the eternal tortures of hell, which include an unquenchable thirst for boiling water like molten copper and pus which tears the bowels [18:29; 38:57]. The resurrection provides for only these two extremes of water experience, but the situation on earth is more varied, which requires further guidance from the life of the Prophet.

The hadith*s on water for animals*
As noted earlier, the *Sunna* of the Prophet includes sayings and actions ascribed to him by his companions, which are known as *hadith*s or "traditions." The *hadith*s deal with water for humans more than animals, but animals are mentioned in several well-known cases. One of the most moving accounts is transmitted as follows:

> While a man was walking, he became thirsty. He went to a well and drank from it. Afterwards he noticed a dog sniffing at the sand because of thirst. The man said to himself: "This dog is suffering what I have suffered," so he filled his shoes with water and held it for the dog to drink. He then thanked God who bestowed upon him forgiveness for his sins. The Prophet's companions asked: "Are we also rewarded for (kindness to) animals?" He answered: "There is a reward for (kindness to) every living thing."[32]

A similar event related by Abu Haraira presents a striking example of the redemptive power of providing water for animals:

> There was a dog moving around a well whom thirst would have killed. Suddenly a prostitute from the prostitutes of Bani Isra'il happened to see it and she drew water in her shoe and made it drink, and she was pardoned because of this.[33]

Withholding water from animals is a mortal sin, as the following example makes clear: "A woman was punished because she had kept a cat tied until it died and (as a punishment for this offence) she was thrown into Hell. She had not provided it with food, or drink."[34] These accounts underscore the obligation to care for animals as fellow sufferers in the creation.

Legal rights are also articulated in the *hadith*s. For example, people must not prevent animals from having access to available water, including waterworks constructed by humans: "He who digs a well in the desert when there is pasture around this well and when there is no other water nearby cannot prevent the animals from slaking their thirst at this well."[35]

Specific cases involving animals occur within a broad set of social duties related to water, which are: (1) Not to refuse water when it is asked, especially surplus water;[36] and (2) Not to sell water, especially surplus water.[37] "Jabir ibn 'Abdullah (Allah be pleased with him) reported that Allah's Messenger (may peace be upon him) forbade the sale of excess water."[38] Wilkinson notes that in Oman, "the drawing of water [from a communal channel] for domestic purposes and the watering of animals is free to all."[39] At a time when pressure to create markets for water is growing around the world, it is worth remembering that the right to drinking water, free of charge, has a moral basis in some cultures and not a venal association with "free-riders."

Practical distinctions among schools of law

Schools of Islamic law vary in how they apply the right of thirst in concrete situations.[40] Table 12.1 outlines the major classifications of water supplies and water rights in four major branches of Islamic law. In each case, greater emphasis is given to the right of thirst for humans than other animals, and to domesticated animals over all others.

In a peaceful situation, running water is common property available to all people and their animals, "provided, however, the animals do not exhaust the entire supply of water."[41] Desert wells, "falling within the *dira* [territory] of a tribe may be considered its property, but the water is still free to nomads from other tribes not at war with the possessors."[42] There are customs for establishing protected zones (*harim*) around common wells, springs, and water channels that remain accessible to all, including the poor and their animals.[43] Water does not generally become private property until it is confined in a reservoir or vessel.

Table 12.1 Water rights and thirst in branches of Muslim water law

Sunni Schools	*Large expanses of water (lakes, seas, major rivers, and mountain ice)* The right of thirst is unlimited and unconstrained for non-Muslims as well as Muslims

Appropriated waters

Community waters: The right of thirst is unconstrained provided that any damage to the banks of a watercourse is compensated.

Privately owned waters:

Hanifite school provides for the right of thirst including the right to use force to gain access to surplus water. Malikite school provides for the right of thirst, but those who drink should pay compensation if they have the means to do so. Shafi and Hanbal schools allow the right of thirst to be exercised without compensation under certain conditions.

Other waters (e.g., wells dug on unowned property)

Wells dug for public benefit: Under scarce conditions, animals may take water after humans have taken their share.

Wells dug by nomads: Persons who dug the well have exclusive rights while at the well, though they must provide surplus water to anyone who is thirsty. After they leave, it becomes common property.

The priority of users is as follows:

a) Persons suffering most from lack of water
b) Persons who dug the well
c) Travellers who may also request equipment to draw the water
d) Local inhabitants
e) Animals of the person who dug the well
f) Animals of travellers
g) Animals of local inhabitants

Shi'a Doctrine	The right of thirst applies only to public waters. Anyone who takes private water must return an equal amount.
Ibadite Doctrine (Omani)	The same as Sunni, except for water in skins, jars, and containers which is excluded from the right of thirst.
Ottoman Civil Code (Mejelle)	Everyone including animals has a right of thirst from public and private rivers, provided the animals do not damage the waterway.

Source: Caponera, 1973, 14–15, 39

Purity and pollution

Laws that promote sanitation and ritual cleansing place further constraints on animals' access to water. Because water is the primary means of purification before prayers and vital for human health, pollution of water supplies is a serious concern. Certain waters (for example, running water) have traditionally been considered pure for drinking and ablutions. Water may still be regarded as pure after an edible animal, donkey, mule, bird of prey, or cat has drunk from it.[44] However, dogs and pigs definitely pollute a drinking source, as do dead animals, noxious excrement, and bodily secretions.

Laws of purification prescribe various remedies for water sources polluted by animals. If a well becomes impure, the remedy depends upon the type of pollution. It must be completely emptied if a large animal dies in it or a live or dead pig falls in it. If a small animal dies in the well, a prescribed number of buckets of water (twenty to sixty) must be drawn out and thrown away. But if a very small animal dies in the well, or a small amount of wild bird or livestock droppings fall into the well, the well is not ordinarily rendered impure.[45] If a water pot becomes polluted by dog saliva, it must be washed a prescribed number of times. Requirements of this sort determine which animals are allowed access to which sources of water, though it is understood that all animals have a right to some source of water.

Early historical incidents

These laws took shape during the first century of Islam (seventh to eighth century C.E.). They were reinforced by several historical incidents that continue to have broad salience in Muslim societies today. For example, the second caliph, 'Umar ibn al-Khattab (634–644 C.E.) was known for his kindness to animals. He did not tolerate cruelty even to animals about to be slaughtered. In one account, "he saw a man refusing to give a sheep water to drink before it was slaughtered. He flogged this man too, and made him give water to the sheep."[46] Many Muslims believe that animals must be offered water before they are slaughtered.

An incident of tragic significance for Islam was the murder of Husain, grandson of the Prophet, at Karbela in 680–81 C.E., which contributed to the rupture of Islam into its Shi'a and Sunni branches. Before he was murdered, Husain, his followers, and their animals were denied water for three days and suffered terrible thirst.[47] The person who enforced those orders reportedly suffered an unquenchable thirst and vomiting until he died.[48] Although the plight of the horses is a relatively minor theme, Shi'a passion rituals that annually mourn the death of Husain recall the suffering of his horses in a way that keeps the right of thirst alive in the minds of Muslims.[49] In other Shi'a accounts, 'Ali, the father of Husain, was denied access to water, but he refused to let his followers retaliate in a similar manner.[50] Another account implies, however, that the caliph Uthman was denied water by 'Ali's party.[51]

Principles of kindness to animals are recorded in allegorical tales,[52] and in political writings known as "mirrors for princes" where prospective rulers are told that, "[t]he water of a well that is in excess of [an owner's] needs should not be kept from the livestock of others, and nothing should be taken in exchange for it."[53]

Summary

The "right of thirst" (*haq-i shurb*) is thus a "right" and is "Islamic." As Geertz points out, the Arabic word *haq* is "more comparable to the Western notion of 'right' (*Recht, droit*) than [it is] to 'law' (*Gesetz, loi*)."[54] Although shaped by pre-Islamic Arab, Jewish, and Roman conceptions of rights, and the arid environment of Arabia, the right of thirst was reforged in terms of an Islamic understanding of the creation and of right conduct. It subsequently diffused across centuries and continents under the rubric of Islamic law. As a right divinely conferred upon humans and other animals, it bears comparison with theological arguments for natural rights in the West rather than theories based on individual liberty or social contract.[55] But most students of *shari'a* have little interest in theoretical comparisons of this sort: the main issue is its practical relevance for faith and conduct in societies today.

Relevance for Pakistan

The Islamic Republic of Pakistan has a complex legal landscape in which animal rights have little importance. The modern administrative system dates back to the Mughal period (1526–1858 C.E.), when kings organized massive hunting parties in the semi-arid forests of the Indus plains. Although these warrior-kings occasionally had spiritual experiences while hunting, after which they temporarily renounced killing and eating meat, wholesale slaughters were common practice.[56] The fourth Mughal emperor, Jahangir (reigned 1605–27 C.E.), erected a minaret to commemorate a pet antelope at a tank known as the Hiran Minar. To this day people say, "he loved that antelope like a brother"—an antelope trained to lure other animals to the tank for water, where they were casually observed or shot. Although the Mughals constructed wells for the free use of Muslim and Hindu travelers, their texts are silent on the right of thirst for animals.

Colonial rulers sought to confine Muslim and Hindu law to family and religious matters (marriage, divorce, inheritance, *waqf* [religious endowments of land and property], and so on). However, Kozlowski has shown that colonial magistrates interpreted religious endowment laws more rigidly and dogmatically than their Muslim predecessors.[57] I have found no colonial judicial cases or legislation that dealt with the right of thirst for animals. Instead, colonial administrators amalgamated their conceptions of customary water laws with selected features of common law and Roman law, and set these diverse bits within a broad framework of administrative laws that dealt with canal irrigation, drainage, land revenue, and municipal corporations.[58]

This was the plural context that existed when animal welfare legislation was transplanted from England and America to the Punjab and other regions of India.[59] The first priority of colonial government had been horse breeding and veterinary science for the cavalry.[60] But by the late nineteenth century, Societies for the Prevention of Cruelty to Animals (SPCA) were established in major cities like Lahore and Karachi. As in the West, SPCA officers had quasi-police powers to set standards, make arrests, confiscate animals, build hospitals, and levy fines against abusive owners.[61] Sympathetic patrons built water troughs along public roads for draft animals.[62]

Colonial water laws showed little concern for animals, except as they affected human health and welfare. Irrigation laws forbade livestock watering in canals because it damaged canal banks.[63] Land tenancy laws touched upon rights to pasture, stock ponds, and fishing—with an eye toward revenue generation and administration.[64] Municipal and military laws focused on health hazards associated with wild and domesticated animals in cities and cantonments.[65]

The only colonial analogue with the right of thirst was the "natural water rights" in the 1882 Law of Easements.[66] These easements were based on English interpretations of Roman rural servitudes that allow property owners to herd their animals across neighboring lands for water. They did not extend to wild or untended animals.

Independence for Pakistan in 1947 raised fresh disputes about the role of *shari'a* law relative to other legal traditions.[67] To the judicial system of district courts, high courts, and supreme court were added separate *shariat* courts that have a somewhat ambiguous scope and jurisdiction. Not surprisingly, animal welfare law receives almost no attention in these struggles between legal systems.

Recently, Muslim scholars have argued compellingly about the relevance of Islamic law for environmental protection.[68] But others fear a long-term erosion by Islamic political movements of science, welfare, and freedom.[69] The past two decades have witnessed a divergence of these views.

Independent organizations like the SPCA have adapted their colonial heritage to the new situation. The SPCA convenes committees of Islamic scholars to help it address social and moral issues of animal protection. In 1963, for example, a committee on conditions in slaughterhouses led to raids and draft legislation supported by Islamic principles.[70]

Environmental legislation at the provincial and national levels has established standards, agencies, and programs that have yet to be tested. The new laws give more attention to ambient environmental monitoring than to animal or habitat protection. A research project on water law in India is breaking new ground on the access of low-caste social groups to water, but it has neglected animal water needs.[71] Municipal laws in Lahore recently banned livestock holding from the urban center for sanitary reasons, and it is only incidental that the animals now have greater access to water in the rural outskirts.

In many respects, the welfare of draft animals in cities like Lahore and

Karachi is declining. Colonial water troughs have been displaced by shops and houses. Overworked, overloaded, and underfed horses, donkeys, and bullocks are forced to compete with a burgeoning fleet of aggressive motorized transport. In polluted forty-five degree centigrade summer heat, animals collapse on the congested urban roads, desperately gasping for air.

In more remote areas, nongovernmental environmental organizations have made some gains in wildlife and aquatic habitat protection. They have also drawn more heavily upon international environmental norms than Islamic traditions, for reasons that will be discussed later.[72]

With this brief sketch of the situation in Pakistan, we can return to the question, what relevance does the "Islamic right of thirst" have in a plural syncretic context like Pakistan? Pakistan is an Islamic state and one might presume that it would embrace the "right of thirst" and related norms. But from its creation, Pakistan has had difficulty determining the role of *shari'a* law relative to other sources of law. Islamic political parties demand a complete shift to *shari'a* law, while mainstream political parties, religious minorities, and progressive movements oppose it or seek to limit it to explicitly religious affairs.[73] In November 1994, for example, Islamic groups protested in Islamabad, Mingora, and Saidu Sharif for the adoption of *Shari'a* law in the Northwest Frontier Province of Pakistan. Comparable situations exist in other post-colonial Muslim societies.[74]

The suffering and treatment of animals plays no part in these struggles, which revolve instead around the treatment of human beings, political-economic relations, and religious duties. Although one expects the "right of thirst" to have direct relevance for Islamic states, several factors work against it.

Because Islam constitutes a complete system, the acceptance of some legal principles, such as the right of thirst, without the whole system would be, for many devout Muslims, worse than meaningless. In formerly colonial societies like Pakistan *shari'a* law does not in fact constitute the complete system. It exists in tension with other legal systems that have varying degrees of usage and legitimacy. The politics of Pakistani *shari'a* law movements have become so heavily associated with patriarchy, archaic criminal laws, Sunni hegemony, and opportunistic political parties that few progressive, cosmopolitan, or minority groups would be inclined to open a legal door to *shari'a* laws concerning animals if these ends could be accomplished through other means. Last and most important, animal welfare groups in Pakistan recognize that massive problems of human poverty and suffering contribute to the suffering of animals and diminish public concern for them.

Thus, despite Pakistan's status as an Islamic state, legal recognition of the right of thirst seems unlikely—with one important exception. In its deepest sense, *shari'a* law governs the inner relations of heart and mind and their outward expression in public conduct. In these private and public realms, some individuals have told me that their commitment to animal welfare has been motivated by the religious beliefs outlined above.[75] For such persons, every passage on animals in the *Qur'an* and *hadith*s is a precious guide. They

Figure 12.3   Boys in the Punjab who are responsible for watering their animals

respect the right of thirst for animals and seek to inspire others to do the same. If they witness abuse in public, they feel obliged as Muslims to intervene. When they organize with effective nongovernmental organizations, like the IUCN or SPCA, their influence is significant.

Relevance for the American West

If the right of thirst has limited relevance for water law in a Muslim society, what possible relevance could it have in places with small Muslim minorities like the American West? Even if it is granted that Islamic water law indirectly influenced early irrigation institutions in the American West through Spanish colonization, does it have any additional relevance for animal water rights in the West? I will address this question using the water laws of my home state of Colorado as an example.

In Colorado, the doctrine of prior appropriation rests on the constitutional right of citizens to divert the unappropriated waters of the state for a beneficial use.[76] Seniority of use determines who gets the water in times of scarcity. The courts asserted long ago that the "natural" riparian rights at common law never obtained in Colorado. Until recently, there were no provisions for recognizing the water remaining in the streams as beneficial for society or required for animals.[77]

The closest analogue to the right of thirst in Colorado law is the preference given to human domestic water uses (including water for one's domestic animals) over other types of use.[78] Human beings may take water from public water bodies for domestic needs without asking permission, subject to restrictions of trespass and nuisance. Livestock ponds to collect runoff have also been treated somewhat preferentially.

But animals themselves have no rights of any kind to water. Even in states that follow the riparian doctrine, animals have no constitutional rights. Instead, their owners have rights in and for them. Owners also have duties toward them. States often exercise their police powers to stop cruelty to animals. They do this not in recognition of animal rights, but rather to "conserve public morals."[79] Similarly, under state criminal codes, failure to provide water and other necessities for domestic animals constitutes neglect, which is a form of cruelty.[80]

Federal statutes and regulations have comparable provisions for confined and wild animals. A short list of relevant laws includes the Animal Welfare Act, which requires humane treatment in research facilities, stores, and transportation facilities; the Endangered Species Act, which requires protection of critical habitats; and numerous laws establishing sanctuaries and preserves where animals presumably have better access to water than in unprotected areas.

This panoply of state and federal laws probably provides greater protection for animals than do laws in Pakistan today; but it is no less fragmented, piecemeal, and inconsistent with policies that permit abuse, eradication, and habitat destruction. More to the point of this chapter, Western water law does not acknowledge that animals have any inherent right to water at all, other than what society chooses to provide in the interest of animal owners and public morals.

So what is the relevance, if any, of the Islamic right of thirst for Colorado water law? This question has three answers. The first is that it has no direct relevance. The structure of Islamic law does not encourage legal transfers, except through the conversion of individuals or the political transformation of states. As noted in the Pakistan case study, Islam denies the legitimacy of partial acceptance. To understand the right of thirst for animals in Islam, one must understand the broader legal and moral context in which it is framed, as well as the theistic principles on which it is founded. From a Muslim perspective, the adoption of selected principles, like the right of thirst, represents a serious error. In light of modern American (mis)perceptions of Islam, such errors seem unlikely.

However, there is another way of looking at the right of thirst from a non-Muslim perspective. Granted that Coloradans are unlikely to adopt *shari'a* law, they are also less likely to feel obliged to maintain the connections between the whole and its parts. The "right of thirst," taken alone, asserts a moral imperative that stands in sharp contrast with Western water law. This contrast may accentuate concerns about the moral adequacy of Coloradan, and more generally Western, water law. This possibility may then initiate a chain of inquiry, stimulated by comparison, that does not conclude with an "import," "transfer," or "transplant" of Islamic law to the West. Instead, comparative inquiry has several possible outcomes. Most conservatively, it may suggest ways to elaborate on existing duties to provide water or access to water for animals. It may stimulate a search for comparable customs in Western history that are sensitive to issues of thirst, for example, customary practices of providing access to water for animals on rangelands, farms, and in urban areas.

More boldly, comparison may stimulate analogies and extensions between human rights and animal rights. Many passages cited earlier in this paper involved analogies between the right of thirst for humans and the right of thirst for animals in Islamic water law. Analogy is a rigorous, and highly constrained, "source" of Islamic law.[81] Animal rights advocates in the West also use analogies to extend human rights to animals.[82]

In the case of water rights, however, the rights to be extended to animals may not be secure for humans either.[83] In the American West, there is not a great deal of difference between the situations of some animals and those of the homeless and poor.[84] Human rights to water for domestic needs have received minimal attention in legal treatises. Those rights have been taken for granted, despite the fact that access to public streams, baths, washrooms, and drinking water may have less public salience than earlier in this century. Hundreds of families have their water cut off every year in cities such as Denver, and municipalities that have pricing policies known as "lifeline rates." Is ensuring access for the poor and homeless merely a matter of conserving public morals or a more fundamental right to the means of survival? Some readers might find it surprising that comparisons with Islamic law lead to concern about human as well as animal rights in the West.

Conclusion: Comparing Pakistan and the US

Our final task is to compare the relevance of Islamic water law in Pakistan and the US. Far from what was presumed at the outset of this paper—that Islamic law has clear applicability in Pakistan and little or none in the USA—the results are in each case more nuanced.

In Pakistan, a constitutional foundation exists for *shari'a* law, and a religious foundation exists for the right of thirst. But the struggles surrounding *shari'a* law, and the seemingly insignificant role of animal welfare in those struggles, diminishes the likelihood that the right of thirst would be formally recognized and enforced. In the US, there is no such constitutional basis for recognizing animal water rights. There are instead legal duties to provide water for animals in certain situations (for example, domesticated animals, instream flows, and endangered species habitats). Expanding these duties would be far less difficult than expanding the definition and scope of water rights holders (as Native American tribes know well).

Formal adoption of Islamic water law is thus unlikely in both regions, for different reasons, but that does not mean that it is entirely irrelevant in either. In Pakistan, *shari'a* law has inspired personal commitment to animal rights and support for organizations and actions that stop abuse. In the US, persons informed about the Islamic right of thirst might begin to wonder, as I have, why comparable water rights are not recognized here and whether existing or even expanded duties constitute an adequate alternative.

When we consider the relevance of Islamic law for generating alternatives in the two regions, we find that analogy and extension have important roles in both cases. For Pakistan, Islamic law already makes clear analogies between human and animal rights to water. In principle, all animals are included; but in practice it is usually domesticated animals that benefit. It should be possible in some situations to extend practices toward the boundaries of principle. "New" analogies, by contrast, are severely constrained in Islamic jurisprudence; but they are not really needed to extend the scope of established analogies to a larger number of animals. Given the current controversies surrounding *shari'a* law, informal analogies that transfer *shari'a* values to other branches of civil law also have promise, though syncretism of this sort raises other problems.

For Colorado, and the US more broadly, the relevance of Islamic water law depends almost entirely on analogy. If previous experiments in comparative water law are any indicator, cross-cultural analogies would be most influential in the early stages of reform. As inquiry proceeds, societies tend to turn from comparisons with foreign models toward a reworking of existing laws or framing of new ones.[85]

The Islamic right of thirst reminds us that access to water is by no means a new moral imperative, but is instead an ancient one that is all too often denied or unenforced. It insists that all "communities" have rights to the water necessary for their survival, and it challenges every culture to find ways—through personal commitment, resource sharing, and legal force—to ensure access to water even in situations of extreme scarcity, where it matters most.

Acknowledgements

I would like to thank Syed Wali Waheed of WAPDA, Salman Husain, and Obaidullah Beg of IUCN–Pakistan for discussing this topic with me and directing me to helpful persons, publications, and organizations. The Smithsonian Institution, Arthur M. Sackler Gallery, and US Environmental Protection Agency have supported my research in Pakistan. Jody Emel, Gerry Pratt, Fred Denny, Dale Jamieson, and anonymous reviewers provided constructive criticism of earlier drafts. Pion Limited, publishers of *Environment and Planning D, Society and Space*, vol. 13, no. 6, December 1995, in which a fuller version of this essay appeared (at pp. 637–54), have generously granted permission to reprint. I take responsibility for remaining errors and deficiencies.

Notes

1. All passages from the *Qur'an* are from *The Meaning of the Glorious Qu'ran*, M. Pickthall, trans, Karachi: Taj Company, n.d. Citations are set within square brackets and follow the convention of chapter (*sura*) number followed by verse number. Dates are given in the Common Era calendar (C.E.).

2. B. Rolls and E. Rolls, *Thirst*, Cambridge: Cambridge University Press, 1982, pp. 137–51.

3. D. Caponera, *Water Laws in Moslem Countries*, revised edition, 2 vols, Irrigation and Drainage Paper No. 20, Rome: Food and Agriculture Organization, 1973.

4. D. Macdonald, *"Hakk" Encyclopedia of Islam*, new edition, vol. 3, Leiden: E. J. Brill, 1979, pp. 82–3.

5. Specific examples are drawn from the Punjab province of Pakistan and the state of Colorado in the US. These two regions have a significant record of historical interaction and comparison in water management: J. Wescoat, "L'acqua nei giardini islamici: religione, rappresentazione e realta," in A. Petruccioli, ed., *Il Giardino Islamico: Architettura, Natura, Paesaggio*, Milan: Electa, 1994, pp. 109–26; J. Wescoat, R. Smith, and D. Schaad, "Visits to the US Bureau of Reclamation from South Asia and the Middle East, 1946–1990: An Indicator of Changing International Programs and Politics," *Irrigation and Drainage Systems* 6, 1992, pp. 55–67.

6. A. Weiss, ed., *Islamic Reassertion in Pakistan: The Application of Islamic Laws in a Modern State*, Syracuse, N.Y.: Syracuse University Press, 1986.

7. D. Forte, "Islamic Law in American Courts," *Suffolk Transnational Law Journal* 7, 1983, pp. 1–33.

8. See P. Crone, *Roman, Provincial and Islamic Law: The Origins of the Islamic Patronate*, Cambridge: Cambridge University Press, 1987; G. Makdisi, *The Rise of Humanism in Classical Islam and the Christian West*, Edinburgh: Edinburgh University Press, 1990; G. Radosevich, *Proceedings: International Conference on Global Water Law Systems*, Valencia, Spain, 1975, Fort Collins: Colorado State University, 1976.

9. For example, N. Blomley, "Text and Context: Rethinking the Law-Space Nexus," *Progress in Human Geography* 13, 1989, pp. 512–34; G. Clark, "The Geography of Law," in R. Peet and N. Thrift, eds, *New Models in Human Geography*, London: Unwin Hyman, 1989, pp. 310–37; J. Emel, R. Roberts, and D. Sauri, "Ideology, Property, and Groundwater Resources: An Exploration of Relations," *Political Geography* 11, 1992, pp. 37–54; O. Matthews, *Water Resources, Geography and Law*, Washington, D.C.: Association of American Geographers, 1984; J. Wescoat, "Water Rights in South Asia and the United States: Comparative Perspectives, 1873–1993," Paper presented to Comparative Examination of Landed Property Rights Project, Stowe, Vermont, 19–22 August 1994, copy with author.

10. L. Teclaff, *Abstraction and Use of Water: A Comparison of Legal Regimes*, UN/ST/ECA 154, New York: United Nations, 1972; United Nations, Economic Commission for Asia and the Far East, *Water Legislation in Asia and the Far East. Part 1*, Water Resource Service No. 1, Bangkok: UN ECAFE, 1967.

11. G. Frankenburg, "Critical Comparisons: Re-thinking Comparative Law," *Harvard International Law Journal* 26, 1985, p. 411; O. Kahn-Freund, "On the Uses and Misuses of Comparative Law," *Modern Law Review* 37, 1974, pp. 1–27; B. Markesinis, "Comparative Law—A Subject in Search of an Audience," *The Modern Law Review* 53, 1990, pp. 1–21.

12. J. Hill, "Comparative Law, Law Reform and Legal Theory," *Oxford Journal of Legal Studies* 9, 1989, pp. 101–15.

13. U. Baxi, "Understanding the Traffic of Ideas in Law between America and India," in R. Crunden, *Traffic of Ideas between America and India*, Delhi: Chanakya Publications, 1985, pp. 319–42; L. Beer, ed., *Constitutionalism in Asia: Asian Views of the American Influence*, Berkeley: University of California Press, 1979; B. Cohn, "Law and the Colonial State of India," in J. Starr and J. Collier, eds, *History and Power in the Study of Law: New Directions in Legal Anthropology*, Ithaca, N.Y.: Cornell University Press, 1989, pp. 131–52; R. Dhavan, "Borrowed Ideas: On the Impact of American Scholarship on Indian Law," in Crunden, ed., *Traffic of Ideas*, pp. 291–318; M. Galanter, *Law and Society in Modern India*, Delhi: Oxford University Press, 1992; M. Hooker, *Legal Pluralism: An Introduction to Colonial and Neo-colonial Laws*, Oxford: Clarendon Press, 1975; D. Kolff, "The Indian and British Law Machines: Some Remarks on Law and Society in British India," in W. Mommsen and J. De Moor, eds, *European Expansion and Law: The Encounter of European and Indigenous Law in 19th and 20th Century Africa and Asia*, Oxford: Berg, 1992, pp. 201–36; A. Watson, *Legal Transplants: An Approach to Comparative Law*, Charlottesville: University Press of Virginia, 1974; Watson, "Legal Transplants and Law Reforms," in *Legal Origins and Legal Change*, London: Hambledon Press, 1991, pp. 293–8.

14. J. Dewey, *Art as Experience*, New York: Capricorn Books, 1934.

15. R. Afshari, "An Essay on Islamic Cultural Relativism in the Discourse of Human Rights," *Human Rights Quarterly* 16, 1994, pp. 235–76; A. Naim, ed., *Human Rights in Cross-cultural Perspectives*, Philadelphia: University of Pennsylvania Press, 1992; B. Tibi, "Islamic Law/shari'a, Human Rights, Universal Morality, and International Relations," *Human Rights Quarterly* 16, 1994, pp. 277–99.

16. A. al-Masri, "Animal Experimentation: the Muslim Viewpoint," *Animal Sacrifices: Religious Perspectives on the Use of Animals in Science*, T. Regan, ed, Philadelphia: Temple University Press, 1986; G. Bousquet, "Des animaux et de leur traitement selon le Judaisme, le Christianisme et l'Islam," *Studia Islamica* IX, 1958, pp. 31–48; L. Regenstein, *Replenish the Earth: A History of Organized Religion's Treatment of Animals and Nature—Including the Bible's Message of Conservation and Kindness toward Animals*, New York: Crossroad, 1991.

17. M. Deen, "Islamic Environmental Ethics, Law and Society," in J. Engel and J. Engel, eds, *Ethics of Environment and Development: Global Challenge, International Response*, Tucson: University of Arizona Press, 1990; S. Nasr, "Islam and the Environmental Crisis," in S. Rockefeller and J. Elder, eds, *Spirit and Nature: Why the Environment Is a Religious Issue*, Boston: Beacon Press, 1992, pp. 83–108; Nasr, "The Ecological Problem in the Light of Sufism: The Conquest of Nature and the Teachings of Eastern Science," *Sufi Essays*, 2nd edition, Albany: State University of New York, 1991, pp. 152–63; Regenstein, *Replenish the Earth*, pp. 249–60.

18. Caponera, *Water Laws*, pp. i, 1–9; F. Rahman, *Islam*, 2nd edition, Chicago: University of Chicago Press, 1979, p. 100.

19. F. Rahman, *Islamic Methodology in History*, Islamabad: Islamic Research Institute, 1965.

20. N. Aghnides, *An Introduction to Mohammedan Law*, Lahore: Sang-e-Meel, 1981; J. Schacht, *"Fikh" Encyclopedia of Islam*, new edition, vol. 1, Leiden: E. J. Brill, 1983, pp. 886–91.

21. H. Kassis, *A Concordance of the Qur'an*, Berkeley: University of California Press, 1983.

22. A. Wensinck, *A Handbook of Early Muhammadan Traditions*, Leiden: E. J. Brill, 1960; Muslim, *Sahih Muslim (al-Jami'-us-Sahih)*, A. Siddiqi, trans., 4 vols, Lahore: Sheikh Muhammad Ashraf, 1990.

23. H. Gatje, *The Qur'an and Its Exegesis: Selected Texts with Classical and Modern Muslim Interpretations*, A. Welch, trans., London: Routledge & Kegan Paul, 1976, pp. 30–44.

24. Kassis, *Concordance of the Qur'an*, p. 363. Compare A. Abel, *"Dabba" Encyclopedia of Islam*, new edition, vol. ii, Leiden: E. J. Brill, 1983, p. 71.

25. al-Masri, "Animal Experimentation: the Muslim Viewpoint."

26. L. Goodman, trans., *The Case of the Animals Versus Man before the King of the Jinn, A Tenth-Century Ecological Fable of the Pure Brethren of Basra*, Boston: Twayne Publishers, 1978; S. Nasr, *An Introduction to Islamic Cosmological Doctrines*, Boulder, Colo.: Shambala, 1978, pp. 93–104, 249–50.

27. Nasr, "Islam and the Environmental Crisis."

28. A. al-Masri, *Islamic Concern for Animals*, The Athene Trust, 5A Charles St, Petersfield, Hants GU32 3EH, no date, and "Animal Experimentation: the Muslim Viewpoint."

29. Caponera, *Water Laws*.

30. al-Masri, *Islamic Concern for Animals*, p. 181.

31. A. Schimmel, "The Celestial Garden of Islam," in E. Macdougall and R. Ettinghausen, eds, *The Islamic Garden*, Washington, D.C.: Dumbarton Oaks, 1976; Wescoat, "L'acqua nei giardini islamici"; J. Wescoat, "From the Gardens of the Qur'an to the Gardens of Lahore," *Landscape Research*, 20, 1995, pp. 19–29.

32. M. Haleem, "Water in the Qur'an," *The Islamic Quarterly* 33, 1989, pp. 34–50; al-Bukhari, ii, 106, cited in Caponera, *Water Laws*, p. 11.

33. Muslim, *Sahih Muslim*, vol. 4, p. 1216.

34. Muslim, *Sahih Muslim*, vol. 4, p. 1381.

35. al-Bukhari, ii, 103, as quoted in Caponera, *Water Laws*, p. 12.

36. Wensinck, *Handbook of Early Muhammadan Traditions*, p. 249; Caponera, *Water Laws*, p. 11.

37. Wensinck, *Handbook of Early Muhammadan Traditions*, p. 249.

38. Muslim, *Sahih Muslim*, vol. 3, p. 824.

39. J. Wilkinson, *Water and Tribal Settlement in South-east Arabia*, Oxford: Oxford University Press, 1977, p. 948.

40. Caponera, *Water Laws*, pp. 14–15.

41. Aghnides, *Introduction to Mohammedan Law*, pp. 518–19.

42. G. Rentz, *"Bi'r" Encyclopedia of Islam*, new edition, vol. 2, Leiden: E. J. Brill, 1979, p. 1231.

43. Deen, "Islamic Environmental Ethics," p. 196; Caponera, *Water Laws*.

44. M. Dabas and J. Zarabozo, eds, *Fiqh us-Sunnah at-Tahara and as-Salah*, vol. 1, Indianapolis, Ind.: American Trust Publications, 1989, pp. 5–6.

45. Kamal, 1975, pp. 54–7.

46. Reported in Regenstein, *Replenish the Earth*, p. 244.

47. al-Tabari, *The History of al-Tabari (Ta'rikh al-rasul wa'l muluk). Volume XIX. The Caliphate of Yazid b. Mu'awiyah*, I. Howard, trans., Albany: State University of New York Press, 1990, pp. 92–197.

48. al-Tabari, *The History of al-Tabari*, p. 107.

49. S. Shams al-Din, *The Rising of al-Husayn: Its Impact on the Consciousness of Muslim Society*, I. Howard, trans., London: The Muhammadi Trust, 1985, esp. pp. 129–30, 164.

50. Ali ibn Abu Talib, *Nahjul Balaga: Peak of Eloquence*, S. Reza, trans., Elmhurst, N.Y.: Tahrike Tarsile Qur'an, Inc., 1984, pp. 182–3.

51. al-Tabari, *The History of al-Tabari*, p. 107.

52. M. al-Damiri, *Al-Damiri's Hayat al-Hayawan (A Zoological Lexicon)*, A. Jayakar, trans., London: Luzac & Co., 1908; Goodman, *The Case of the Animals*.

53. Anonymous, *The Sea of Precious Virtues (Bahr al-Fava'id): A Medieval Islamic Mirror for Princes*, J. Meisami, trans., 12th century; reprint, Salt Lake City: University of Utah Press.

54. C. Geertz, "Local Knowledge: Fact and Law in Comparative Perspective," *Local Knowledge: Further Essays in Interpretive Anthropology*, New York: Basic Books, 1983, p. 187.

55. S. Buckle, *Natural Law and the Theory of Property: Grotius to Hume*, Oxford: Clarendon Press, 1991.

56. E. Findly, "Jahangir's Vow of Non-violence," *Journal of the American Oriental Society*, vol. 107, no. 2, 1987, pp. 245–56.

57. G. Kozlowski, *Muslim Endowments and Society in British India*, Cambridge: Cambridge University Press, 1985.

58. Hooker, *Legal Pluralism*.

59. D. Favre and V. Tsang, "The Development of Anti-cruelty Laws during the 1800s," *Detroit College of Law Review*, 1993, pp. 1–35; S. Goodkin, "The Evolution of Animal Rights," *Columbia Human Rights Law Review* 18, 1987, pp. 259–88.

60. I. Ali, *Punjab under Imperialism, 1885–1947*, New Delhi: Oxford University Press, 1989.

61. Society for the Prevention of Cruelty to Animals (SPCA), "Report on the Conditions of Slaughter Animals in Lahore," Lahore: n.p., 1972; SPCA, Interviews with SPCA staff, Punjab branch, December 1993, with S. Wali Waheed.

62 A. Hasan, *Analogical Reasoning in Islamic Jurisprudence: A Study of the Juridical Principle of Qiyas*, Islamabad: Islamic Research Institute, 1986; compare in England, P. Davies, *Troughs & Drinking Fountains: Fountains of Life*, London: Chatto & Windus, 1989.

63. S. Masood ul Hassan, *The Manual of Canal and Drainage Laws*, Lahore: Irfan Book House, 1992.

64. O. Aggarwala, *The Manual of Tenancy Laws (in Pakistan)*, Lahore: Khyber Law Publishers, 1991; Ali, *Punjab under Imperialism*.

65. J. Hume, "Colonialism and Sanitary Medicine: The Development of Preventive Health Policy in the Punjab, 1860–1900," *Modern Asian Studies* 20, 1986, pp. 703–24.

66. S. Abbas, *Law of Easement (V of 1882)*, Lahore: Kausar Brothers, 1992–3.

67. E. Ahmed, "Islam and Politics," in A. Khan, ed., *Islam, Politics and the State: The Pakistan Experience*, London: Zed, 1985, pp. 13–30; A. Weiss, *Islamic Reassertion in Pakistan: The Application of Islamic Laws in a Modern State*, Syracuse, N.Y.: Syracuse University Press, 1986.

68. Deen, "Islamic Environmental Ethics"; M. Helmy, *Islam and Environment 2—Animal Life*, Kuwait City: Environmental Protection Council, 1989; IUCN-Pakistan, *The Pakistan National Conservation Strategy*, Karachi: IUCN, 1992; A. Kader, A. al Sabbagh, M. al Glenid, and M. Izzidien, *Basic Paper on the Islamic Principles for the Conservation of the Natural Environment*, Gland, Switzerland: IUCN and the Kingdom of Saudi Arabia, 1983; Nasr, "Islam and the Environmental Crisis"; Nasr, "The Ecological Problem"; Nasr, *Introduction to Islamic Cosmological Doctrines*.

69. P. Hoodbhoy, *Muslims and Science: Religious Orthodoxy and the Struggle for Rationality*, Lahore, 1991.

70. SPCA, "Report on the Conditions of Slaughter Animals."

71. C. Singh, *Water Rights and Principles of Water Resources Management*, Water Project Series, Indian Law Institute, with "Research Programme" by U. Baxi, Bombay: N.M. Tripathi Pvt Ltd., 1991; Singh, ed., *Water Law in India*, New Delhi: Indian Law Institute, 1992.

72. IUCN-Pakistan, *Pakistan National Conservation Strategy*.

73. M. Ahmad, "Islamic Fundamentalism in South Asia: The Jamaat-i-Islami and the Tablighi Jamaat," in M. Marty and R. Appleby, *Fundamentalisms Observed*, Chicago: University of Chicago Press, 1991, pp. 457–530.

74. R. Braibanti, ed., *Asian Bureaucratic System Emergent from the British Imperial Tradition*, Durham, N.Y.: Duke University Press, 1966.

75. SPCA, Interviews with SPCA staff.

76. J. Wescoat, "On Water Conservation and Reform of the Prior Appropriation Doctrine in Colorado," *Economic Geography* 61, 1985, pp. 3–24.

77. Colorado Revised Statutes (CRS), 37–92–102(3).

78. CRS, 35–49–101, 37–92–602.

79. 3A Corpus Juris Secundum, "Animals," pp. 585–96.

80. CRS, 18–9–201; CRS, 35–42–105; and *Waters v. People* 23 Colo 33, 46, P. 112 [1896].

81. Hasan, *Analogical Reasoning in Islamic Jurisprudence*.

82. P. Cavalieri and P. Singer, eds, *The Great Ape Project: Equality beyond Humanity*, New York: St Martin's Press, 1993.

83. S. McCaffrey, "A Human Right to Water: Domestic and International Implications," *The Georgetown International Environmental Law Review* 5, 1992, pp. 1–24.

84. F. Brown and H. Ingram, *Water and Poverty in the Southwest*, Tucson, University of Arizona Press, 1987.

85. Wescoat, "L'acqua nei giardini islamici."

Animals, Ethics, and Geography

William S. Lynn

> You ought to follow the example of shunk-Tokecha (wolf).
> Even when he is surprised and runs for his life, he will pause
> to take one more look at you before he enters his final retreat.
> So you must take a second look at everything you see.
>
> Ohiyesa[1]

Introduction

For many of us the moral status of animals seems abundantly clear. For thousands of years most people believed animals to be resources that lay beyond the boundaries of moral community. Like the rest of the earth, animals were said to exist solely for the benefit of humans. Aristotle put the matter nicely:

> Plants exist for the sake of animals, and brute beasts for the sake of man—domestic animals for his use and food, wild ones (or at any rate, most of them) for food and other accessories of life, such as clothing and various tools. Since nature does nothing purposeless or in vain, it is undeniably true that she has made all animals for the sake of man.[2]

If the moral status of animals is so clear, why am I bothering with this topic? The reason is that, like shunk-Tokecha, geographers and other scholars across the academy are taking a second (and clear-eyed) look at animals and animal ethics. As a consequence of our inquiries, we are remapping the moral landscape of animal-human relations, revealing a diverse world of ethically relevant nonhuman beings. Moral value is the keystone concept for remapping this world and locating animals in our moral landscape. My intention in this chapter is to center our attention on the subject of moral value, and present a geographically informed argument on the moral status of animals. This avowedly normative project is indispensable, for it holds the key to reconfiguring how humans (including geographers) understand and relate to the animal world.

The chapter begins with a brief discussion of "geoethics," a geographically informed theory of moral understanding that positions context at the centre of our moral concerns. Geoethics' contextual emphasis on geographical being and community serves as the starting point for our exploration of moral value. Next, I examine anthropocentric and non-anthropocentric value paradigms (including one I call geocentrism) to explain the case for including animals in our moral community. I conclude with a set of principles to help guide our thought and action toward the animal world.

Geoethics

Ethics and geography

Like any complex tradition of scholarship, ethics has a blizzard of concepts—definitions, distinctions, principles, theories, and so on—and espying order in this haze can be difficult. Yet Socrates's definition provides an excellent place to start. Socrates reminds us that when we discuss ethics, "we are discussing no small question, but how we ought to live."[3] In other words, ethics concerns how we should live our lives, what ends we should seek, and what means we should use in pursuit of our ends. Through moral discourse we develop norms to serve as guidelines for evaluating and directing our conduct toward animals and people, nature and society. Moral reflection does this by generating justifications for our actions using principles about what is good, right, just, or of value. Perhaps most importantly, ethical norms not only reflect who we are, but simultaneously condition how we think and act, and thereby who we may become.[4]

Yet what does it mean to seek justifiable principles for action? Answers to this question vary with one's ethical tradition. My answer is rooted in the hermeneutic tradition of moral understanding, where "justifiable" means good and defensible reasons, "principles" are rules-of-thumb applicable in a wide range of circumstances, and "actions" are statements or behaviors that affect the well-being of others. Ethicians are therefore in the business of using principles that are reasonable, widely applicable, and practicable to adjudicate (and perhaps solve) moral problems.[5]

Ethics is sometimes dismissed as an ineffective word-game in the face of unequal power relations. Such a dismissal is ill-advised, as ethics is both a critique and source of power. As a critique of power, moral norms (however they may be named or generated) are the necessary concepts we use to identify and analyze oppressive power relations. Ethics is also the creative ground from which we envision a non-exploitative future. The importance of ethics, therefore, does not diminish in the face of indifference or malevolence: the harsher the circumstances, the more we need recourse to sources of moral critique and renewal. As a source of power, ethics constitutes (in part) how we understand (describe, explain, and evaluate) whether we are in "right relationship" with the world, that is, whether our individual and collective lives are morally worthwhile and defensible. Ethics delivers a power of insight that unmasks previously unproblematized power relations. Just as importantly, it deeply affects human motivations and actions in light of that insight. People routinely struggle against injustice and compassionately help others, often at great personal risk or self-sacrifice. They do so not only because their moral outlook helps identify wrong-doing or suffering, but because it helps motivate them to act and make ethically directed change a reality. This is why ethical debate is so important, contentious, and consequential—a moral "victory" will influence not only our vision but our actions in the world, and this certainly applies to the world of animals.[6]

Unfortunately, geographers commonly set their moral interests and intentions in the background. Perhaps this is because many of us think of ourselves as disinterested physical or social scientists seeking universal explanations for spatial phenomena. In a similar vein, we often marginalize the ethical relevance of geography itself, regarding moral inquiry as external to the discipline, something best left to philosophy, a stance that does not reflect the character of ethics itself, nor the ethical inquires of other disciplines alongside philosophy.[7] Both these postures contribute to a moral lassitude: we do not ask moral questions of ourselves in an ongoing or deep manner, nor do we fully appreciate the sources of moral insight latent in the discipline.[8]

As importantly, geography is not a value-free or value-neutral inquiry. There is a wide-ranging consensus that all human understanding is at least value-laden, if not rife with moral implications. The moral experience—a conscientious reflection on the ends and means of life, an emotional disposition to nurture, a desire to know what is right or good as opposed to expedient or profitable, a sense of injustice over a state of affairs—is ubiquitous to human life. Humanity's continual engagement with ethical debate is one of our species's special competencies. The reasons for this competence are intricate, and whether it is explained by an innate sociality or cultural dispositions, the phenomenon is ever-present in human beings.[9] Geographers are like other people is this regard. Whether we know it or not, we bring moral presuppositions to our work, and whether we like it or not, our work has moral ramifications. Indeed, many of us became geographers because we care about the world and want to make a positive difference.

Geoethics

What is geoethics, and what does it add to our moral understanding of animals? Geography has much to contribute to ethical theory, discourse, and action. The contribution emphasized by geoethics is the importance of context. Geoethics develops this insight into a distinct kind of contextual ethics, one which generates situated understandings of moral problems. With respect to animals, geoethics directs our attention to the shared contexts of all life-forms, contexts which inform our moral understanding and relationship to animals.

Geoethics develops the contextual insight of geography in the following way. Geographical knowledge is more often than not appreciative of the natural, social, spatial, and temporal circumstances of phenomena. These circumstances are what we call contexts, and when taken together, what I call *geographic contexts*. Geography is a contextualizing tradition of scholarship: geographers commonly contextualize cultural and natural phenomena by emphasizing the interrelations between sites and situations, humans and nature, values and social actions. We indirectly refer to contextuality in many ways: space, place, location, positionality, networks, linkages, scale.

All human activity, including moral conflict, occurs at *sites* embedded in *situations*, making geographic context a constitutive element of all ethical problems. As the site and situation change, so too (to greater or lesser degree) does the moral problem, the interpretations, intentions, and actions regarding that problem, as well as the intended or unintended consequences of those actions. In other words, changes in the geographic context of ethical problems can change the problems themselves, as well as our understanding of and response to those problems. If we wish to find appropriate moral guidance, we must take context into account.

With respect to shared contexts, there is a continuity between humans and animals. From distant evolutionary lineages to intimate loving relationships, animals are both "familiar and extraordinary."[10] Humans have often articulated a radical separation from other species using cultural, linguistic, cognitive, social, technological, and theological criteria (to name a few). These differences do exist and we should not make light of them. But the differences remain distinctions, not dichotomies. We are simultaneously part of and distinct from the natural world, related to yet different from the other species with whom we inhabit the earth.

As *geographic beings* we are necessarily embodied as individuals, and ennatured, that is, situated in nature's rich web of life-forms and life-forming processes. The natural world is a precondition for our individual and species existence. What this means for human and nonhuman animals alike is that our consciousness manifests differing kinds of cognitive and perceptual faculties that are consistent with our particular species's traits. At one end of the spectrum are sophisticated biological "machines," life-forms akin to automatons, such as single-celled organisms. Their consciousness is nil, but their relation to us is both evolutionary and ecological. At the other end are species like ourselves that manifest a high degree of consciousness. Humans, for example, excel at cognition, language, social organization, and technology. No other species manifests these characteristics so abundantly (for better or worse). The power thus enabled over the natural world is astounding. George Perkins Marsh was correct when he termed humans "geographic agents" of environmental change.[11] Some groups of animals—canines, cetaceans, felines, and primates—share analogous characteristics, the emergent properties of well-developed nervous systems, comparatively large brains, and complex social groupings. They have a commensurate (if different) degree of consciousness and emotional authenticity that parallels, and in some cases exceeds, our own.[12] More striking still is the communicative competence manifest by human and some nonhuman animals. Humans are supremely capable of empathizing and communicating with our own kind. Yet we are also capable, to varying degrees, of a similar understanding with others of different species. This is so because they too, in their own ways, are sentient, sapient, social, and communicative, emotional, and social beings.[13]

Animals and humans share more than an individualizing embodiment and a contextualizing ennaturement. Wild and domestic, in wilderness or the countryside or the city, animals and humans share geographic environments—reciprocally constituting natural, social, and artifactual contexts. We humans live amongst a host of other social creatures, from companion and work animals, to adaptable yet wild neighbors such as squirrels, skunks, raccoons, and coyotes, to wild and seldom seen bears, caribou, cougars, pine martens, and wolves. This is the *geographic community*—multiple and overlapping communities of humans, domestic animals, and wild creatures.[14]

Moral Value and Geocentrism

Moral value

How scholars conceptualize and identify moral value varies significantly.[15] Some claim it is an intuitively known, objective truth. Others believe it reports nothing more than emotional states or aggregate social preferences. A more sophisticated and contextual account is developed by Holmes Rolston. Without denying the element that feelings and preferences can play in moral valuations, Rolston also points out how moral value names distinctive and/or intrinsic properties of a being or thing, properties that are necessary for cogent ethical thinking. In this view, while humans are "valuers" generating "valuations," the "value" can be a property intrinsic or distinctive to the being or thing itself. A moral value in this sense is both objective and subjective. People recognize something in the world external to themselves that is a critical element for moral understanding, and they call this a "moral value."[16]

An example may help to clarify this point. Sapience (self-awareness) is a distinctive and intrinsic property of *Homo sapiens*. It is an emergent property of our physical natures and cultural milieu. This is not to say that it is exclusively or uniquely human, but it is a defining feature of our species nonetheless. To reflect on human morality without taking sapience into account is impossible. Indeed, it is our sapient nature that is responsible for our species's capacity for ethical thought and conduct. In a similar way, sapience is a defining feature of certain families of animals—primates, canines, felines, and cetaceans (to name a few). While their self-awareness differs in degree from that of humans, there is a continuity between related kinds of animals and the forms of consciousness they manifest. More importantly, while humans may recognize and name this sapience, it exists independent of our recognition and naming. It is an objective (nonsubjective) feature of some animals themselves. So my friend Copper (an exuberant Weaton terrier) is cognizant of his own life-world, irrespective of whether I or other humans acknowledge that fact. Reflecting on the moral value and considerability of animals, like Copper, without taking their sapience into account, would miss a feature as critical to them as it is to ourselves.

Geographic context is an important, if implicit, element in Rolston's system of recognizing and naming moral value. Rolston implicitly uses the characteristics of geographic context when providing case-study vignettes to explore the nuances of a particular moral value. What distinctive properties exist, as well as which of these we use to generate moral values, will depend on the natural, social, spatial, and temporal circumstances of a particular case. The moral values identified by human valuers may therefore differ from place to place, and from time to time (even in the same place). Yet values are not simply relative to social norms and personal desires. While contextual differences produce a diversity of values, some of these values (like sapience) are real properties of the world. In animal ethics, subjectivity (sapience and sentience) is often emphasized. We tend to direct our attention (and affections) to "charismatic megafauna" such as wolves, apes, and elephants, subjective creatures like ourselves. And, in my experience, it is easier for people to appreciate the moral value of a highly subjective creature than a disembodied social or ecological relationship. Yet subjectivity is not the only real property of the world from which we generate a moral value. The mutual attachments between people and companion animals, the ability people have to empathize with the personal and collective well-being of wild or distant animals, the effects that social structures and policies have on animal well-being, and the essential role animals play in the integrity of ecological relationships (reproduction, adaptation, and speciation) are all real features of animal-human relations. Social relationships and ecological processes are, therefore, generative sources of moral value as well. Alongside the emphasis on subjectivity then, our moral values may legitimately emphasize care and integrity. Whether we stress subjectivity, care, or integrity in ethical deliberation will depend on the characteristic features of the animals themselves, as well as the context in which they exist.

A common means of identifying the presence of subjective, social, or ecological values is through two interrelated distinctions: intrinsic versus extrinsic value, and direct versus indirect duties.[17] Intrinsic value refers to a being having moral value in and of itself, while extrinsic value (also called instrumental value) refers to the usefulness of a being for someone else. Because we regard beings with intrinsic value as ends in themselves, they are owed direct duties, meaning we have direct moral responsibilities to these creatures. Beings with extrinsic value, however, are not regarded as ends in themselves. They are things, means to another's ends, and we can only have indirect duties to them, meaning ancillary duties that derive from our direct duties to others. To claim that animals have intrinsic value, then, is to say that they are ends in themselves and that humans have direct duties

to nonhuman creatures. Alternatively, to claim that animals have only
extrinsic value is to say that they are only means to our ends, and we have
only indirect duties to these creatures.

Moral community and boundary transgressions
Having distinguished between intrinsic and extrinsic, we can now appreciate
why the concept of moral value is so critically important: it determines who
has intrinsic value and is the direct recipient of moral duty. Beings with
intrinsic value are said to be within our "moral community." A moral
community is composed of all beings having moral standing, where *standing*
means that one's well-being can be considered for moral reasons. Moral
value is the criterion (often unrecognized) by which we determine who has
standing within our moral community. Without moral value one is left
outside the boundaries of moral community.[18] In effect, moral value marks
a boundary between ethically considerable persons and inconsiderable
things. In human affairs, these boundaries are most obvious in extreme
cases. For example, racism and ethnocentrism attach greater significance to
the moral value of some human beings than others. Thus, Nazi ideology
regarded Jews as subhuman and homosexuals as pathologically abnormal.
Both were consequently placed outside the Nazi's moral community. The
consequence of this kind of boundary marking was tragic—genocide. This
is a pattern of boundary marking with which we are all too familiar, from
the massacres in Cambodia and Rwanda to ethnic cleansing in Bosnia.

   This kind of boundary marking is also at work with regard to animals. Just
as it does for people, the placing of animals outside humanity's moral
community justifies the most brutal and exploitative of power relations. But
unlike humans, animals cannot organize and challenge the practice for
themselves: they require human interlocutors to speak and act in their
interests. When we speak out for the moral value of animals, we are
engaging in boundary transgressions, that is, transgressing the boundaries
of our human-centered moral community by demanding the inclusion of
animals.[19] Boundary transgressions elicit great alarm amongst anthropo-
centrists, and eventuate several objections. Rooted in claims about theology,
agency, and species loyalty, each objection tends to be acontextual and
categorical, predicating its recognition of moral value on one or more
human characteristics. Because these criteria are self-referential, they have
the effect of creating, then reinforcing, specious moral boundaries between
animals and humans.

   One theological objection holds that God made men and women in His
[*sic*] image, gave dominion over the earth to human beings, and did not
endow animals with a soul. For these reasons animals have only extrinsic
value. While this represents the oldest and dominant tradition of Judeo-
Christian argument, theological interpretations of creation have undergone
a tremendous shift in recent decades. For Judeo-Christian ecotheologians,
all of nature is in the image of God (with humans representing only one
aspect of that image), dominion is not to be interpreted as exploitation but

as stewardship, and God pronounced the creation and all its animals good on each of the six "days" before the creation of humans. For these reasons animals have intrinsic value, with or without a soul. Given our distinct species capacities, our earthly role is to act as guardians of creation, to protect and restore the earth for the well-being of ourselves, future generations, animals, and the rest of nature.[20]

According to the agency objection, animals lack the sentience to be self-aware political subjects, the linguistic skill to understand moral rights and obligations, and the capacity to reciprocate a moral regard for human beings. They are not, therefore, agents of their own moral lives. Because moral agency is the "test" of moral value, only humans are within the boundary defining the moral community. Critics of the agency argument challenge its veracity and relevancy. First, as we have noted before, many animals are self-aware subjects living complex psychological lives. Denying moral value to these creatures, while justifying such for ourselves using similar criteria, is contradictory. Second, while animals are generally incapable of linguistic production (chimpanzee and ape exceptions notwith-standing), they are quite capable of communicative expression and compre-hension, as anyone who has been charged by a moose in rut or been comforted by a companion animal in a moment of emotional need can tell you. It is certainly true that animals cannot speak to us about their needs or intentions, for like human infants they are not the sort of creatures who are capable of doing so. Yet by attending to their communications and behaviors we can interpret their needs, and thus what interests they have that require our respect. Third, the ability to reciprocate rights and duties is hardly a determinative criterion for human membership in a moral community. Infants, persons with severe developmental disabilities, and otherwise unconscious or asleep adults are in no position to reciprocate anything. Yet we continue to recognize moral value in them, and they retain their standing in our moral community.[21]

A nonreductive resolution of this issue is to recognize that animals and humans are distinct kinds of moral beings. Animals are moral "recipients," incapable of ethical actions but legitimate recipients of human moral consideration nonetheless. In contrast, humans are moral "agents"—"the moral primate," in the words of Mary Midgley—beings capable of moral thought and conduct, obligated to consider the consequences of their actions for other moral beings, human and nonhuman alike.[22]

The final objection is the assertion that humans have (or should have) a species loyalty that overrides all moral relationships to other animals. Species loyalists worry that we lose our moral concern for humans when we become concerned about animals. There are powerful social and biological reasons why humans do have a partiality for our own kind. Although we should not overstate the case, there are human relationships that are impossible, difficult, or inappropriate to share with animals. Yet there are interspecies relationships that are as deeply (if differently) satisfying, and we should not discount these either. Moreover, there is no reason to suppose that a moral regard for animals will diminish our moral regard for people. Indeed, the practice of moral concern across a range of beings and issues may strengthen our ethical insights and commitments. In addition, moral boundary-marking based on species membership is potentially malicious. It replicates, in the worst possible ways, the identity-based arguments that legitimate prejudice, injustice, violence, and genocide against other humans. Indeed, many moralists have noted the continuities that exist between the exploitation of the human and animal worlds.[23]

Value paradigms

We are now in a position to specify the value paradigms that structure our individual and collective understandings of the moral status of animals. A value paradigm is a conceptual map of moral value. Value paradigms are configurations of concepts regarding moral value that are used to orient and guide our actions in the social and natural world. They help us navigate ethical and social space by locating where we stand in *moral* relation to others. Most crucially, they lay down the boundaries of intrinsic value, and in so doing map the extent to which our moral community overlaps the geographic community.

Anthropocentrism claims that moral value is centerd in *Homo sapiens* alone. Humans are the center of all intrinsic value: we are ends in ourselves, alone within the boundary of moral community, and owe consideration only to other human beings. We are, in short, the only beings to have moral standing. Nonhuman animals have only extrinsic value: they are either means to human ends or instrumentally valuable for the continuation of ecosystemic functions. They are not morally considerable as they exist outside the moral community, and they can consequently have no moral standing or significance within that community. In anthropocentrism, the boundaries of the moral community are a subset of the geographic community. Anthropocentrism has been the dominant value paradigm in Western discourse on animals and nature. As a taken-for-granted and generally uncontested norm, it is embedded in virtually all laws, policies, and regulations about animals, society, and nature.[24]

Despite its dominance in our public discourse, anthropocentrism is not the only value paradigm about animals. In opposition to anthropocentrism are multiple value paradigms collectively termed non-anthropocentrism. As a whole, non-anthropocentrism claims that moral value is not centred on human beings. It extends beyond human beings to include parts or all of the natural world, including animals as individuals, species, or ecosystems. Non-anthropocentrism does not deny the moral value of human beings, nor does it set the moral value of animals above that of humans. Rather, it expands the boundary of moral concern to include nonhuman animals.

Where non-anthropocentric value-paradigms do differ is in their scale of analysis. From these different scales of analysis come different locations of moral value. Biocentrism emphasizes the parts of nature, that is, animals at an individual scale of analysis, and centers moral value in individual creatures themselves. Biocentrists see a continuity between humans and animals, a continuum of subjectivity that helps us recognize the moral value in ourselves as akin to that of other animals. Biocentrism is the predominant value paradigm of the traditional humane movements, as well as the more radical animal rights movement. Ecocentrism emphasizes wholes in nature, that is, collectivities of animals at a systemic scale of analysis (for example, populations, species, ecological communities), and centers moral value in the ecological functions of species and ecosystems. Ecocentrists claim that the moral value of these wholes outweighs the significance of their constitutive units, that is, individual animals. Ecocentrism is a widely held doctrine in both mainstream environmentalism and the values-oriented wing of disciplines like conservation biology.[25] In biocentrism and ecocentrism, the boundary of intrinsic value overlaps both the moral and geographic communities, but only at a certain scale of analysis.

Geocentrism

I have no argument with anthropocentrism's strong moral valuation of human beings, and I appreciate the equally strong claims of biocentrism and ecocentrism that moral value extends beyond the sphere of humans. Yet I am equally uncomfortable with biocentrism and ecocentrism's a priori location of moral value in individual or collective life-forms. Because most animal and environmental ethicists are contending with anthropocentrism, the differing scales of analysis in biocentrism and ecocentrism remain undertheorized. Unfortunately, this undertheorization has led animal ethicists into dead ends. Tom Regan's overinterpretation of ecocentrism as a form of "environmental fascism" is a case in point. Because Regan is pre-committed to an "individualistic" (biocentric) scale of analysis, he rejects the efforts of other ethicists to think "holistically" (ecocentrically) about species or ecosystems.[26] Geographers too have appropriated these paradigms as fixed categories, deploying them in a manner insensitive to geographic context.[27] This rigid and acontextual approach to the moral value of animals is unnecessarily restrictive, and positions us between binary choices—to value either parts or wholes, individuals or species and ecosystems, but not both. This binary logic creates a false dichotomy and constrains our ability to think clearly and creatively.

As a value paradigm, geocentrism avoids false dichotomies. It emphasizes both the parts and wholes of the earth. This includes animals and humans as individuals, species, and ecosystemic components of the geosphere. Geocentrism is similar to other forms of non-anthropocentrism because it does not locate moral value in humans alone. It is dissimilar to biocentrism and ecocentrism, however, because it recognizes plural centers of moral value in both parts *and* wholes. The vast diversity of life creates a multiplicity of diverse life-forms with distinct properties. From these we generate moral values, the diversity of which reflect this continuum of parts and wholes. To account for the shifting scales and circumstances of moral values, we need to contextually adjust our understanding to fit the properties of the being(s) and situations we are examining. Recognizing the simultaneous and overlapping importance of subjectivity, social relationships, and ecological function in moral understanding, geocentrism regards all animals (including humans) as ends in themselves, as well as a means to other ends; we are variable mixtures of both intrinsic and extrinsic values. In geocentrism, the maps of the moral and geographic communities are isomorphic at whatever scale of analysis we choose.

Practicing Solidarity in the Geographic Community

> The whales turn and glisten, plunge
>   and sound and rise again,
> Hanging over subtly darkening deeps
> Flowing like breathing planets
> in the sparkling whorls of
>   living light—
>
> Gary Snyder, "Mother Earth: Her Whales"[28]

Solidarity

Gary Snyder is a geographer. Not a professional geographer, but a geographer at heart, a poet with a profound sense of place, and an acute understanding of the intricate symbolic, social, and natural connections between animals, humans, and nature. He has a talent for opening our hearts and minds to the wondrous geodiversity of this world. He can also strike a sadder, even angry, note, unveiling how we live at the expense of ourselves and our neighbors—wild and domestic animals, species and ecosystems, indigenous peoples, the poor, the oppressed, and the marginalized—all the "others" with whom we share this planet.

> Brazil says "sovereign use of Natural Resources"
> Thirty thousand kinds of unknown plants.
> The living actual people of the jungle
>   sold and tortured—
> And a robot in a suit who peddles a delusion called "Brazil"
>   can speak for *them*?[29]

Penned at the United Nations Environment Conference (Stockholm) in 1972, "Mother Earth: Her Whales" portrays our home as resplendent—a wondrous "breathing planet," a cybernetic ecosystem transforming matter and radiant energy into "living light," a geome (global biome) packed with an ever-evolving diversity of life-forms and life-ways. Our home is also materially and spiritually degraded—raped of resources, its creatures (human and nonhuman alike) enslaved or exploited for profit, the richness of life "parceled out" like commodities with nary a concern for the well-being of ourselves, our neighbors, or future generations. Thus,

The robots argue how to parcel out our Mother Earth
To last a little longer
      like vultures flapping
Belching, gurgling,
      near a dying Doe.[30]

Central to Snyder's writing is a moral valuation of animals and the rest of nature.[31] Snyder envisions a moral community to which all life belongs, a global geographical community in the language of geoethics. This moral valuation of animals and nature destabilizes the routine view of animals as resources, as instrumental means to human ends. The whales, for example, are not fungible commodities for humans to use at our pleasure, but members of an extended moral community. They are morally considerable in and of themselves; they have intrinsic value; they are the property of no one—except the web of life we call Mother Earth.

For these reasons, Snyder would have us stand in solidarity with each member of the geographical community against the forces that ravage and despoil animals, nature and ourselves.

Solidarity. The People.
Standing Tree People!
Flying Bird People!
Swimming Sea People!
Four-legged, two-legged, people![32]

Thus for Snyder our solidarity should not be restricted to ourselves, human beings, the "two-legged." Our solidarity should extend to the standing, flying, swimming, and four-legged "people," that is, with the geographical community of life in which we are pragmatically and morally interlaced. But what does solidarity *mean*?

Solidarity is a condition of relative unity binding members of a group into a fellowship of rights and responsibilities. Amongst human beings the basis for unity is multi-faceted. It has a social edge with respect to material interests or shared identities; a locational edge regarding affective ties to place; a political edge with regard to contested and common purposes. There is a moral edge as well. Solidarity presupposes a felt recognition that we are part of a moral community of beings whose welfare is not only important to our own well-being, but is important in its own right, a value for which we are willing to struggle and to sacrifice.

The question remains, however, whether solidarity is possible between humans and the other animals. Because of our distinct capabilities, humans practice solidarity with one another through flexible strategies of mutual aid: fair trade, social networks and alliances, political negotiations and protests and direct actions, appeals to good sense, moral arguments, and the like. Basic to all these strategies is a context of linguistically constituted personal and social interactions. We argue, bargain, promise, scheme, organize, and evaluate our actions. Structuralist accounts of social forces notwithstanding, we act with a high degree of intentionality. We are, after all, *Homo sapiens*, the wise ones. Yet the gifts of animals are different from our own. Animals are not the sort of creatures who can linguistically conceptualize and strategize mutual aid. How we may practice solidarity in our relationship to them is, therefore, on the face of things, not at all clear. Moreover, acting on behalf of animal well-being is obviously a large and complex undertaking. We have individuals and species, domestic and wild, microfauna and megafauna, human needs and social justice, as just some of the contextual considerations to take into account.

Principles for solidarity
A moral principle is a rule of thumb used to help guide our thought and conduct in the world. Principles are developed from moral concepts into prescriptive or proscriptive statements. Thus, the moral concept of equality is developed into the principle "treat like cases alike; different cases differently." Because they inform our ethical outlooks and deliberations, principles are very helpful (if insufficient in themselves) to the task of changing or concretizing our moral relations in the world. Good principles are guidelines, not axiomatic truths, and have four characteristics that we should keep in mind. First, they should be general, that is, applicable in a wide variety of contexts. Second, they should be practicable, by which I do not mean narrowly pragmatic or immediately attainable, but possible in the long run for agents of average intelligence, good will, and social organiz-ation. Third, they should be interpretable, subject to a legitimate range of interpretation and innovative use. Fourth, they should be limited, meaning we should be suspicious of simplistic formulas and give due regard to a principle's intended use(s) and limitations.

Geoethics can help us actualize solidarity by providing principles directed at animal-human relations. So I conclude this chapter with four principles—geocentrism, equal consideration, hard cases, and moral carrying capacity. These are not the only principles we might (or should) consult, but I offer these as a geographically informed place for geographers to begin their moral and policy reflections.[33]

1. Principle of Geocentrism—*Recognize the moral value of animals, humans, and the rest of nature.* This is our "first principle" because it values animals in our moral landscape, encourages humans to acknowledge their member-ship in the geographic community, and implies duties to respect and protect the community's human and nonhuman members. It also emphasizes the importance of considering "soft" (nontangible) values in the public policy process, and as such helps reveal the implicit value paradigms that pervade such discussions. The words "respect and protect" are intentionally meant to make this principle a strong regulative ideal, that is, we not only have duties to refrain from causing harm (a duty of nonmalevolence), but duties to act to promote the well-being of the geographic community as a whole (a duty of benevolence). This principle does not deny or diminish our

parallel concerns for human rights, interests, or justice. It does, however, reject the automatic privileging of human over animal well-being.

2. Principle of Equal Consideration—*Give equal consideration to the well-being of all creatures affected by our actions.*[34] This principle's intent is to help us identify and then balance the well-being of animals and humans. If we refuse to transgress anthropocentrism's moral boundary, the matter is comparatively simple: we restrict moral value to human beings, pass humane legislation to protect our emotional sensibilities about the treatment of individual animals, and conserve the biological resources of species and ecosystems efficiently. The matter becomes complex when we recognize animals as members of our moral community, and is much more so when we realize there are several scales of analysis and associated varieties of moral value.

Differences in moral value are important when weighing competing ethical concerns, and thus the *moral significance* we accord a claim is a critical feature of moral deliberation.[35] The well-being of certain beings can be more significant than that of others, such as when the well-being of a human being outweighs that of an insect because of the person's sapience and social relationships. But this does not mean that we never face inevitably difficult choices about whose well-being to favour. Weighing multiple kinds of moral values (subjectivity, social relationships, ecological relations) is complex and fraught with uncertainty. An example is the conundrum of animal experimentation. There are reasonable arguments that justify certain experiments as necessary for the benefit of both human and nonhuman animals. Yet there is no justification beyond anthropocentrism to forcibly use apes and chimpanzees in medical experiments we are unwilling to perform upon ourselves.[36] In each case, the significance of different ethical claims must be weighed carefully.

Most importantly, this principle does not justify outlandish comparisons. Humans are not microbes, sapient creatures are not machines, and species are not individuals. Each of these will manifest differing kinds of moral value as we change the scale of our analysis. Nor does it require us to treat different animals in exactly the same manner. "'Equality of consideration' does not imply 'sameness of treatment.' Where beings differ, then equality of consideration will positively require appropriate differences of treat-ment."[37] A contextual recognition of moral value encourages us to pay close attention to the characteristics of the being(s) themselves, as well as to the circumstances of the moral problem. Finally, what this principle does require is for humans to give due consideration to the well-being of other creatures, and to do so without prejudice.

3. Principle of Hard Cases—*When faced with hard cases pitting animals against humans, solve the problem, look for alternatives, or choose a geographic compromise that defends the well-being of animals.* Cases of win/lose conflict are a fact of life, and when these cases involve multiple values, they can be hard to judge. This is clear to all of us who value predators as well as prey and must therefore accept (even embrace) the suffering and death that comes

with predation. Of course, there is no choice in predation, for it is a natural given of the geosphere, an intrinsic property necessary to the survival of all geographic beings. So too all geographic beings require habitat, viable populations, food sources, and so on. Our universal need for "space" makes conflicts over resource use, land-use planning, habitat change or destruction, and destructive or dangerous animals inevitable features of our lives. Inevitably, we will face hard cases where the well-being of humans and animals are in conflict, for geographically it is impossible to maximize the well-being of all members of the geographic community in exactly the same place. These conflicts are especially acute at the macro-scale—between the growing urban and rural landscapes of the humanitat, and the shrinking wildlands necessary to sustain large carnivores, herds of herbivores, and the full range of biodiversity. Yet they are as troublesome at the micro-scale, as when suburban homeowners refuse to share their habitat with adaptable wild creatures like raccoons.

When we face such difficult choices, the principle of hard cases asks that we take the following steps. First, we should resolve the underlying conflict in order to eliminate the problem and prevent its recurrence in the future. Second, if no resolution to the conflict is possible, we should look for the best possible alternative, a course of action that does the least harm while optimizing animal and human well-beings. Third, if no clear alternative is available (or agreeable), we should seek a geographic compromise that maximizes animal and human well-being in different places. One caveat, however, is that the reality of hard cases should not be used as a justification for "mitigation" planning or other subterfuges to acting on behalf of animals. In the words of Rolston, while we should emphasize "non-rival values," we should be careful of compromise, especially when the balance of land-use practices and habitat change is already skewed against the interests of animals.[38]

4. Principle of Moral Carrying Capacity—*Humans should live within a carrying capacity that preserves the integrity of the entire geographical community.* Viewed historically and geographically, there is no "essential" carrying capacity for the earth, meaning we cannot specify a single, quantifiable value for the human population. How many people nature can sustain depends on our numbers, consumption, negative externalities, social organization and conflict, technology, and the like. This is not to imply there are no limits, whether to resource use or population growth or environmental degradation, but these limits do vary with other factors.[39] Embedded in this discourse of carrying capacity, however, is an anthropocentric bias. We often speak of carrying capacity as if it were solely a matter of matching economic resources to human demands, as if the conflict over carrying capacity only pertains to human haves and have-nots. So too we often fail to consider the implications for animal well-being and diversity, not to mention aboriginal cultures and their ways of life.[40]

This is especially important at the global scale; there is nowhere else to go, after all. Yet it is as important in certain biogeographical regions and locales, some of which are more sensitive to human disruption, others of which harbor animals especially threatened by human activities. So the carrying capacity of a locale, a region, or the globe is partially constituted by the value we recognize in the various kinds of animals, biomes, and societies living or capable of living in these areas. Different forms of human ways-of-life are more (or less) compatible with the existence of diverse life-forms and cultures, if only because of variable resources and habitat requirements. The use of these resources and habitats inextricably affects the well-being of nonhuman animals, as well as of other (especially indigenous) cultures. A comparison of the cultural diversity and biological fecundity of the Serengeti plain with the cultural and biological impoverishment of North America's prairie drives this point home. When we talk about carrying capacity then, we must justify the burden our society places on the natural world, and justify that burden according to ethical principles. If we wish to live morally in concert with a diversity of human and nonhuman animals, we must adjust our population, consumption, and interactions accordingly.

Acknowledgements

I would like to thank Helga Leitner, Fred Lukermann, Debra Martin, Roger Miller, Laura Pulido, and Eric Sheppard for their helpful critiques. I am also grateful to Jody Emel and Jennifer Wolch for their patience and editing. I owe a special debt to Nancy Krawetz, who inspired this manuscript. All shortcomings are, of course, my own.

Notes

1. As quoted in Joseph Epes Brown, *Animals of the Soul: Sacred Animals of the Oglala Sioux*, Rockport, Mass: 1992, p. 1.

2. Aristotle, *Politics*, London 1916, 1256b.

3. James Rachels, *The Elements of Moral Philosophy*, New York 1986, p. 1. I will use the words "ethics," "moral," and their cognates interchangeably. There are scholars who assign these terms mutually exclusive definitions, but these are stipulative not descriptive definitions, and do not affect my arguments. See J. O. Urmson and Jonathan Rée, eds, *The Concise Encyclopedia of Western Philosophy and Philosophers*, revised edition, Cambridge 1989, pp. 100–101.

4. Mary Midgley, *Can't We Make Moral Judgements?*, New York 1993; Peter Singer, ed., *A Companion to Ethics*, Cambridge 1993; Paul W. Taylor, *Problems of Moral Philosophy: An Introduction to Ethics*, Belmont, Calif. 1967.

5. Aristotle, *Nicomachean Ethics*, Terence Irwin, trans., Indianapolis, Ind. 1985, 441; David Cooper, *Value Pluralism and Ethical Choice*, New York 1993; Alisdair MacIntyre, *A Short History of Ethics*, New York 1966; Stephen Toulmin and Albert R. Jonsen, *The Abuse of Casuistry: A History of Moral Reasoning*, Berkeley, Calif 1988; Michael Walzer, *Spheres of Justice: A Defense of Pluralism and Equality*, New York 1983.

6. See Terence Ball, "Deadly Hermeneutics; Or, Sinn and the Social Scientist," in Terence Ball, ed., *Idioms of Inquiry: Critique and Renewal in Political Science*, Albany 1987, pp. 96–112.

7. Daniel Callahan and Bruce Jennings, eds, *Ethics, the Social Sciences, and Policy Analysis*, New York 1983; James M. Gustafson, "Ethics: An American Growth Industry," *Key Reporter* 56, 1991, pp. 1–5; Allan J. Kimmel, *Ethics and Values in Applied Social Research*, Newbury Park, Calif. 1988, p. 160; Joan E. Sieber, *Planning Ethically Responsible Research: A Guide for*

Students and Internal Review Boards, Newbury Park, Calif. 1992; Daniel E. Wueste, ed., *Professional Ethics and Social Responsibility*, Lanham, Md. 1994.

8. But ethics is no stranger in our midst. The geographical tradition already incorporates a long history of moral discourse, from ancient Greek and Islamic cosmology, early-modern writing on natural history and teleology, turn-of-the-century debates over environmental determinism, to contemporary research on social and environmental justice. See Clarence J. Glacken, *Traces on the Rhodian Shore: Nature and Culture in Western Thought from Ancient Times to the End of the Eighteenth Century*, Berkeley, Calif. 1967; David N. Livingstone, *The Geographical Tradition: Episodes in the History of a Contested Discipline*, Oxford 1992; William S. Lynn, "Geography, Value Paradigms, and Environmental Justice," *Newsletter of the Society for Philosophy and Geography* 1, 1995, pp. 2–4; David M. Smith, *Geography and Social Justice*, Oxford 1994.

9. Mary Midgley, *The Moral Primate*, New York 1995.

10. Paul Shepard, *The Others: How Animals Made Us Human*, Covelo, Calif. 1996, p. 9.

11. George Perkins Marsh, *Man and Nature, or, Physical Geography as Modified by Human Action*, 1864; reprint Cambridge 1965.

12. I do not mean these comments to privilege human characteristics. Animals have a wide range of skills (sensory acuity, physical grace, and power), many of which exceed our own.

13. Donald R. Griffin, *Animal Minds*, Chicago 1992; Jeffrey Moussarett Masson and Susan McCarthy, *When Elephants Weep: The Emotional Lives of Animals*, New York 1995. As I write these pages, Casey (a sweet and energetic Chow-Shepherd) is dropping a tennis ball in my lap, cocking her ears, and wagging her tail, communicating to me in no uncertain terms what she desires at this moment!

14. See Barry Holstun Lopez, *Wolves and Humans*, New York 1986; Mary Midgley, *Animals and Why They Matter*, Athens, Georgia, 1984; Arne Naess, "Self-Realization in Mixed Communities of Humans, Bears, Sheep, and Wolves," *Inquiry* 22, 1974, pp. 231–41.

15. While this discussion is directed towards animals, much of its content reflects a parallel debate about the moral value of nature.

16. Holmes Rolston III, *Environmental Ethics: Duties to and Values in the Natural World*, Philadelphia 1988.

17. An ethical duty is simply an action we "should" take for defensible moral reasons.

18. Warwick Fox, *Toward a Transpersonal Ecology: Developing New Foundations for Environmentalism*, Boston 1990; I. G. Simmons, *Interpreting Nature: Cultural Constructions of the Environment*, London 1993, pp. 124–5.

19. These are not just crossings, but transgressions. Humans are the beneficiaries of excluding animals from the sphere of moral concern, and we can be quite callous or reactionary about deconstructing such boundaries.

20. Thomas Berry, *The Dream of the Earth*, San Francisco 1988; David Kinsley, *Ecology and Religion: Ecological Spirituality in Cross-Cultural Perspective*, Englewood Cliffs, N.J. 1995; Lynn White, "The Historical Roots of Our Ecologic Crisis," in Lynn White, ed., *Machina Ex Deo: Essays in the Dynamism of Western Culture*, Boston 1968, pp.75–94.

21. The clearest discussion of these issues is by Midgley, *Animals and Why They Matter*. See also Joel Feinberg, "The Rights of Animals and Unborn Generations," in Ernest Partridge, ed., *Responsibilities to Future Generations: Environmental Ethics*, Buffalo, N.Y. 1981, pp. 139–50; Michael Fox, "'Animal Liberation': A Critique," in Kristin S. Shrader-Frechette, ed., *Environmental Ethics*, Pacific Grove, Calif. 1981, pp. 113–24; Kenneth E. Goodpaster, "On Being Morally Considerable," *Journal of Philosophy* 75, 1978, pp. 308–25.

22. See Midgley, *Moral Primate*, as well as Ted Benton, "Animal Rights and Social Relations," in Andrew Dobson and Paul Lucardie, eds, *The Politics of Nature: Explorations in Green Political Theory*, New York 1993, pp. 161–76.

23. Ted Benton, *Natural Relations: Ecology, Animal Rights and Social Justice*, London 1993; Greta Gaard, ed., *Ecofeminism: Women, Animals, and Nature*, Philadelphia 1993; Arne Naess, *Ecology, Community and Lifestyle: Outline of an Ecosophy*, Cambridge 1989; Marjorie Spiegel, *The Dreaded Comparison: Human and Animal Slavery*, Philadelphia 1988; Karen J. Warren, "The Power and the Promise of Ecological Feminism," *Environmental Ethics* 12, 1990, pp. 125–46.

24. See Daniel J. Decker and Gary R. Goff, eds, *Valuing Wildlife: Economic and Social Perspectives*, Boulder, Colo. 1987; Thomas Dunlap, *Saving America's Wildlife: Ecology and the American Mind, 1850–1990*, Princeton, N.J. 1988; Gill Langley, ed., *Animal Experimentation: The Consensus Changes*, New York 1989; Lisa Mighetto, *Wild Animals and American Environmental Ethics*, Tucson 1991; US Fish and Wildlife Service, *Refuges 2003: A Plan for the Future*, Washington, D.C. 1993; Frederic H. Wagner et al., *Wildlife Policies in the U.S. National Parks*, Covelo, Calif. 1995.

25. J. Baird Callicott, *In Defense of the Land Ethic: Essays in Environmental Philosophy*, Albany, N.Y. 1989; Arne Naess, "Intrinsic Value: Will the Defenders of Nature Please Rise?" in Michael E. Soulé, ed., *Conservation Biology: The Science of Scarcity and Diversity*, Sunderland 1986, pp. 504–15; Rolston III, *Environmental Ethics*; Paul W. Taylor, *Respect for Nature: A Theory of Environmental Ethics*, Princeton, N.J. 1986. Naess and Rolston are the philosophers most associated with "biocentrism" and "ecocentrism," respectively. They use these terms to identify their non-anthropocentric alternatives, and neither imposes a bifurcation between the moral value of individuals or collectives. My intention in specifying a narrower meaning to biocentrism and ecocentrism is not to turn them into fixed categories, but to enunciate a distinction to clarify our scales of analysis.

26. Tom Regan, *The Case for Animal Rights*, Berkeley, Calif. 1983, pp. 361–3.

27. For example, see Michael Dear, "Who's Afraid of Postmodernism? Reflections on Symanski and Cosgrove," *Annals of the Association of American Geographers* 84, 1994, pp. 295–300; William S. Lynn, "Contested Moralities: Animals and Moral Value in the Dear/Symanski Debate," Manuscript, 1996; Richard Symanski, "Contested Realities: Feral Horses in Outback Australia," *Annals of the Association of American Geographers* 84, 1994, pp. 251–69; Symanski, "Why We Should Fear Postmodernists," *Annals of the Association of American Geographers* 84, 1994, pp. 301–4.

28. Gary Snyder, "Mother Earth: Her Whales," in *Turtle Island*, New York 1974, pp. 47–9.

29. Snyder, "Mother Earth: Her Whales," p. 47.

30. Snyder, "Mother Earth: Her Whales," p. 48.

31. Gary Snyder, *The Practice of the Wild: Essays*, San Francisco 1990; Snyder, *Place and Space: Ethics, Aesthetics, and Watersheds*, Washington, D.C. 1995.

32. Snyder, "Mother Earth: Her Whales," p. 48.

33. Equally helpful principles can be found in the "Code of Environmental Ethics & Conduct" of the Canadian Environmental Network (Http://nceet.snre.umich.edu/green-life/canamsn.html); the "platform of the deep ecology movement" in Naess, *Ecology, Community, and Lifestyle*, p. 29; and "rules for balancing natural and cultural values" in Holmes Rolston III, *Conserving Natural Value*, New York 1994, pp. 26–33.

34. Peter Singer is the author of this principle, and phrases it as follows: "We [should] give equal weight in our moral deliberations to the like interests of all those affected by our actions." Peter Singer, *Practical Ethics*, second edn, Cambridge 1993, p. 21.

35. *Moral significance* means that one's well-being counts in comparison with another's well-being for ethical reasons.

36. Paola Cavalieri and Peter Singer, eds, *The Great Ape Project: Equality Beyond Humanity*, New York 1994.

37. Ted Benton, "Animal Rights and Social Relations," in Dobson and Lucardie, eds, *The Politics of Nature*, p. 163.

38. Rolston III, *Conserving Natural Value*, pp. 27–9.

39. William R. Catton, *Overshoot: The Ecological Basis of Revolutionary Change*, Urbana, Ill. 1982; William Ophuls, *Ecology and the Politics of Scarcity: Prologue to a Political Theory of the Steady State*, San Francisco 1977; Billy Lee Turner, ed, *The Earth as Transformed by Human Action: Global and Regional Changes in the Biosphere over the Past 300 Years*, Cambridge 1990.

40. Arne Naess, "Sustainable Development and Deep Ecology," in J. Ronald Engel and Joan Gibb Engel, eds, *Ethics of Environment and Development: Global Challenge, International Response*, Tucson, Ariz. 1990, pp. 87–96.

Notes on Contributors

Kay Anderson is Associate Professor of Geography at University College, University of New South Wales. Her recent research has centered on the nexus of race/culture/nature with publications on geographies of civility and wildness (in *Progress in Human Geography*), "the savage" (in *Ecumene*), and animality constructs (forthcoming in *Society and Space*). These interests form the basis for the forthcoming *Domesticating the Wild: Rethinking the Colonial Encounter in Australia*, with Oxford University Press.

Glen Elder is Assistant Professor of Geography and Director of African Studies at the University of Vermont. His recent publications focus on the politics of racialized sexuality and the state sanctioning of black homosexuality in the mining communities of South Africa during Apartheid, and appear in the journals *Jewish Affairs* and the *Journal of the Geography of Higher Education*.

Jody (Jacque) Emel is Associate Professor of Geography and Director of Women's Studies at Clark University Graduate School of Geography. She teaches courses on the economy and environment in history, social theory and resources, and feminism and nature. Prior to coming to Clark, she worked as a water planner in Arizona and as a resource analyst in the coalfields of Pennsylvania. Her co-authored book, *The Southern High Plains: Rise and Decline of a Modern Irrigation Culture*, will be published in 1998 by U.N. University Press. She is a long-standing animal rights and environmental activist.

Andrea Gullo holds a Masters in Urban Planning from the University of California, Los Angeles. She is currently employed as a land use planner with the California Department of Parks and Recreation. Her district (Los Lagos) includes Chino Hills, where mountain lions have been seen.

Unna Lassiter holds degrees in art and geography, and is presently a doctoral student in geography at the University of Southern California. Her previous work focused on the creation of place in a deaf community. Currently, she is exploring the relationship between cultural diversity, attitudes towards animals, and non-profit animal-related organizations in Los Angeles County.

William S. Lynn received his doctorate in geography at the University of Minnesota in 1998, with a dissertation entitled *Geoethics: Ethics, Geography and Moral Understanding*. He is an assistant professor at Green Mountain College, review editor for the journal *Ethics, Place and Environment*, and a consultant on ethics and values in ecosystem management. His research merges ethics and hermeneutics into the intellectual horizon of geography, contributing a distinctly contextual, casuistic, and geocentric approach to practical ethics.

Suzanne M. Michel is a doctoral student in the Department of Geography at the University of Colorado. She is the author of *Raptors and Rhetoric: How Communicative Actions Shape Golden Eagle Habitat in Ramona, CA,* and her dissertation concerns problems of transnational environmental justice in the Tijuana River Valley of the US–Mexico border region.

Chris Philo is Professor of Geography at the Department of Geography and Topographic Science, University of Glasgow. His research is on the historical geographies of "madness" and "asylums," and his wider interests encompass the sociocultural geographies of "outsider" social groups (human and non-human). He is the co-author of *Approaching Human Geography*, the compiler of *New Words, New Worlds* (1991), the co-editor of *Selling Places* (1993), and the editor of *Off the Map* (1995).

James D. Proctor is an assistant professor at UC Santa Barbara, where he teaches courses on human and environmental geography. His research concerns overlapping questions in cultural and environmental geography, especially those pertaining to environmental ethics. He is the co-editor with David M. Smith of the forthcoming *Geography and Ethics: Journeys in a Moral Terrain.*

Paul Robbins is an assistant professor in the Department of Geography at Ohio State University. His work focuses on changes in community, authority and agro-pastoral ecology in arid regions, concentrating on the desert grassland of India, and his research currently involves competing accounts of change in flora and fauna in the desert state of Rajasthan.

Frances M. Ufkes received her doctorate from the University of Iowa. She conducts research into North American and Asian agro-food systems, and her current work deals with the changing geographies of food consumption in Southeast Asia as they relate to democratic space and state legitimacy.

James L. Wescoat, Jr is an associate professor at the University of Colorado, Boulder, where he conducts research into the historical geography of water problems in South Asia and the US. From 1986 to 1995 he organized collaborative research with scholars and agencies in Pakistan, the Smithsonian Institution and the US Environmental Protection Agency.

Jennifer Wolch is Professor of Geography at the University of Southern California, where she has taught courses on urban theory, social policy, and animals and ethics since 1979. She is the co-author of *Malign Neglect: Homelessness in an American City* and *Landscapes of Despair: From Deinstitution-alization to Homelessness*, author of *The Shadow State: Government & Voluntary Sector in Transition*, and co-editor of *The Power of Geography: How Territory Shapes Social Life*. Her research on human-animal relations has appeared in many journals, including *Landscape, Environment and Planning D: Society and Space, Capitalism, Nature, Socialism*, and *Society & Animals*.

Index

301

LES SEPT PÉC

Figure 13.1   Les sept péchés capitaux (The Seven Deadly Sins), by M. Jadin

M. Jadin.

Please remember that this is a library book,
and that it belongs only temporarily to each
person who uses it. Be considerate. Do
not write in this, or any, library book.